MATERNAL EFFECTS IN MAMMALS

MATERNAL EFFECTS IN MAMMALS

EDITED BY

DARIO MAESTRIPIERI

AND

JILL M. MATEO

THE UNIVERSITY OF CHICAGO PRESS • CHICAGO AND LONDON

Dario Maestripieri is professor
of comparative human
development, neurobiology,
and evolutionary biology at
the University of Chicago and
author of *Macachiavellian
Intelligence: How Rhesus
Macaques and Humans Have
Conquered the World.* Jill M.
Mateo is assistant professor of
comparative human development
and evolutionary biology at
the University of Chicago.

The University of Chicago Press, Chicago 60637
The University of Chicago Press, Ltd., London
© 2009 by The University of Chicago
All rights reserved. Published 2009
Printed in the United States of America
18 17 16 15 14 13 12 11 10 09   1 2 3 4 5

ISBN-13: 978-0-226-50119-2   (cloth)
ISBN-13: 978-0-226-50120-8   (paper)
ISBN-10: 0-226-50119-1   (cloth)
ISBN-10: 0-226-50120-5   (paper)

Library of Congress Cataloging-in-Publication Data

Maternal effects in mammals / edited by Dario Maestripieri and
Jill M. Mateo.
     p. cm.
Includes bibliographical references and index.
ISBN-13: 978-0-226-50119-2 (cloth : alk. paper)
ISBN-13: 978-0-226-50120-8 (pbk. : alk. paper)
ISBN-10: 0-226-50119-1 (cloth : alk. paper)
ISBN-10: 0-226-50120-5 (pbk. : alk. paper) 1. Mammals—
Genetics. 2. Parental behavior in animals. 3. Mother and child.
I. Maestripieri, Dario. II. Mateo, Jill M.
  QL738.5.M38   2009
599.135—dc22

                                         2008028726

CONTENTS

CONTRIBUTORS

DAVID F. BJORKLUND
Department of Psychology, Florida Atlantic University, Boca Raton, FL 33431 USA, dbjorklu@fau.edu

W. DON BOWEN
Population Ecology Division, Bedford Institute of Oceanography, Department of Fisheries and Oceans, Dartmouth, Nova Scotia, B2Y 4A2 CANADA, BowenD@mar.dfo-mpo.gc.ca

FRANCES A. CHAMPAGNE
Department of Psychology, Columbia University, New York, NY 10027 USA, fac2105@columbia.edu

JAMES M. CHEVERUD
Department of Anatomy & Neurobiology, Washington University School of Medicine, St. Louis, MO 63110 USA, cheverud@pcg.wustl.edu

TAMI E. CRUICKSHANK
Department of Biology, Indiana University, Bloomington, IN 47405 USA, tcruicks@indiana.edu

ADRIANA CSINADY
Department of Psychology, Florida Atlantic University, Boca Raton, FL 33431 USA, csinady_adriana@hotmail.com

JAMES P. CURLEY
Department of Psychology, Columbia University, New York, NY 10027 USA, jc3181@columbia.edu

STEPHANIE M. DLONIAK
Department of Zoology, Michigan State University, East Lansing, MI 48824 USA, and Masai Mara Predator Research, PO Box 86, Karen 00502 KENYA, smdloniak@gmail.com

MARCO FESTA-BIANCHET
Département de Biologie, Université de Sherbrooke, Sherbrooke, Québec J1K 2R1 CANADA
Marco.Festa-Bianchet@USherbrooke.ca

BENNETT G. GALEF, JR.
Department of Psychology, Neuroscience and Behaviour, McMaster University, Hamilton, Ontario L8S 4K1
CANADA, galef@mcmaster.ca

JASON GROTUSS
Department of Psychology, Florida Atlantic University, Boca Raton, FL 33431 USA, grotuss@msn.com

KAY E. HOLEKAMP
Department of Zoology, Michigan State University, East Lansing, MI 48824 USA, holekamp@msu.edu

DARIO MAESTRIPIERI
Department of Comparative Human Development, and Committee on Evolutionary Biology, The University
of Chicago, Chicago, IL 60637 USA, dario@uchicago.edu

JILL M. MATEO
Department of Comparative Human Development, and Committee on Evolutionary Biology, The University
of Chicago, Chicago, IL 60637 USA, jmateo@uchicago.edu

ANDREW G. MCADAM
Department of Integrative Biology, University of Guelph, Guelph, Ontario N1G 2W1 CANADA,
amcadam@uoguelph.ca

NICHOLAS K. PRIEST
Department of Biology & Biochemistry, University of Bath, Bath, BA2 7AY UK, nkpriest@indiana.edu

JOHN G. VANDENBERGH
Department of Biology, North Carolina State University, Raleigh, NC 27695 USA, vandenbergh@ncsu.edu

MICHAEL J. WADE
Department of Biology, Indiana University, Bloomington, IN 47405 USA, mjwade@indiana.edu

ALASTAIR J. WILSON
Institute of Evolutionary Biology, University of Edinburgh, Edinburgh, EH9 3JT UK, Alastair.Wilson@ed.ac.uk

JASON B. WOLF
Faculty of Life Sciences, University of Manchester, Manchester, M13 9PT UK,
Jason@EvolutionaryGenetics.org

1

The Role of Maternal Effects in Mammalian Evolution and Adaptation

DARIO MAESTRIPIERI AND JILL M. MATEO

THE ROLE OF MATERNAL EFFECTS IN EVOLUTION

Offspring phenotype is, in part, the result of the genes they inherit from their parents. In sexually reproducing animal species, offspring inherit half of their genes from their mothers and half from their fathers. Therefore, both parents are equally likely to affect their offspring's phenotype through direct genetic effects. Offspring phenotype can also be affected by the parental phenotype. These parental phenotypic effects on the offspring's phenotype can have a significant impact on the offspring's survival and reproduction and therefore have important consequences for evolution and adaptation. In evolutionary biology, they are referred to as "maternal effects," perhaps because in many of the organisms in which they have been demonstrated they are more likely to originate from mothers than from fathers.

Maternal effects can result in phenotypic similarities between mothers and offspring. For example, in many mammalian species, large mothers in good body condition can produce large offspring by virtue of the fact that they can transfer to them large amounts of nutrients during pregnancy and lactation. Cross-fostering studies in rodents and nonhuman primates have shown that there is a strong resemblance between the parental care patterns of mothers and those of their adopted daughters (Francis et al. 1999; Maestripieri et al. 2007). When maternal effects result in phenotypic similarities between mother and offspring, they essentially represent a mechanism for

1

nongenetic transmission of traits from one generation to the next. Correlations of traits between mothers and offspring due to maternal effects can significantly bias estimates of additive genetic variance and genetic correlations. Therefore, knowledge of maternal effects is required to understand the genetic basis of traits and their potential for evolutionary change.

Although it has long been known that maternal effects can complicate our ability to estimate the genetic basis of traits, the adaptive ecological and evolutionary significance of maternal effects has only recently been appreciated (Cheverud 1984; Kirkpatrick & Lande 1989; Cheverud & Moore 1994; Bernardo 1996; Mousseau & Fox 1998a, b; Wolf et al. 1997; 1998; 2002; Wolf & Wade 2001). When individual variation in the maternal phenotypic traits that affect the offspring's phenotype has a significant genetic basis, this variation can be subject to natural selection. In other words, natural selection can favor the evolution of maternal genes whose effects are expressed in the offspring's phenotype, i.e., "maternal-effect genes." Genetic maternal effects can play an important role in evolutionary dynamics. For example, theoretical studies have demonstrated that maternal effects can dramatically affect the strength and direction of evolution in response to selection (e.g., Kirkpatrick & Lande 1989; Wade 1998). They can slow down or accelerate the rate of evolution of a character and, in some cases, also lead to evolution in the opposite direction to selection (Kirkpatrick & Lande 1989; Qvarnström & Price 2001). Therefore, knowledge of maternal effects is required to fully understand the evolution of traits by natural selection. Accordingly, maternal effects have been the focus of intense scrutiny in recent evolutionary-biology research.

The source of variation in some maternal phenotypic traits that have the potential to affect offspring phenotype may be largely environmental rather than genetic. Given the high degree of phenotypic plasticity in many vertebrate taxa, mothers can adjust their phenotype in response to their environment and shape their offspring's phenotype accordingly. By doing so, mothers are in effect transferring information about the environment to their offspring. If the mothers' adjustments to the environment are adaptive and if the environment is stable across generations, that is, if the cues from the mother's environment are a good predictor of the environment in which offspring will find themselves, then the offspring's phenotypic adjustments are also adaptive. Therefore, nongenetic maternal effects provide a mechanism for cross-generational phenotypic plasticity and make a significant contribution to an organism's fit with the environment (Bernardo 1996; Mousseau & Fox 1998a, b). By modifying the offspring's phenotype or inducing the expression of new phenotypic traits, nongenetic maternal effects can also allow offspring to colonize new ecological niches and be exposed to new selective pressures. This, in turn, may result in the expression of previously unexpressed genes in the offspring that have significant phenotypic effects on their fitness. Therefore, nongenetic maternal effects can play an important role not only in promoting adaptation to local environmental conditions but, more generally, in evolutionary change in the population. A high degree of phenotypic plasticity may also imply that some individual responses to the environment can be maladaptive. If a mother's responses to the environment are maladaptive, maternal modifications of the offspring phenotype may be maladaptive as well. Therefore, maternal effects may also provide a mechanism by which maladaptive phenotypic traits are transmitted across generations (e.g., Maestripieri 2005a).

MATERNAL EFFECTS IN MAMMALS

Maternal effects have been reported in a wide range of taxa but, until recently, most empirical research on adaptive maternal effects concentrated on plants and insects, and to a lesser extent fish, amphibians, reptiles, and birds (e.g., Mousseau & Fox 1998b). Maternal effects have long been known by breeders of domestic mammals (e.g., Bradford 1972; Koch 1972). Indeed, the covariation between maternal nutrition and offspring size and growth in domestic mammals is the system in which maternal effects were first recognized and studied (McLaren 1981; Bernardo 1996). Until recently, maternal effects in wild mammals received little systematic attention by evolutionary biologists. Yet maternal effects arguably play a larger role in the evolutionary dynamics and adaptation of mammals than in any other animal taxa (Reinhold 2002). Mammals are unique among vertebrates in that mothers and offspring have an intimate and extended association during gestation and lactation. This provides the opportunity for offspring size and growth to be influenced by maternal body condition. Aside from these nutritional maternal effects, other maternal effects involving physiological and behavioral mechanisms are also likely to be common. This is because of the extended period of maternal care and offspring dependence characteristic of many mammals, and therefore the many opportunities for interactions between mothers and offspring throughout their lifetime. For example, in some social mammals, offspring remain associated with their mothers through their entire lifetime and mothers continue to invest in them and influence their behavior and reproduction for many years after weaning.

In recent years there has been a dramatic increase in evolutionary studies of maternal effects in mammals. Maternal effects have been especially studied in wild populations of rodents (e.g., several genera and species of lemmings and voles, subfamily *Arvicolinae*: Boonstra & Hochachka 1997; Inchausti & Ginzburg 1998; Ergon et al. 2001; Oksanen et al. 2003; Ylonen et al. 2004; wild house mice, *Mus domesticus*: Nespolo et al. 2003; leaf-eared mice, *Phyllotis darwini*: Banks & Powell 2004; North American red squirrels, *Tamiasciurus hudsonicus*: McAdam et al. 2002; McAdam & Boutin 2003a, b) and wild ungulates (red deer, *Cervus elaphus*: Schmidt et al. 2001; mountain goats, *Oreamnos americanus*: Gendreau et al. 2005; Côté & Festa-Bianchet 2001; bighorn sheep, *Ovis canadensis*: Festa-Bianchet et al. 2000; Soay sheep, *Ovis aries*: Wilson et al. 2005). Many of these studies used data on nutritional maternal effects on offspring size and growth to test the predictions of evolutionary models. Nutritional maternal effects have also recently been studied in other groups of wild mammals such as pinnipeds and nonhuman primates (e.g., Ellis et al. 2000; Bowen et al. 2001; Altmann & Alberts 2005).

Nonnutritional maternal effects involving behavioral or physiological mechanisms are probably widespread in mammals but have only been investigated in a few mammalian groups, most notably rodents and primates. In these taxa, studies have also considered maternal effects on offspring phenotypic traits other than body size and growth rate. For example, field and laboratory studies of rodents have uncovered effects of maternal presence and behavior on a wide range of offspring phenotypic traits, including habitat selection, food preferences, antipredator behavior, and social preferences (e.g., Yoerg & Shier 1997; Mateo & Holmes 1997; 1999; Holmes & Mateo 1998; Galef 2002; Davis & Stamps 2004). Recent research with laboratory rodents has also elucidated the neuroendocrine and molecular mechanisms through which naturally occurring variations in maternal care affect the responsiveness to stress, and the behavior and reproduction of offspring, including the intergenerational transmission of maternal care (e.g., Francis et al. 1999; Meaney 2001; Champagne et al. 2003; Cameron et al. 2005). Maternal dominance rank in cercopithecine monkeys has long been known to affect a wide range of offspring phenotypic traits, including sex, growth rate, timing of first reproduction, and behavior (e.g., Silk 1983; Holekamp & Smale 1991; Berman & Kapsalis 1999; Altmann & Alberts 2005). Studies of primates have also addressed some of the physiological mechanisms underlying the influence of maternal behavior on offspring's reactivity to the environment as well as the intergenerational transmission of maternal behavior from mothers to daughters (Fairbanks 1989; Berman 1990; Chapais

& Gauthier 1993; Maestripieri 2005b; Maestripieri et al. 2007). Finally, interest in maternal effects is also growing among researchers studying human behavior, development, and cognitive evolution (e.g., Bjorklund & Pellegrini 2002; Bjorklund 2006).

The study of maternal effects in mammals is currently a highly heterogeneous area of research. In part, this is because these effects are of interest to and have been studied not only by evolutionary biologists but also by behavioral ecologists, biological psychologists, and anthropologists. As a result, there is notable heterogeneity in the use of concepts and definitions of maternal effects by different scientists. Research is also heterogeneous with regard to whether the focus is on the analysis at the level of the population versus the individual, whether the studies focus on genetic versus environmental maternal effects, whether these effects are quantified versus simply described, and whether the emphasis is on their evolutionary consequences versus their underlying mechanisms. For example, whereas studies focusing on maternal effects in wild rodents and ungulates have mostly focused on genetic maternal effects and their evolutionary consequences at the population level of analysis, studies of carnivores and nonhuman primates have mostly focused on environmental maternal effects and their underlying mechanisms at the level of the individual. Because of this heterogeneity, current research on maternal effects in mammals is not framed within a common body of theories and there is little communication between researchers working with different taxonomic groups. This volume reflects the variation in approaches to maternal effects that researchers have taken. We believe that it is premature to attempt a definitive synthesis of all of the research represented in this volume; rather, our goal is to integrate the conceptual and empirical studies of maternal effects in mammals, and hope that this will stimulate and direct the growth of this area of research. Moreover, a comprehensive and integrated elucidation of the role of maternal effects in the evolution and adaptation of mammals could enhance the understanding of these effects in other taxonomic groups as well.

CONTENT OF THE VOLUME

Maternal Effects in Mammals aims to provide a comprehensive representation of maternal-effects research in different mammalian groups, with a balanced emphasis between theory and data, genetic and environmental effects, evolutionary approaches and studies of mechanisms, field and laboratory approaches, and analyses at the populational, organismal, and molecular level.

The first two invited chapters provide the theoretical background necessary for those unfamiliar with the study of genetic maternal effects. The following chapters summarize the relevant work in mammals, sometimes in a topic-oriented perspective and sometimes in a taxon-oriented perspective.

Cheverud and Wolf (Chapter 2) and Wade et al. (Chapter 3) review recent theoretical advances in our understanding of maternal effects and their role in evolution. Their chapters mainly focus on genetic maternal effects and illustrate some experimental approaches with which these effects can be studied in the laboratory and in the field. McAdam (Chapter 4) and Wilson and Festa-Bianchet (Chapter 5) also focus on the evolutionary significance of genetic maternal effects, but the emphasis of their chapters is on the measurement of these effects, particularly nutritional maternal effects on offspring growth, in wild mammalian populations. McAdam draws examples from his research with wild red squirrels while Wilson and Festa-Bianchet review studies of wild ungulates. Chapters 6–13 largely focus on the study of individuals and their behavior rather than on population-level processes, and address both the ecological-evolutionary significance of maternal effects and the social, physiological, molecular, and cognitive mechanisms through which these effects may operate. Bowen (Chapter 6) discusses nutritional maternal effects on offspring size and development in pinnipeds, within the context of the ecological adaptations and the maternal investment strategies of these animals. In Chapter 7, Mateo provides an overview of behavioral and physiological maternal effects on a wide range of offspring fitness-related traits such as social relationships, reproduction, habitat use and dispersal, antipredator behavior, and reactivity to the environment in wild rodents and other mammals, with a special emphasis on seasonally breeding animals. Galef (Chapter 8) discusses the behavioral and cognitive mechanisms through which mammalian mothers may influence the development of food preferences and feeding behavior of their offspring, drawing from experimental studies of laboratory rodents as well as observations of free-ranging animals. Champagne and Curley (Chapter 9) review a large body of recent research showing that variations in maternal care patterns of laboratory rats and mice can alter the behavior, reproductive development, neuroendocrine responsiveness to stress, and maternal care patterns of the offspring, as mediated by alterations in gene expression. Vandenbergh (Chapter 10) reviews evidence that hormones from the mother, from exogenous hormone mimics, and from adjacent fetuses in the uterus can have profound effects on the behavior and physiological development of offspring. The most compelling evidence of these prenatal hormone effects results from studies of rodents in which the postnatal development of fetuses adjacent to other males and exposed to their testosterone is compared to that of fetuses that lack this androgenic exposure. In Chapter 11, Holekamp and Dloniak describe a wide range of maternal effects on offspring development in fissiped carnivores, with particular emphasis on the influence of maternal social rank on offspring status, play behavior, reproductive development, association patterns, growth, survivorship, and dispersal patterns in spotted hyenas, *Crocuta crocuta*. The mechanisms underlying some of these effects are also discussed, including androgenic hormones, insulin-like growth factors, stress hormones, nutritional variables, and differential maternal care. Well-known examples of maternal effects in nonhuman primates, reviewed by Maestripieri in Chapter 12, include the maternal inheritance of dominance rank and its effects on reproduction in cercopithecine monkeys, the influence of maternal social networks and behavior on the social development, kin preferences, and social networks of the offspring, the effects of maternal behavior on the offspring's fearfulness, tendency to explore, and responsiveness to stress, the effects of maternal behavior on female reproductive maturation, and the nongenetic cross-generational transmission of maternal styles. Bjorklund et al. (Chapter 13) discuss the many ways in which maternal characteristics in humans can affect children's social, emotional, and cognitive development. Emphasizing the plasticity of human development and the contribution of variation in maternal behavior to epigenetic inheritance, Bjorklund et al. suggest that developmental processes driven by maternal effects may have played an important role in the evolution of human intelligence. Finally, in Chapter 14, Mateo and Maestripieri synthesize and integrate the conceptual and empirical information provided in the other chapters and discuss future directions for research on maternal effects in mammals and other taxa.

REFERENCES

Altmann, J. & Alberts, S. C. 2005. Growth rates in a wild primate population: ecological influences and maternal effects. *Behavioral Ecology and Sociobiology*, 57, 490–501.

Banks, P. B & Powell, F. 2004. Does maternal condition or predation risk influence small mammal population dynamics? *Oikos*, 106, 176–184.

Berman, C. M. 1990. Intergenerational transmission of maternal rejection rates among free-ranging rhesus monkeys. *Animal Behaviour*, 39, 329–337.

Berman, C. M. & Kapsalis, E. 1999. Development of kin bias among rhesus monkeys: maternal transmission or individual learning? *Animal Behaviour*, 58, 883–894.

Bernardo, J. 1996. Maternal effects in animal ecology. *American Zoologist*, 36, 83–105.

Bradford, G. E. 1972. The role of maternal effects in animal breeding. VII. Maternal effects in sheep. *Journal of Animal Science*, 35, 1324–1334.

Bjorklund, D. F. 2006. Mother knows best: epigenetic inheritance, maternal effects, and the evolution of human intelligence. *Developmental Review*, 26, 213–242.

Bjorklund, D. F. & Pellegrini, A. 2002. *The Origins of Human Nature: Evolutionary Developmental Psychology*. Washington, DC: APA Press.

Bowen, W. D., Ellis, S. L., Iverson, S. J. & Boness, D. J. 2001. Maternal effects on offspring growth rate and weaning mass in harbour seals. *Canadian Journal of Zoology*, 79, 1088–1101.

Boonstra, R. & Hochachka, W. M. 1997. Maternal effects and additive genetic inheritance in the collared lemming, *Dicrostonyx groenlandicus*. *Evolutionary Ecology*, 11, 169–182.

Cameron, N. M., Champagne, F. A., Parent, C., Fish, E. W., Ozaki-Kuroda, K. & Meaney, M. J. 2005. The programming of individual differences in defensive responses and reproductive strategies in the rat through variations in maternal care. *Neuroscience and Biobehavioral Reviews*, 29, 843–865.

Champagne, F. A., Francis, D., Mar, A. & Meaney, M. J. 2003. Variations in maternal care in the rat as a mediating influence for the effects of environment on development. *Physiology & Behavior*, 79, 359–371.

Chapais, B. & Gauthier, C. 1993. Early agonistic experience and the onset of matrilineal rank acquisition in Japanese macaques. In: *Juvenile Primates: Life History, Development, and Behavior* (Ed. by M. E. Pereira & L. A. Fairbanks), pp. 245–258. Oxford: Oxford University Press.

Cheverud, J. M. 1984. Evolution by kin selection: a quantitative genetic model illustrated by maternal performance in mice. *Evolution*, 38, 766–777.

Cheverud, J. M. & Moore, A. J. 1994. Quantitative genetics and the role of environment provided by relatives in behavioral evolution. In: *Quantitative Genetic Studies of Behavioral Evolution* (Ed. by C. R. B. Boake), pp. 67–100. Chicago: University of Chicago Press.

Côté, S. D. & Festa-Bianchet, M. 2001. Birthdate, mass and survival in mountain goat kids: effects of maternal characteristics and forage quality. *Oecologia*, 127, 230–238.

Davis, J. M. & Stamps, J. A. 2004. The effect of natal experience on habitat preferences. *Trends in Ecology and Evolution*, 19, 411–416.

Ellis, S. L., Bowen, W. D., Boness, D. J. & Iverson, S. J. 2000. Maternal effects on offspring mass and stage of development at birth in the harbor seal, *Phoca vitulina*. *Journal of Mammalogy*, 81, 1143–1156.

Ergon, T., Lambin, X. & Stenseth, N. C. 2001. Life-history traits of voles in a fluctuating population respond to the immediate environment. *Nature*, 411, 1043–1045.

Fairbanks, L. A. 1989. Early experience and cross-generational continuity of mother-infant contact in vervet monkeys. *Developmental Psychobiology*, 22, 669–681.

Festa-Bianchet, M., Jorgenson, J. T. & Réale, D. 2000. Early development, adult mass, and reproductive success in bighorn sheep. *Behavioral Ecology*, 11, 633–639.

Francis, D., Diorio, J., Liu, D. & Meaney, M. J. 1999. Nongenomic transmission across generations of maternal behavior and stress responses in the rat. *Science*, 286, 1155–1158.

Galef, B. G. 2002. Social influences on food choices of Norway rats and mate choices of Japanese quail. *Appetite*, 39, 179–180.

Gendreau, Y., Côté, S. D. & Festa-Bianchet, M. 2005. Maternal effects on post-weaning physical and social development in juvenile mountain goats (*Oreamnos americanus*). *Behavioral Ecology and Sociobiology*, 58, 237–246.

Holekamp, K. E. & L. Smale, L. 1991. Dominance acquisition during mammalian social development: the "inheritance" of maternal rank. *American Zoologist*, 31, 306–317.

Holmes, W. G. & Mateo, J. M. 1998. How mothers influence the development of litter-mate preferences in Belding's ground squirrels. *Animal Behaviour*, 55, 1555–1570.

Inchausti, P. & Ginzburg, L. R. 1998. Small mammal cycles in northern Europe: patterns and evidence for a maternal effect hypothesis. *Journal of Animal Ecology*, 67, 180–194.

Kirkpatrick, M. & Lande, R. 1989. The evolution of maternal characters. *Evolution*, 43, 485–503.

Koch, R. M. 1972. The role of maternal effects in animal breeding. VI. Maternal effects in beef cattle. *Journal of Animal Science*, 35, 1316–1323.

Maestripieri, D. 2005a. Early experience affects the intergenerational transmission of infant abuse in rhesus monkeys. *Proceedings of the National Academy of Sciences USA*, 102, 9726–9729.

Maestripieri, D. 2005b. Effects of early experience on female behavioural and reproductive development in rhesus macaques. *Proceedings of the Royal Society of London, Series B*, 272, 1243–1248.

Maestripieri, D., Lindell, S. G. & Higley, J. D. 2007. Intergenerational transmission of maternal behavior in rhesus monkeys and its underlying mechanisms. *Developmental Psychobiology*, 49, 165–171.

Mateo, J. M. & Holmes, W. G. 1997. Development of alarm-call responses of Belding's ground squirrels: the role of dams. *Animal Behaviour*, 54, 509–524.

Mateo, J. M. & Holmes, W. G. 1999. How rearing history affects alarm-call responses of Belding's ground squirrels (*Spermophilus beldingi*, Sciuridae). *Ethology*, 105, 207–222.

McAdam, A. G. & Boutin, S. 2003a. Maternal effects and the response to selection in red squirrels. *Proceedings of the Royal Society of London, Series B*, 271, 75–79.

McAdam, A. G. & Boutin, S. 2003b. Effects of food abundance on genetic and maternal variation in the growth rate of juvenile red squirrels. *Journal of Evolutionary Biology*, 16, 1249–1256.

McAdam, A. G., Boutin, S., Réale, D. & Berteaux, D. 2002. Maternal effects and the potential for evolution in a natural population of animals. *Evolution*, 56, 846–851.

McLaren, A. 1981. Analysis of maternal effects on development in mammals. *Journal of Reproduction and Fertility*, 62, 591–596.

Meaney, M. J. 2001. Maternal care, gene expression, and the transmission of individual differences in stress reactivity across generations. *Annual Review of Neuroscience*, 24, 1161–1192.

Mousseau, T. A. & Fox, C. W. 1998a. The adaptive significance of maternal effects. *Trends in Ecology and Evolution*, 13, 403–407.

Mousseau, T. A. & Fox, C. W. (Eds.). 1998b. *Maternal Effects as Adaptations*. Oxford: Oxford University Press.

Nespolo, R. F., Bacigalupe, L. D. & Bozinovic, F. 2003. Heritability of energetics in a wild mammal, the leaf-eared mouse (*Phyllotis darwini*). *Evolution*, 57, 1679–1688.

Oksanen, T. A., Jokinen, I., Koskela, E., Mappes, T. & Vilpas, H. 2003. Manipulation of offspring number and size: benefits of large body size at birth depend upon the rearing environment. *Journal of Animal Ecology*, 72, 321–330.

Qvarnström, A. & Price, T. D. 2001. Maternal effects, paternal effects, and sexual selection. *Trends in Ecology and Evolution*, 16, 95–100.

Reinhold, K. 2002. Maternal effects and the evolution of behavioral and morphological characters: a literature review indicates the importance of extended maternal care. *Journal of Heredity*, 93, 400–405.

Schmidt, K. T., Stien, A., Albon, S. D. & Guinness, F. E. 2001. Antler length of yearling red deer is determined by population density, weather, and early life-history. *Oecologia*, 127, 191–197.

Silk, J. B. 1983. Local resource competition and facultative adjustment of sex ratios in relation to competitive abilities. *American Naturalist*, 121, 56–66.

Wade, M. J. 1998. The evolutionary genetics of maternal effects. In: *Maternal Effects as Adaptations* (Ed. by T. A. Mousseau & C. W. Fox), pp. 5–21. Oxford: Oxford University Press.

Wilson, A. J., Pilkington, J. G., Pemberton, J. M., Coltman, D. W., Overall, A. D. J., Byrne, K. A. & Kruuk, L. E. B. 2005. Selection on mothers and offspring: whose phenotype is it and does it matter? *Evolution*, 59, 451–463.

Wolf, J. B. & Wade, M. J. 2001. On the assignment of fitness to parents and offspring: whose fitness is it and when does it matter? *Journal of Evolutionary Biology*, 14, 347–356.

Wolf, J. B., Moore, A. J. & Brodie, E. D., III. 1997. The evolution of indicator traits for parental quality: the role of maternal and paternal effects. *American Naturalist*, 150, 639–649.

Wolf, J. B., Brodie, E. D., Cheverud, J. M., Moore, A. J. & Wade, M. J. 1998. Evolutionary consequences of indirect genetic effects. *Trends in Ecology and Evolution*, 13, 64–69.

Wolf, J. B., Vaughn, T. T., Pletscher, L. S. & Cheverud, J. M. 2002. Contribution of maternal effect QTL to genetic architecture of early growth in mice. *Heredity*, 89, 300–310

Ylonen, H., Horne, T. J. & Luukkonen, M. 2004. Effect of birth and weaning mass on growth, survival and reproduction in the bank vole. *Evolutionary Ecology Research*, 6, 433–442.

Yoerg, S. I. & Shier, D. M. 1997. Maternal presence and rearing condition affect responses to a live predator in kangaroo rats (*Dipodomys heermanni arenae*) *Journal of Comparative Psychology*, 111, 362–369.

2

The Genetics and Evolutionary Consequences
of Maternal Effects

JAMES M. CHEVERUD AND JASON B. WOLF

INTRODUCTION

Maternal effects are most generally defined as the influence of the maternal phenotype on the phenotype of her offspring. In most cases, they can be thought of as the influence that the environment provided by the mother has on the development of her offspring (using the term "environment" in a very general sense). In considering maternal effects, it is important to distinguish the effects of maternally generated environments on the development of her offspring from the direct effects of the offspring genes and environment on offspring development. Maternal effects may arise from either environmental or genetic influences (or their interaction) on maternal traits that, in turn, ultimately influence the expression of traits in their offspring. The former are referred to as maternal environmental effects to indicate that they arise as a result of the environment experienced by the mother and cannot be attributed to the maternal genotype. The latter are referred to as maternal genetic effects because they ultimately map back to the maternal genome. These maternal genetic and maternal environmental effects are contrasted with "direct effects," where the environment experienced by an individual or the individual's genotype directly affects its own phenotype. These are referred to as direct environmental effects and direct genetic effects, respectively. Although we refer to maternal effects throughout, it is important to keep in mind that there are opportunities in some mammals for fathers to affect the expression of traits in their offspring, and consequently, all of our discussions of maternal effects apply equally to such paternal effects.

Maternal effects have been defined in a number of ways, some more restricted (e.g., Ginzburg 1998) than the definition presented above and some encompassing additional phenomena. Our very broad definition of maternal effects encompasses effects that do not fit into traditional definitions. For instance, maternal effects have often been defined as the effect of the maternal phenotype on the offspring phenotype independent of the offspring genotype (e.g., Mousseau & Fox 1998). However, such independence is not necessary, and this definition leaves out the potentially common and important cases of maternal-offspring interaction, in which the effect of the maternal phenotype differs depending on the offspring genotype. These effects can be conceptualized as a genotype-by-environment interaction, where the effect of the maternally provided environment on offspring traits differs depending on offspring genotype and vice versa (see Rossiter 1998; Wolf 2000a, b). For example, embryo transfer experiments in mammals have found that the effect of the maternal uterine environment on offspring growth and development depends on offspring genotype (e.g., Cowley et al. 1989) and is, therefore, not independent of offspring genotype.

There have also been discussions of maternal effects that have included all situations in which there is an increased resemblance of relatives through the maternal lineage compared to the paternal lineage (e.g., Kirkpatrick & Lande 1989). Such a broad definition confounds maternal effects with the more general phenomenon of maternal inheritance (as in Kirkpatrick & Lande 1989), the latter of which includes cytoplasmic inheritance (see below). This definition of maternal effects is invoked, for example, when maternal-offspring resemblance is stronger than paternal-offspring resemblance and the difference is attributed to maternal effects (with the opposite pattern suggesting paternal effects). Similarly, this broad definition is implied in studies that compare the resemblance of sibs that share a mother with sibs that share a father and attribute the increased resemblance of the former as evidence of maternal effects. However, in both cases many different kinds of effects may contribute to the difference in resemblance between offspring and parents and these are confounded with maternal effects sensu stricto. For example, the difference between paternal and maternal lineages confounds maternal effects, paternal effects, cytoplasmic inheritance, sex chromosome effects, and genomic imprinting (i.e., parent-of-origin-dependent expression of alleles). It is important to differentiate these varied effects because they all have very different effects on the genotype-phenotype relationship and all have different evolutionary consequences.

The inheritance of mitochondria, for example, should not be considered a maternal effect since the effect of the mitochondrial genome on offspring traits is a direct consequence of the expression of their own mitochondrial genotype, not an effect operating through their mother (Lacey 1998). That is, the effect of the mitochondrial genes on an individual's phenotype is a direct genetic effect, comparable to the direct effects of its nuclear genes. One would not consider the effect of a nuclear allele inherited from the mother to be a maternal effect. Similarly, the effects of sex chromosomes, despite being inherited from a particular parent, are a direct consequence of the expression of genes within an individual and thus, there is no effect acting through the parent from whom that chromosome was received.

In contrast to the direct effects of mitochondria and sex chromosomes, which are very clearly not maternal or paternal effects, one may be tempted to consider genomic imprinting a maternal or paternal effect since parents affect the expression of genes in their progeny (e.g., Lacey 1998). As a result, one may consider the maternal genotype to affect the trait of gene expression in the offspring, implying that imprinting is analogous to a maternal effect. However, this is very different from our definition of maternal effects. This is clearest if one considers the ordered genotypes of the offspring, where we designate the genotypes of offspring according to the parent of origin of alleles. For example, if we have a single locus with two alleles (A and a) we would have four ordered genotypes, AA, Aa, aA, and aa (with the paternal allele first and the maternal allele second). If this locus is imprinted we should be able to account for all genetic variation associated with this locus by examining the association of the ordered genotypes with the individual phenotype (see Spencer 2002). That is, if we know the parent of origin of alleles at a locus, we can assign all variation associated with an imprinted locus to the individual genotype, leaving no variation to be accounted for by factors acting through the mother or father.

IMPORTANCE OF MATERNAL EFFECTS IN MAMMALS

Maternal effects are particularly important for mammals where the maternal-offspring interaction is prolonged and intimate. Indeed, the very name of the class Mammalia refers to a specialized organ by which the mother feeds her offspring after birth. In mammals, the mother provides an all-encompassing environment early in life, from conception to birth. While very early life is supported by diffusion of nutrients from the uterine lining and egg constituents, the mammalian embryo soon implants in the wall of the uterus and induces the formation of a specialized organ formed from both maternal and embryonic tissues, the placenta. From the time of implantation to birth, the fetus will draw nutrients, oxygen, and hormones and dispel wastes through the placenta, rather than processing them through its own immature organs.

The mammalian placenta is quite variable in structure in ways that affect transfer of materials between mother and offspring. For example, prosimian primates have appositional, noninvasive placentas with no direct contact between maternal blood flow and fetal tissues. These are referred to as epitheliochorial placentas. Anthropoids have highly invasive, hemochorial placentas in which the fetal tissues erode the uterine wall so that they come into direct contact with maternal blood (Le Gros Clark 1959; Rinkenberger & Werb 2000). It also appears that the hemochorial placenta with its greater intimacy of connection between maternal and fetal tissues has evolved repeatedly in mammals, being found in flying lemurs, anthropoid primates, tenrecs, rodents, insectivores, hyenas, bats, hyraxes, armadillos, anteaters, and elephant shrews (Carter & Enders 2004). Even after birth most mammals are nearly entirely dependent on their mothers for nutrition and care until weaning. This includes providing milk for the growing offspring, protection from predators or other interlopers, and a warm nest to promote growth.

The period prior to weaning provides some of the greatest opportunity for survival selection (Crow & Kimura 1970) on the offspring during life, as loss before weaning can approach 50% in some mammals. While some life history and behavioral ecology models consider offspring survival as a feature of the mother, doing so conflates and confuses the fitness of two different individuals, the mother and her offspring (see Wolf & Wade 2001). If maternal effects were the only factor affecting offspring survival, this would be a sensible approach. However, experimental studies show that offspring characteristics also depend on both the direct effects of offspring genes and nonmaternal environments. In maternal effects models, maternal reproductive success is complete with her production of offspring. Offspring survival affects offspring fitness, not maternal fitness. However, the model allows the mother to affect her offspring's survival through the environment she provides. Indeed, it is the environment that the mother provides for her offspring that is the single most important factor in causing differences among offspring phenotypes and fitness among newborns and weanlings. Thus, the joint evolution of maternal effects and offspring phenotypes plays a critical role in mammalian evolution.

GENETIC AND EVOLUTIONARY MODELS OF MATERNAL EFFECTS

The evolutionary consequences of maternal effects arise largely as a result of the influence that they have on the relationship between the genotype and the phenotype (Wolf et al. 1998). With maternal effects there are two genotype-phenotype (G-P) relationships to consider. First is the relationship between maternal genotype and maternal phenotype—i.e., the relationship between alleles possessed by mothers and their own phenotype caused by the direct effect that the genes have on the expression of maternal traits. If these maternal traits affect the expression of features in their offspring, then variations at these loci are the causal origin of maternal genetic effects. Second is the relationship between the offspring genotype at these maternal-effect loci and the offspring phenotype. The former is generally assumed to be similar to that of other ordinary traits since the maternal genotype directly affects the maternal phenotype (although maternal traits may also themselves be affected by maternal effects; Kirkpatrick & Lande 1989). However, the latter is of particular interest because the relationship between offspring phenotype and offspring genotype at maternal-effect loci is not a consequence of direct gene effects but rather is a consequence of maternal effects. That is, because there is a causal link between the genotype of an individual's mother and the individual's phenotype (due to maternal effects) and because the individual is related to its mother (and therefore shares alleles with her), the individual's own phenotype may reflect the alleles it inherits at maternal-effect loci. This G-P relationship between maternal-effect loci and offspring traits is very different from that associated with "ordinary" direct effects, where there is a direct mapping from individual genotype to individual phenotype. Because of this, most evolutionary models of maternal effects (e.g., Cheverud 1984; Kirkpatrick & Lande 1989) focus on the genetics of offspring phenotypes since the G-P relationship generated by maternal effects can have very different genetic and evolutionary consequences than do ordinary direct effects. Furthermore, this relationship is of particular interest because early mortality often contributes the largest component of variance in offspring fitness, and, as a result, selection acting on offspring phenotypes may be the primary force driving evolution of maternal characters and their associated maternal effects.

Quantitative Genetic Maternal Effects Models

The effects of the environment provided by the mother on the character-istics of her offspring have long been of interest in animal sciences (Dick-erson 1947; Willham 1963; 1972). Researchers were especially interested in selection for early growth and thus were concerned with the environment the mother provided for her growing offspring. What makes this situation unique is that there will be genetic variations between mothers in the quality of the environments they provide for their growing offspring and a poten-tial genetic correlation between the effect of maternal environment and the direct effects of genes on offspring. Researchers found that the response to selection on "offspring" traits could not be predicted by traditional quantita-tive genetic models because of the influence of these maternal effects. Tradi-tional models fail because selection on offspring traits indirectly selects on the quality of the environment provided by mothers.

The models developed in animal sciences to study this situation consoli-dated all aspects of the maternally provided environment into a single factor, maternal performance, and measured this factor as the overall influence of the maternally provided environment on the expression of offspring pheno-types (Legates 1972). Mechanisms by which maternal performance affects offspring traits include uterine factors and postnatal factors such as nest building, supportive care, and milk production. These kinds of maternal traits coalesce to determine overall maternal performance, but performance itself is measured in terms of the offspring phenotype (e.g., grams of growth in offspring body size). As a result, to measure a mother's maternal perfor-mance one need not measure any maternal traits; rather, one can simply estimate maternal performance by measuring the net effect that a mother has on the expression of traits in her offspring.

In the agricultural model the offspring's phenotype (P_o) is composed of several components (see Figure 2-1)

$$P_o = A_o + E_o + P'_m.$$ [1]

where A_o indicates an additive genetic value for the offspring's phenotype, E_o is the environmental deviation (that includes nonadditive genetic effects as well as environmental effects) and P'_m is "maternal performance." The prime (') is used to indicate that the phenotypic value P'_m is a trait possessed by a different individual (the mother) than the individual being considered. Eq. [1] illustrates that, although maternal performance is a characteristic of the mother, it is measured as the effect of the mother on the expression of traits in her offspring. Therefore, the maternal performance phenotype may

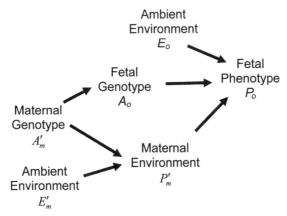

Figure 2-1. Model for maternal effects on offspring phenotypes (after Dickerson 1947). The offspring phenotype (P_o) is impacted by the direct effects of genes expressed by the offspring on its own phenotype (A_o), the ambient environment experienced by the offspring (E_o), and the environment provided by the mother (P'_m). Variations in the environment provided by the mother are due to both genetic factors (A'_m) and ambient environmental factors (E'_m).

be composed of a number of maternal traits each weighted by its effect on the expression of the offspring phenotype. Because maternal performance is composed of traits of the mother, it may have a genetic basis and so we can write a linear equation for components that affect maternal performance:

$$P_m = A_m + E_m,$$ [2]

assuming that maternal performance itself is not affected by maternal effects. Taking this definition for maternal performance (eq. [2]) and substituting into eq. [1] we can decompose the offspring phenotype as:

$$P_o = A_o + E_o + A'_m + E'_m.$$ [3]

Writing it out like this, we can see that the causal genetic components include both the direct effect of the offspring's genes on its own phenotype (A_o) and the indirect effect of the mother's genes for maternal performance on the offspring phenotype (A'_m).

In quantitative genetic theory the response to selection (R_P) is given by

$$R_P = \mathrm{cov}_{AP_o}\beta_o,$$ [4]

where cov_{AP_o} is the covariance of the individuals' additive genetic values (A) with their phenotypic values (P_o) and β_o is the selection gradient on the phenotype (Lande & Arnold 1983). Ordinarily, cov_{AP} is equal to the additive genetic variance (V_A) of the trait (Falconer & Mackay 1997). This result produces the standard breeders' equation of $R_{P_o} = V_{A_o} b_o$, which is more commonly written as $R_{P_o} = h^2 s$, where h^2 is the narrow sense heritability (defined as V_{A_o} / V_{P_o}) and s is the selection differential (NB $b_o = s / V_{P_o}$). However, the standard result is not so when there are maternal effects. This is because the causal genetic component of the phenotype is $A_o + A'_m$ (where A_o is measured in one individual but A'_m is measured in its mother) while the additive genetic value of the selected individual with phenotype P_o is $A_o + A_m$ (where A_o and A_m are now both properties of single individuals, hence no prime on A_m in this equation). Therefore,

$$\text{cov}_{AP_o} = \text{cov}(A_o + A_m, A_o + A'_m)$$
$$= \text{cov}(A_o, A_o) + \text{cov}(A_o, A'_m) + \text{cov}(A_m, A_o) + \text{cov}(A_m, A'_m). \qquad [5]$$

Given that mother's and offspring have a genetic correlation of ($\frac{1}{2}$) due to Mendelian inheritance, we can rewrite eq. [5]:

$$\text{cov}_{AP_o} = V_{A_o} + \frac{3}{2} \text{cov}_{A_{om}} + \frac{1}{2} V_{A_m}, \qquad [6]$$

where V_{A_o} is the direct additive genetic variance for the trait, V_{A_m} is the additive genetic variance for maternal performance, and $\text{cov}_{A_{om}}$ is the additive genetic covariance between the direct effects on P_o and maternal genetic effects on P_o. Then the response of the offspring trait to selection (R_{P_o}) acting directly on the offspring trait (b_o) is defined as:

$$R_{P_o} = \text{cov}_{AP_o} \beta_o$$
$$= (V_{A_o} + \frac{3}{2} \text{cov}_{A_{om}} + \frac{1}{2} W_{A_m}) \beta_o. \qquad [7]$$

The response to selection contains two new terms that do not ordinarily appear in the standard breeders' equation; $\frac{3}{2}\text{cov}_{A_{om}}$, the genetic covariance between direct and maternal effects, and $\frac{1}{2}V_{A_m}$, the heritable variance in maternal performance. The contribution of these components to the evolution of the offspring trait has been described as a consequence of the evolution of the maternally provided environment (Wolf et al. 1998), where the offspring trait evolves because the average maternal performance changes, not because of changes in the direct genetic contribution to the trait. Indeed, with maternal effects, there can be a response to selection even when there are no direct genetic effects on a trait at all (which means that V_{A_o} and $\text{cov}_{A_{om}}$ are both 0) due solely to the evolution of maternal performance. This evolutionary model also has a peculiar characteristic different from standard evolutionary models (Hanrahan 1976) in that the response to selection may be opposite the direction of selection if the genetic covariance between direct and maternal effects is strongly negative. In these circumstances it is possible to have a negative response to positive selection.

Cheverud (1984) showed how this overall response to selection could be divided between the maternal and direct effects and used the model to develop a quantitative genetic version of Hamilton's rule, the conditions under which "altruism" (increased offspring phenotype with direct selection for the offspring trait but direct selection on the mother against maternal performance) will increase in a population. The component of kin selection can be seen in the previous equation, where the term $\frac{1}{2}V_{Am}$ indicates that selection on the offspring trait results in indirect (kin) selection on maternal performance due to the Mendelian correlation between maternal and offspring genotypes (i.e., this $\frac{1}{2}$ is the relatedness term from Hamilton's rule).

Kirkpatrick and Lande (1989; 1992) extended this agricultural model, which lumps all maternal traits into a single maternal performance index, to account for the effects of individual maternal traits contributing to maternal performance, such as milk yield, nest temperature, and hormonal status. While they provide a general multivariate solution to the problem, we will illustrate it by considering only two traits, a single maternal trait affecting a single offspring phenotype. The "offspring trait" is defined as in the performance trait model, except we now keep track of a particular trait expressed in mothers and its effect on the expression of the offspring trait (cf. eq. 4 in Kirkpatrick & Lande 1989):

$$P_o(t + 1) = A_o(t + 1) + E_o(t + 1) + mP'_m(t), \qquad [8]$$

where traits measured in the offspring generation are designated with a $(t + 1)$ while the traits measured in the maternal generation are marked with a (t) to indicate that these values are measured for individuals in two different generations. The new parameter "m" is a measure of the strength of the effect of the mother's phenotype on the phenotype of her offspring (partial regression of offspring phenotype on maternal phenotype holding direct genetic effects constant). The earlier maternal performance models combined the composite term mP'_m to produce a single trait of maternal performance, while the Kirkpatrick and Lande model explicitly decomposes this term into maternal traits (P'_m) and their maternal effects (m). For simplicity, we can again assume that the maternal trait is not affected itself by maternal effects (see Kirkpatrick & Lande 1989, for the example where a trait maternally affects itself):

$$P_m(t) = A_m(t) + E_m(t). \qquad [9]$$

In this situation there is the potential for selection acting directly on both maternal (β_m) and offspring phenotypes (β_o). Under this model, the response of the offspring phenotype to selection in generation "t," $R_{P_o}(t)$, is

$$R_{P_o}(t) = [V_{A_o} + m\,\text{cov}_{A_{mo}}/2]\beta_o(t) + \text{cov}_{A_{mo}}\beta_m(t) + mR_{P_m}(t-1)$$
$$+ mV_m\Delta\beta_m(t-1) + m\,\text{cov}_{P_{mo}}\Delta\beta_o(t-1) \qquad [10]$$

which demonstrates the complexity of the evolutionary response to selection when maternal effects are present, even in a simple two-trait system. The main difference between this equation and the previous response to selection equation (eq. [7]) is that, when multiple generations of selection are considered, additional terms enter the equation representing selection in the previous generation (generation $t-1$). The first of these additional terms keeps track of the response to selection of the maternal trait in the previous generation $R_{P_m}(t-1)$, which has carryover effects onto the next generation and may be considered a form of evolutionary momentum. The last two terms account for changes in the intensity of selection between generations and may be thought of as evolutionary time lags (Kirkpatrick & Lande 1989).

Population Genetic Maternal Effects Models

The effects of genetic architecture for maternal effects can be more closely considered with population genetic models based on the effects of individual loci. We illustrate the interesting and unusual genetic architecture of maternal effects using two simple models (see also Wade 1998 for a population genetic treatment of maternal effects models), a single-locus model and a two-locus model. In the single-locus model, we assume that a locus in the mother affects the development of her offspring and that selection acts on the offspring phenotype. In the two-locus model, we assume that the effect of this "maternal" locus differs depending on the offspring genotype. This interaction is referred to as maternal-offspring epistasis and it plays a role in a number of evolutionary processes (Wolf & Brodie 1998; Wolf 2000a). We contrast these two models with traditional single-locus and two-locus models to demonstrate the differences between the genetic architecture and evolution of maternal effects relative to the expectations for "ordinary" direct effects. Using these models we focus on two complementary issues: first we examine the genotype-phenotype relationship for maternal-effect loci and for maternal-offspring epistasis, and second we examine the evolutionary consequences of these patterns.

Consider a locus A that has two alleles, A_1 and A_2 with frequencies p and q. For simplicity, we focus on the simple additive model for the maternal trait and its maternal effects and present a model that includes dominance elsewhere (Wolf et al., unpublished manuscript). We designate the expected phenotypes for each of the three maternal genotypes (i.e., the "genotypic values") as M_1, M_2, and M_3 for the A_1A_1, A_1A_2, and A_2A_2 genotypes respectively. These genotypic values correspond to those defined for the maternal trait (P_m; eq. [9]) in the quantitative genetic model described above. These three genotypic values for the maternal trait are defined as

$$M_1 = +a_m,$$
$$M_2 = 0,$$
$$M_3 = -a_m, \qquad\qquad [11]$$

where a_m is the additive direct effect of the A locus on the maternal trait. Eq. [11] is analogous to eq. [9] except that we have left out random environmental effects for simplicity.

We assume that the maternal trait has an additive effect on the expression of some offspring trait O. We designate the expected genotypic values of the offspring raised by a mother with the genotypic value M_i as O_i', where the prime indicates, as above, that the trait O is measured in a different generation—i.e., in the mother's offspring (note, however, that we have now reversed the use of the prime in that we now use the prime to indicate the phenotype of a mother's offspring as opposed to designating the phenotype of an offspring's mother). That is, a mother with phenotype M_i for the maternal trait has offspring with the phenotype O_i' due to the maternal effect of her trait on the expression of their phenotype. Thus, it is important to keep in mind that the values of O_i' are not the expected genotypic value of individuals with the three genotypes, but rather are the genotypic values of offspring reared by mothers with each of the three genotypes. The O_i' values represent the indirect effects of the mother's genes on her offspring. As in the quantitative genetic model presented above, the effect of the maternal trait on the expression of her offspring's trait is designated m (cf. Kirkpatrick & Lande 1989), such that the three types of mothers have offspring with the expected genotypic values for trait O defined as $O' = mM_i$, which makes the three genotypic values for trait O equal to $+ma_m$, 0, and $-ma_m$ respectively. This definition of the offspring trait is analogous to that given in eq. [8] for the quantitative genetic model, except we have again left out random environmental effects. The parameter m is a measure of the strength of the maternal effect that trait M expressed in mothers has on the expression of trait O in her offspring and is analogous to the parameter m in the Kirkpatrick and Lande (1989) quantitative genetic model of maternal effects.

If we assume that mothers mate at random in a large population, we can calculate the expected phenotype of offspring that have each of the three genotypes. That is, we can ask what the expected genotypic value for trait O of an individual with a particular A locus genotype would be. These values differ from the genotypic values as defined above (O_i') because this time the value is defined given the offspring genotype rather than for the genotype of its mother, representing the apparent effect of the offspring genotype on its own phenotype. The expected phenotype for trait O (O_i) of the A_1A_1, A_1A_2, and A_2A_2 genotypes in the offspring are $+ma_mp$, $\frac{1}{2}ma_m(p - q)$, and $-ma_mq$ respectively. These expected values point out two important features of loci having maternal effects.

First, it is clear that the maternal effect of the locus also results in a statistical relationship between offspring A locus genotype and offspring genotypic value O_i, although the offspring genotype plays no causal role in its own phenotype. The offspring's A locus maps to their phenotype because the offsprings' and mothers' genotypes are correlated through Mendelian inheritance. Thus, even though the A locus only affects the maternal trait (M), it maps from the offspring A locus genotype to the offspring phenotype (O) through the maternal effect. This indirect mapping from genotype to phenotype means that a gene that is never expressed in offspring nevertheless appears to have an effect on their phenotype. For example, a gene that affects only maternal milk production will appear to map from offspring genotype to offspring growth when milk production affects offspring growth rate, despite the fact that females will not express the active gene until they become mothers themselves and males will never express the gene. This means, for example, that for typical quantitative trait locus (QTL) mapping strategies, the positional candidate genes in the QTL region may not be expressed in the offspring during growth at all, but rather in their mothers. Such a result has an important implication for studies of gene expression, such as microarrays, since there can be genotype-phenotype associations caused by maternal effects that have no causal pattern based on gene expression in the offspring. Since it is gene expression in mothers that affects trait expression in their offspring, one would not find any expression of maternal-effect loci in the offspring even though their genotypes correlate with their phenotypes.

Second, note that the expected phenotype of a given offspring genotype is *frequency dependent* (e.g., $O_1 = +ma_m p$). In models of direct effects the expected phenotype of a particular genotype (i.e., the genotypic value) is defined as a property of the genotype itself and does not change with population allele frequencies (Falconer & Mackay 1997). This is still true for the maternal genotypes and both their direct effects on maternal traits (e.g., $M_1 = +a_m$) and their indirect maternal effects on offspring traits (e.g., $O' = +ma_m$). However, maternal effects result in a frequency-dependent genotype-phenotype relationship in the offspring. This frequency dependence occurs because allele frequencies determine the frequency with which each offspring genotype is produced by each maternal genotype. For example, when the A_1 allele is rare most individuals with the A_1A_1 genotype would have been produced by A_1A_2 mothers. The genotypic value of offspring from heterozygous mothers is 0, and we see that when A_1 is rare, making p very small, the expected phenotype of the A_1A_1 genotype ($+mamp$) approaches 0 as p approaches 0. In contrast, when A_1 is very common, the A_1A_1 individuals would mostly be produced by A_1A_1 mothers and their expected phenotype would approach $+ma_m$ as p approaches 1 (i.e., as A_1 goes to fixation).

The evolutionary consequences of maternal effects can also be illustrated using this one-locus additive model. Under this model, the mean value of the maternal trait ($\overline{M}$) is $a_m(p - q)$ and the mean of the offspring trait ($\overline{O}$) is $ma_m(p - q)$. To understand how selection on the maternal and offspring traits will result in a change in allele frequencies and subsequently in the means of these traits, we can examine the additive genetic variances of the maternal and offspring traits. The direct additive genetic variance of the maternal trait is calculated as it would be for any ordinary trait with direct effects, and in this case has the value $2pqa_m^2$. Likewise, the additive genetic variance for the maternal effect of locus A on trait O' has the value $2pqm^2a_m^2$ (i.e., this is the additive genetic variance of maternal performance for trait O, and therefore is the additive variance among mothers). The additive genetic variance of the offspring trait O, generated by the maternal effect, has the value $[\frac{1}{2}pqm^2a_m^2]$ (i.e., this is apparent additive genetic variance for trait O among offspring). This shows that the maternal effect of M on trait O results in additive genetic variation for trait O generated *by* the maternal effect.

These results highlight a few important points. First, the additive genetic variance equations show that selection on the offspring trait O will result in much slower evolution of allele frequencies at a locus (A) than would selection on trait M itself ($2pqa_m^2 \gg \frac{1}{2}pqm^2a_m^2$). They also demonstrate that maternal effects can generate real additive genetic variation that can contribute to trait evolution and are, therefore, not simply nuisance sources of variation to be controlled for. As a result, selection on the offspring trait will lead to the evolution of the maternal trait and its associated maternal effect, which in turn will result in an evolutionary change in the offspring trait.

We can now add a second locus, B, which directly affects the expression of the offspring trait (see also Wade 1998). Like the A locus, we assume that the B locus has two alleles, B_1 and B_2 with frequencies x and y. We assume that the B locus has an additive effect on offspring trait O, with the three genotypes having the phenotypes $+a_o$, 0, and $-a_o$, for the B_1B_1, B_1B_2, and B_2B_2 genotypes respectively, where a_o is the additive direct effect of the B locus on the offspring trait. However, we will now complicate things and assume that the maternal and offspring traits (M and O) interact to affect the expression of some other trait (like offspring fitness) that we will designate as Q. We assume a simple form for this interaction (more complex forms of interaction are discussed in Wolf 2000a), where the offspring genotypic value, Q_{ij}, is defined as a pure interaction between traits M and O (where the subscript i denotes the mother's genotype and j denotes the offspring's genotype) such that we can define $Q_{ij} = mM_iO_j$, where m is again a measure of the maternal effect, but in this case it is an interaction effect caused by the maternal and offspring genotypes. One way to view this is that the A locus determines the maternally provided environment and the B locus determines the offspring response or adaptation to that maternally provided environment. Mothers may express one trait and offspring another trait, and the two traits interact to determine another offspring phenotype. For example, the maternal locus may affect a component of milk quality while the offspring locus may affect expression of some digestive enzymes and these two may interact to affect offspring growth rate. This interaction of maternal and offspring genomes is called maternal-offspring epistasis (see Wolf 2000a, b).

To understand the genetic and evolutionary consequences of the interaction between the maternal and offspring loci we will focus on its relationship to physiological epistasis and its relationship to the results of the single-locus model above. As with the single-locus case, the "epistatic" maternal effect results in a relationship between the expression of the offspring trait O and the maternal effect locus A, but in this case, the effect of A appears as a form of epistasis. This is illustrated in Figure 2-2, which shows the pattern of interaction between maternal A locus and offspring B locus genotypes (Figure 2-2a) and the within-genome pattern of "apparent" epistasis between

A

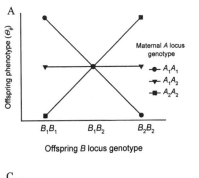

B

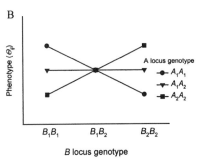

C

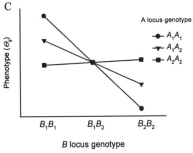

Figure 2-2. Maternal (locus A)–offspring (locus B) epistatic interactions. (a) The pattern of interaction between maternal A locus and offspring B locus genotypes. (b) The within-genome pattern of "apparent" epistasis between A and B locus genotypes at intermediate allele frequencies. (c) The within-genome pattern of "apparent" epistasis between A and B locus genotypes at extreme allele frequencies.

A and B locus genotypes in their offspring (Figure 2-2b). Figure 2-2 clearly demonstrates that the interaction between the maternal and offspring traits generates epistasis between the loci. However, note that Figure 2-2b was drawn for the allele frequencies of $p = q = x = y = 0.5$ (i.e., all alleles at equal frequency), and the actual observed pattern of apparent within-genome epistasis between the A and B loci will depend on allele frequencies and linkage disequilibrium between loci. Figure 2-2c shows the expected pattern of apparent within genome epistasis when the A_1 and B_1 alleles are common. Note that the patterns of phenotypes associated with B locus genotypes appears closer to the additive case, where the B_1 allele has an additive effect on the offspring trait, whereas when all alleles are at equal frequency (Figure 2-2b) neither allele has a net effect across A locus backgrounds. Thus, unlike physiological epistasis, where one can define a phenotype associated with a particular multilocus genotype and consider that phenotype to be a property of that genotype, with maternal-offspring epistasis the expected phenotype of a particular multilocus genotype is frequency dependent. Wolf & Hager (2006) have also shown that this pattern of maternal-offspring epistasis favors the evolution of genomic imprinting.

EMPIRICAL STUDIES OF MATERNAL EFFECTS IN MAMMALS: RESEARCH DESIGNS AND RESULTS

Measurement of maternal effects on mammalian offspring phenotypes is difficult and can require complex breeding designs. A common experimental design used with the mouse is a cross-fostering design (Riska et al. 1986) where mothers are paired with each other and half of each litter is switched between mothers. This results in mothers rearing offspring that are unrelated to them, experimentally breaking the usual link between maternal environment and the effects of the maternal genome passed on to the offspring. While this design has been very successful, it is somewhat limited in that only postnatal maternal environments are measured and accounted for. The uterine prenatal environment undoubtedly also has a very large effect that is confounded with the direct genetic effect in this design. Furthermore, genetic and environmental sources of variation in the environment provided by the mother cannot be distinguished unless sisters are included as mothers in the experiment. Typically, in these postnatal cross-fostering studies the effects of maternal environment are the major contributor to variance in body weight and growth from birth to weaning ($R^2 > 0.40$; compiled in Cheverud 1984).

Quantitative Genetic Analyses

A survey of the agricultural experimental literature based primarily on cross-fostering after birth (Cheverud 1984) showed that maternal effects were responsible for nearly half of the phenotypic variance in size between animals before weaning. After weaning, when offspring are physically separated from their mothers, the maternal effect declines over time, reaching a more modest value of about 10% for many adult traits in rodents (Cheverud et al. 1983a, b; Leamy & Cheverud 1984; Cheverud & Leamy 1985). Studies of multiple traits have indicated that maternal performance for different body measurements are very highly intercorrelated ($r_M \sim 1.00$), indicating a unified effect of maternal environment on various offspring traits (Cheverud et al. 1983a, b).

Heritabilities of maternal performance have been measured in some agricultural studies and are usually about 40% of the phenotypic variance in maternal performance (Cheverud 1984). This moderate heritability is typical of that found for morphological traits and higher than usual for physiology and behavior (Mousseau & Roff 1987). The genetic covariance between maternal performance and the direct effects of the offspring's genes on offspring traits changes over ontogeny, starting at a moderate positive level neonatally, declining to a very low negative level at weaning, and then rising again as the offspring matures (Cheverud 1984; see also Bowen, this volume). The strong negative genetic covariance between maternal and direct effects can sometimes overwhelm the direct and maternal genetic variance (heritabilities), resulting in a negative response to positive selection for weanling weight (Cheverud 1984).

Atchley and colleagues noted that the prenatal maternal environment also should have important effects on offspring characters (Cowley et al. 1989; Ernst et al. 2000). They performed a very interesting and difficult experiment based on a cross-fostering design by performing the cross-fostering before birth, using embryo-transfer technology in mice. They compared three offspring genotypes, inbred C3HeB/FeJ, inbred SWR/J, and the F_1 hybrids between them and cross-fostered them as embryos into three uterine genotypes (again C3HeB/FeJ, SWR/J, and their F_1 hybrid). Since the differences between mothers in this experiment are due to genes, the experiment only considers genetic sources of variation in maternal effects. At birth, all pups were fostered to BALB/cJ × C57BL/6J F_1 hybrid mothers who had given birth on the same day as the experimental mother. This design allows the separation of uterine maternal effects from both direct genetic and postnatal maternal effects. Uterine genotype had significant effects on body weights and tail lengths from birth to 9 weeks and was especially strong for the first postnatal growth period (3–12 days). Regardless of genotype, pups that developed in the strain with larger body size (C3HeB/FeJ) were always bigger than those developed in the smaller strain (SWR/J). These effects were about 10% of the mean body weight from birth to 12 days and remained at 5% of the mean even up to 63 days. They also discovered significant maternal-fetal interactions for body weight, indicating that the response of the offspring genotypes varied depending on which kind of mother they developed in (suggesting the presence of maternal-offspring epistasis). These experiments are among the few to successfully separately quantify the effects of prenatal environment on offspring traits.

Jarvis et al. (2005) performed another style of experiment designed to measure pre- and postnatal maternal effects jointly through reciprocal crosses between inbred mouse lines. This design results in genetically identical offspring being born from genetically diverse mothers and allows genetic maternal effects to be measured using the differences between genetically identical F_1 animals reared by genetically different mothers. This design includes a more complete assessment of the strength of maternal effects, as

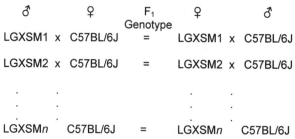

Figure 2-3. Reciprocal breeding design to isolate genetic variation in maternal environments from their genetic contributions to the offspring genotype by mating a set of recombinant inbred strains (LGXSMi) with a standard strain (C57BL/6J). In a reciprocal cross the male from the first strain is mated to the female from the second strain and vice versa. F$_1$ offspring born of these matings are genetically identical to each other so that the only difference between reciprocal litters is the genotype of the mother. In this design, the effect of having a C57BL/6J mother is compared to the average effect of having a Recombinant Inbred (RI) strain mother (Maternity) and genetic variations in maternal effects among RI lines are measured by variations among RI strains relative to a standard strain (RI strain by Maternity interaction).

both pre- and postnatal effects are included, but again is restricted to genetic sources of variation in maternal effects.

In this experiment, the inbred strain C57BL/6J was reciprocally crossed with a set of 10 LGXSM recombinant inbred strains. Differences between reciprocal litters (RI ♀ by C57BL/6J ♂ compared to C57BL/6J ♀ by RI ♂) were used to measure maternal genetic effects on offspring phenotypes (see Figure 2-3). There are two different sources of genetic variation in maternal effects included in this design, the difference between C57BL/6J mothers and the LGXSM RI mothers as a group (referred to as the maternity effect) and the genetic variation in maternal effects among the RI strain mothers (RI strain by maternity interaction effects). Pups born from a C57BL/6J mother were smaller than those born from the LGXSM RI strain mothers and remained smaller throughout life. However, animals born from C57BL/6J mothers grew at a faster rate than those born from LGXSM mothers, although they never caught up with regard to body size. Significant genetic variation in maternal performance was also documented among the different LGXSM RI strains for growth and body size with effects declining with age from 33% of the variance at week 2 to about 10% of the variance at week 20. Interestingly, many adult obesity- and diabetes-related traits also showed genetic variation due to maternal environment, including substantial effects (10%–35% of the phenotypic variance) on adult organ weights (heart, kidney, liver), fat depot weights, responses to glucose challenge, and serum insulin, cholesterol, and leptin levels (Jarvis et al. 2005).

Epidemiological studies in humans (Barker 1998), followed by confirmatory experimental studies in rodents and other mammals (Gluckman & Hanson 2004), have shown that the effect of the pre- and postnatal environment provided by mammalian mothers for their offspring extends throughout life, having an important impact on levels of disease-related traits for several adult diseases, including hypertension, obesity, diabetes, and heart disease. These effects are thought to be mediated through "fetal programming." Individuals who experience a relatively nutrient-poor environment pre- and neonatally set their physiological and metabolic processes at a conservative level, necessary for successful maturation in an energy-poor environment, but potentially pathological if the individual is unexpectedly exposed to an energy-rich environment after birth. It is possible that the persistent maternal effects documented by Jarvis et al. (2005) exert their influence through pre- and neonatal programming.

This programming can occur through many different mechanisms. The structure of many internal organs is set early in life, for example, the number of functional units in the kidney (Manalich et al. 2000). Later growth can increase the size of these units but not produce additional modules, thereby limiting kidney capacity as an adult with consequent effects on hypertension. Another mechanism proposed is genetic imprinting. Early environment can result in imprinting of genes, affecting gene expression throughout an animal's lifetime. Meaney and colleagues have found that increased levels of pup grooming and arched-back nursing by rat mothers early in postnatal life alters the epigenome, DNA methylation pattern, at a glucocorticoid-receptor locus promoter in the hippocampus of their pups (Meaney & Szyf 2005; Szyf et al. 2005). This change in methylation at the promoter persists into adulthood and results in altered histone acylation and transcription factor (NGFI-A) binding to the promoter. The persistent change in methylation alters the pup's physiological response to stress as an adult (for more see Champagne & Curley, this volume).

Although these quantitative genetic studies have done much to clarify the importance of the environment provided by mothers for their offspring, several important issues of the genetic architecture of maternal performance remain. How many genes and of what effect size cause variation in maternal performance? Is there dominance for maternal effects? Are there epistatic interactions between genes affecting maternal performance and even more interestingly, are maternal-fetal interactions, such as those demonstrated by Cowley et al. (1989), common? What are the pleiotropic effects of maternal performance genes on other aspects of the phenotype? The answers to these questions are critically important for an understanding of the evolution of both maternal and offspring characters and have only rarely been considered.

Genetic Architecture of Maternal Effects at Individual Gene Loci

The features of the population genetics of maternal effects described above can be used to derive methods for the empirical analysis of these effects. To examine the relative contribution of direct and maternal-effect loci to genetic architecture we (Wolf et al. 2002) used data from a multigeneration intercross of the Large (LG/J) and Small (SM/J) inbred strains of mice to map direct and maternal effect QTL. We use this study to illustrate the analysis of the molecular quantitative genetics of maternal effects and only briefly describe the details of the experimental design herein (complete details are given in Wolf et al. 2002). To generate our experimental population, we mated SM/J strain male mice to LG/J strain female mice to produce an F1 generation of genetically identical animals each carrying one unrecombined chromosome from SM/J and one from LG/J. These mice were intercrossed to generate an F2 population of approximately 500 mice that each had a genome composed of recombinant chromosomes derived from the parental strains. All animals had genetically identical F1 mothers. As a result, there is no opportunity for maternal genetic effects to contribute to variation among these animals, thereby eliminating the relationship between individual genotype and maternal-effect loci discussed above. This is a convenient feature of the genetics of this population if one wishes to identify direct-effect loci without the potential confounding effects of maternal-effect loci, but also leaves out a potentially very important component of genetic variance.

To examine genetic maternal effects, these F2 animals were randomly mated to produce an F3 population of ca. 1600 animals in ca. 200 full-sib families. Most litters of F3 pups were split and half of each litter was reciprocally cross-fostered to another F2 mother who gave birth on the same day. These F3 mice have a similar genome structure to the F2 animals, albeit with more recombinations, but they are very different since maternal genetic effects can now contribute to genetic differences between animals.

To understand the patterns of variation in the LG/JxSM/J intercross we first did a quantitative genetic analysis of early growth traits using the family structure of the F3 generation (Wolf et al. 2002). We found that growth from week 1 to 2 had a broad-sense heritability (the ratio of genetic variance to phenotypic variance) of 0.12, suggesting that only about 12% of the variation in early growth could be attributed to the effects of the individuals' own genome (although some of this variation may also be attributable to prenatal maternal effects). Maternal effects, estimated as the covariance of unrelated individuals raised by the same mother, accounted for 33% of the variance in pup growth from week 1 to 2.

To examine the mapping of direct and maternal effects, the F2 animals were genotyped at 96 microsatellite loci distributed across the genome (additional markers have been done on the X chromosome but have not shown effects and were not discussed in Wolf et al. 2002). They provided a ca. 20 cM map of the genome. We focused on mapping effects on early growth (weight gain from week 1 to week 2) because it is the earliest recorded growth trait, and unlike week 1 weight itself, it is expected to show less of an influence of prenatal maternal effects. We used an interval mapping approach (Haley & Knott 1992) to scan the genome for QTL. To detect direct effect loci we mapped from the phenotypes of the F2 animals to their own genotypes.

To detect maternal-effect loci we mapped from the genotypes of the F2 mothers to the average growth of the cross-fostered pups they reared. By only mapping from maternal genotype to the growth of cross-fostered pups we eliminated the confounding effects of direct-effect loci. That is, if we were to map from maternal genotype to the phenotype of the mothers' own offspring, any locus showing a strong direct effect on early growth would map to the maternal genotype since mothers and offspring share alleles. These would not be maternal effects at all, but rather simply a consequence of Mendelian inheritance.

We found five QTL having direct effects on early growth and four QTL having maternal effects on early growth, but no loci appeared to have both direct and maternal effects (i.e., we found no evidence of pleiotropy). The direct-effect loci together accounted for only 11.8% of the variance in early growth, a value similar to the broad-sense heritability of the trait. In contrast, the maternal-effect loci account for 31.5% of the among-litter variance in early growth. This analysis demonstrated that maternal-effect loci are an important part of the genetic architecture of early growth, accounting for more than twice the among-individual variance accounted for by direct effects of the genes expressed in the offspring.

Because we did not have comprehensive genotypes for the F3 animals at the time of these analyses, we were unable to look at more complex patterns in the analyses presented in Wolf et al. (2002). However, to examine whether

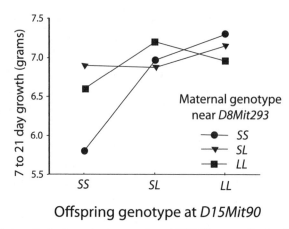

Offspring genotype at *D15Mit90*

Figure 2-4. Mother-offspring interaction between a locus (*D15Mit90*) expressed in the offspring and directly affecting early offspring growth (from day 7 to 21 postnatal) and a locus expressed in the mother (near *D8Mit293*) having a maternal genetic effect on offspring growth. Genotypes are designated with allele "S" from the SM/J parental line and "L" from the LG/J parental line. This interaction is characterized by a strong affect of offspring genotypes at *D15Mit90* on growth when the animals are reared by a nurse with the SS genotype at *D8Mit293* while the effect of offspring genotypes at *D15Mit90* on growth is attenuated when the animals are reared by SL or LL nurses.

the direct effects of loci depended on the genotype of the mother, we later genotyped the F3 animals at loci linked to the direct-effect QTL. Here we will use the case of the largest early growth locus, which is on chromosome 15, to illustrate the sorts of patterns that one might discover in a genetic analysis. This chromosome 15 locus had a primarily additive effect on early growth. We genotyped F3 animals at the microsatellite marker *D15Mit90*, which was located near the estimated most likely location of the QTL. We then tested whether the effect of this locus on early growth (in this case considering growth from week 1 to 3, which is up until weaning) depended on the genotype of the animals' foster mothers (using only cross-fostered pups) by scanning across the maternal genome for loci that interacted with *D15Mit90* in their fostered offspring. This analysis identified a locus on chromosome 8 that showed a significant interaction with *D15Mit90* (the p value generated for the test of this interaction was < 0.0005). The pattern of interaction between the offspring locus (*D15Mit90*) and the maternal locus (*D8Mit293*) is shown in Figure 2-4 (where the offspring locus is measured in the F3 pups and the maternal locus is measured in their foster mothers). The analysis illustrated in Figure 2-4 is directly analogous to a genotype-by-environment interaction analysis in that we are testing whether the effect of the chromosome 15 locus depends on whether the pups experience the environment provided by foster mothers with each of the possible genotypes at the chromosome 8 locus. In future work we will use markers that cover the entire genome of the F3 pups to look in more detail at the patterns and consequences of these types of maternal-offspring epistasis.

In summary, our QTL studies have shown that several genes of relatively small effect can contribute to variation in maternal performance, similar to the results for direct gene effects on early growth (Cheverud & Routman 1995; Cheverud et al. 1996; Wolf et al. 2002). There are epistatic interactions among genes affecting maternal performance and among direct-effects genes. Preliminary analysis also suggests the presence of specific maternal-offspring epistatic interactions. Furthermore, our analyses provided no evidence for pleiotropic effects directly on offspring growth and on maternal performance.

MATERNAL-OFFSPRING RELATIONSHIPS AS A SUBSTRATE FOR EVOLUTION OF SOCIAL INTERACTIONS

The intimate relationship between mothers and their offspring is a central theme in mammalian evolution. However, this close relationship has extended beyond its original role in successfully rearing young to encompass intraspecific social interactions among members of mammalian social groups more generally. Mechanisms that have originally evolved as part of the mother-offspring bond have been coopted as the basis for social interactions with other relatives and, indeed, unrelated individuals. For example, this can be seen clearly in the various roles played by oxytocin in maternal-offspring bonding and in social bonding generally. Oxytocin acts both as a hormone and as a neurotransmitter in the central nervous system. In addition to its role in mother-offspring bonding (Nelson & Panksepp 1998), oxytocin and temporal and spatial variations in the placement of its receptors in the brain have been found to affect the quality of social interactions (Witt et al. 1992), aggression (Winslow et al. 1993), social recognition (Dantzer et al. 1987; Popik & van Ree 1991; Ferguson et al. 2000; 2001), and male-female pair-bonding in prairie voles (Nair & Young 2006).

Other parent-offspring behavioral interactions have been transferred to more general social contexts. Adult tamarins have been known to utilize infant distress vocalizations under severe stress. In some nonhuman primates, individuals who are not the mother help care for the offspring of others un-related to themselves with consequences for the adult's social relationships (West-Eberhard 1975). An extreme version of this is adoption, which has been documented in several social species, when an infant loses its mother and others step in to fill the maternal role.

Cheverud (1984) described how the genetic models developed for mater-nal effects and described here can be generalized to all situations in which animals provide an environment that affects the phenotype of others. Mater-nal effects can therefore serve as a model for understanding social evolution more generally.

CONCLUSIONS

We have defined maternal effects as the effects of the environment provided by the mother on the development of her offspring and described how variation in these maternal environments can be caused by maternal genes and ambient environments. Quantitative and population genetic models have been developed over the years to describe genotype-phenotype rela-tionships and expected evolutionary response in the presence of maternal effects. These models indicate that evolution can be expected to proceed in the opposite direction of selection when the genetic covariance between maternal and direct effects is strongly negative, evolution will develop in-ertia so that it proceeds even after selection stops, and that some of the apparent effects of offspring genes on offspring phenotypes may actually be due to maternal genetic causes when traits are impacted by maternal environment.

Genetic studies of maternal effects on offspring phenotypes indicate that maternal effects are the single most important source of variation among newborns, that the impact of maternal effects declines after weaning when offspring are physically separated from their mothers, but that they can persist throughout life, being important factors for adult disease-related traits, and that the heritability of maternal performance is usually moderate (~40%). Gene mapping studies in mice show that there are many maternal effects loci each of relatively small effect, that these loci epistatically interact with each other, and that they are distinct from direct-effect loci. Finally, both quantitative genetic and gene mapping experiments provide evidence for maternal-offspring epistatic interaction, where the effect of maternal en-vironment on offspring traits varies depending on genotypes present in the offspring.

REFERENCES

Barker, D. J. P. 1998. *Mothers, Babies and Health in Later Life*. London: Churchill Livingstone.

Bowen, W. D. 2009. Maternal effects on offspring size and development in pinnipeds. In: *Maternal Effects in Mammals* (Ed. by D. Maestripieri & J. M. Mateo), pp. 104–132. Chicago: University of Chicago Press.

Carter, A. M. & Enders, A. C. 2004. Comparative aspects of trophoblast development and placentation. *Reproductive Biology and Endocrinology*, 2, 46 (doi:10.1186/1477-7827-2-46).

Champagne, F. A. & Curley, J. P. 2009. The trans-generational influence of maternal care on offspring gene expression and behavior in rodents, pp. 182–202. In: *Maternal Effects in Mammals* (Ed. by D. Maestripieri & J. M. Mateo). Chicago: University of Chicago Press.

Cheverud, J. 1984. Evolution by kin selection: a quantitative genetic model illustrated by maternal performance in mice. *Evolution*, 38, 766–777.

Cheverud, J. & Leamy, L. 1985. Quantitative genetics and the evolution of ontogeny. III. Ontogenetic changes in correlation structure among live-body traits in randombred mice. *Genetical Research*, 46, 325–335.

Cheverud, J. M. & Routman, E. J. 1995. Epistasis and its contribution to genetic variance components. *Genetics*, 139, 1455–1461.

Cheverud, J., Leamy, L., Rutledge, J. & Atchley, W. 1983a. Quantitative genetics and the evolution of ontogeny. I. Ontogenetic changes in quantitative genetic variance components in random-bred mice. *Genetical Research*, 42, 65–75.

Cheverud, J., Rutledge, J. & Atchley, W. 1983b. Quantitative genetics of development: genetic correlations among age-specific trait values and the evolution of ontogeny. *Evolution*, 37, 895–905.

Cheverud, J., Routman, E. J., Duarte, F. M., van Swinderen, B., Cothran, K. & Perel, C. 1996. Quantitative trait loci for murine growth. *Genetics*, 142, 1305–1319.

Cowley, D. E., Pomp, D., Atchley, W. R., Eisen, E. J. & Hawkins-Brown, D. 1989. The impact of maternal uterine genotype on postnatal growth and adult body size in mice. *Genetics*, 122, 193–203.

Crow, J. & Kimura, M. 1970. *An Introduction to Population Genetics Theory*. Minneapolis: Burgess Publishing.

Dantzer, R., Bluthe, R. M., Koob, G. F. & Le Moal, M. 1987. Modulation of social memory in male rats by neurohypophyseal peptides. *Psychopharmacology*, 91, 363–368.

Dickerson, G. 1947. Composition of hog carcasses as influenced by heritable differences in rate and economy of gain. *Iowa Agricultural Experiment Station Research Bulletin*, 354, 489–524.

Ernst, C. A., Rhees, B. K., Miao, C. H. & Atchley, W. R. 2000. Effect of long-term selection for early postnatal growth on survival and prenatal development of transferred mouse embryos. *Journal of Reproduction and Fertility*, 118, 205–210.

Falconer, D. S. & Mackay, T. 1997. *Introduction to Quantitative Genetics*, 4th ed. New York: Longman Press.

Ferguson, J. N., Young, L. J., Hearn, E. F., Matzuk, M. M., Insel, T. R. & Winslow, J. T. 2000. Social amnesia in mice lacking the oxytocin gene. *Nature Genetics*, 25, 284–288.

Ferguson, J. N., Aldag, J. M., Insel, T. R. & Young, L. J. 2001. Oxytocin in the medial amygdala is essential for social recognition in the mouse. *Journal of Neuroscience*, 21, 8278–8285.

Ginzburg, L. R. 1998. Inertial growth: population dynamics based on maternal effects. In: *Maternal Effects as Adaptations* (Ed. by T. A. Mousseau & C. W. Fox), pp. 42–53. Oxford: Oxford University Press.

Gluckman, P. & Hanson, M. 2004. *The Fetal Matrix: Evolution, Development, and Disease*. Cambridge: Cambridge University Press.

Haley, C. S. & Knott, S. A. 1992. A simple regression method for mapping quantitative trait loci in line crosses using flanking markers. *Heredity*, 69, 315–324.

Hanrahan, J. 1976. Maternal effects and selection response with application to sheep data. *Animal Production*, 22, 359–369.

Jarvis, J. P., Kenney-Hunt, J., Ehrich, T. H., Pletscher, L. S., Semenkovich, C. F. & Cheverud, J. M. 2005. Maternal genotype affects adult offspring lipid, obesity, and diabetes phenotypes in LGXSM recombinant inbred strains. *Journal of Lipid Research*, 46, 1692–1702.

Kirkpatrick, M. & Lande, R. 1989. The evolution of maternal characters. *Evolution*, 43, 485–503.

Kirkpatrick, M. & Lande, R. 1992. The evolution of maternal characters: errata. *Evolution*, 46, 284.

Lacey, E. P. 1998. What is an adaptive environmentally induced parental effect? In: *Maternal Effects as Adaptations* (Ed. by T. A. Mousseau & C. W. Fox), pp. 54–66. Oxford: Oxford University Press.

Lande, R. & Arnold, S. 1983. The measurement of selection on correlated characters. *Evolution*, 37, 1210–1226.

Leamy, L. & Cheverud, J. 1984. Quantitative genetics and the evolution of ontogeny. II. Genetic and environmental correlations among age-specific characters in randombred house mice. *Growth*, 48, 339–353.

Legates, J. 1972. The role of maternal effects in animal breeding. IV. Maternal effects in laboratory species. *Journal of Animal Science*, 35, 1294–1302.

Le Gros Clark, W. E. 1959. *The Antecedents of Man: An Introduction to the Evolution of the Primates*. New York: Harper & Row.

Manalich, R., Reeves, L., Herrera, M., Melendi, C. & Fundora, I. 2000. Relationship between weight at birth and the number and size of renal glomeruli in humans: a histomorphometric study. *Kidney International*, 58, 770–773.

Meaney, M. J. & Szyf, M. 2005. Maternal care as a model for experience-dependent chromatin plasticity? *Trends in Neuroscience*, 28, 456–463.

Mousseau, T. A. & Fox, C. W. 1998. The adaptive significance of maternal effects: moms do more than make babies. *Trends in Ecology and Evolution*, 13, 403–407.

Mousseau, T. A. & Roff, D. A. 1987. Natural selection and the heritability of fitness components. *Heredity*, 29, 181–197.

Nair, H. P. & Young, L. J. 2006. Vasopressin and pair-bond formation: genes to brain to behavior. *Physiology*, 21, 146–152.

Nelson, E. E. & Panksepp, J. 1998. Brain substrates of infant-mother attachment: contributions of opioids, oxytocin, and norepinephrine. *Neuroscience and Biobehavioral Reviews*, 22, 437–452.

Popik, P. & van Ree, J. M. 1991. Oxytocin but not vasopressin facilitates social recognition following injection into the medial preoptic area of the rat brain. *European Neuropsychopharmacology*, 1, 555–560.

Rinkenberger, J. & Werb, Z. 2000. The labyrinthine placenta. *Nature Genetics*, 25, 248–250.

Riska, B., Rutledge, J. J. & Atchley, W. R. 1986. Genetic analysis of cross-fostering data with sire and dam records. *Journal of Heredity*, 76, 247–250.

Rossiter, M. 1998. The role of environmental variation in parental effects expression. In: *Maternal Effects as Adaptations* (Ed. by T. A. Mousseau & C. W. Fox), pp. 112–134. Oxford: Oxford University Press.

Spencer, H. G. 2002. The correlation between relatives on the supposition of genomic imprinting. *Genetics*, 161, 411–417.

Szyf, M., Weaver, I. C., Champagne, F. A., Diorio, J. & Meaney, M. J. 2005. Maternal programming of steroid receptor expression and phenotype through DNA methylation in the rat. *Frontiers in Neuroendocrinology*, 26, 139–162.

Wade, M. J. 1998. The evolutionary genetics of maternal effects. In: *Maternal Effects as Adaptations* (Ed. by T. A. Mousseau & C. W. Fox), pp. 5–21. Oxford: Oxford University Press.

West-Eberhard, M. J. 1975. The evolution of social behavior by kin selection. *Quarterly Review of Biology*, 50, 1–33.

Willham, R. 1963. The covariance between relatives for characters composed of components contributed by related individuals. *Biometrics*, 19, 18–27.

Willham, R. 1972. The role of maternal effects in animal breeding. III. Biometrical aspects of maternal effects in animals. *Journal of Animal Science*, 35, 1288–1293.

Winslow, J. T., Shapiro, L., Carter, C. S. & Insel, T. R. 1993. Oxytocin and complex social behavior: species comparisons. *Psychopharmacology Bulletin*, 29, 409–414.

Witt, D. M., Winslow, J. T. & Insel, T. R. 1992. Enhanced social interactions in rats following chronic, centrally infused oxytocin. *Pharmacology, Biochemistry and Behavior*, 43, 855–861.

Wolf, J. B. 2000a. Gene interactions from maternal effects. *Evolution*, 54, 1882–1898.

Wolf, J. B. 2000b. Indirect genetic effects and gene interactions. In: *Epistasis and the Evolutionary Process* (Ed. by J. B. Wolf, E. D. Brodie III & M. J. Wade), pp. 158–176. Oxford: Oxford University Press.

Wolf, J. B. & Brodie, E. D., III. 1998. Coadaptation of parental and offspring characters. *Evolution*, 52, 535–544.

Wolf, J. B. & Hager, R. 2006. A maternal-offspring coadaptation theory for the evolution of genomic imprinting. *PLoS Biology*, 4, 2238–2243.

Wolf, J. B. & Wade, M. J. 2001. On the assignment of fitness to parents and offspring: whose fitness is it and when does it matter? *Journal of Evolutionary Biology*, 14, 347–356.

Wolf, J. B., Brodie, E. D. III, Cheverud, J. M., Moore, A. J. & Wade, M. J. 1998. Evolutionary consequences of indirect genetic effects. *Trends in Ecology and Evolution*, 13, 64–69.

Wolf, J. B., Vaughn, T. T., Pletscher, L. S. & Cheverud, J. M. 2002. Contribution of maternal effect QTL to genetic architecture of early growth in mice. *Heredity*, 89, 300–310.

A Theoretical Overview of Genetic Maternal Effects: Evolutionary Predictions and Empirical Tests with Mammalian Data

MICHAEL J. WADE, NICHOLAS K. PRIEST, AND
TAMI E. CRUICKSHANK

INTRODUCTION

In this chapter, we review the unusual features of the evolution of maternal-effect genes from both a theoretical and an experimental perspective. Maternal care and postnatal provisioning of young are not only ubiquitous features of mammalian life histories but also a defining trait of the class Mammalia. The degree of investment in caregiving by mothers depends on maternal genotype, age, condition, parity, offspring number, or litter size, as well as on a complex and reciprocal series of behavioral interactions between mother and offspring. Together, these result in the dual, negotiated control of offspring provisioning by both the maternal and zygotic genomes (Champagne & Curley, this volume). The standard partitioning of phenotypic variation into genetic and environmental components is inadequate whenever some influences on the phenotype are both genetic and environmental, as is the case with genes of maternal effect.

The first reported maternal genetic effect in animals was that snail shell chirality was determined by genes expressed in the mother, not in the offspring (Boycott & Diver 1923; Sturtevant 1923). The within-population variation in chirality implied that maternal genetic effects can generate and maintain genetic diversity, a finding that has been supported in contemporary work (Schilthuizen & Davison 2005). Recent developments in population genetic theory have led to a fuller understanding of the evolutionary genetics of maternal effects. The theory of relaxed selective constraint (RSC) argues that maternal-effect genes, and other genes with sex-specific expression, should have greater standing polymorphism within populations and should evolve more rapidly among populations or taxa than zygotically expressed genes. The relaxed constraint is the result of selection being weaker on variants of genes with sex-specific expression than on genes with zygotic expression. With RSC theory, we can make specific predictions about the expected patterns of sequence diversity within and across taxa for maternal-effect genes relative to their zygotically expressed homologues and test these predictions with sequence and genomic data from mammals. Testing these patterns in many species will be easier than carrying out the cross-fostering necessary to experimentally separate maternal from zygotic effects.

Our goals are to review recent theoretical advances in the maternal-effect field, especially RSC theory, and to illustrate how to test the theoretical predictions using sequence data. We discuss how genetic questions on the evolution of maternal effects can be approached using a combination of microarray expression data and DNA sequence data. The predicted patterns of sequence diversity within taxa and divergence among taxa result from the different ways that natural selection acts on zygotic, maternal, and maternal-zygotic gene-expression patterns.

To illustrate how to test the theory, we examine patterns of polymorphism and divergence in two maternal-effect genes in mammals, *Mater* and *RHD*. We also discuss the possible pitfalls of conducting this kind of analysis using GenBank sequences and how these pitfalls might be diminished or avoided with better preplanned sampling from mammalian populations. Nevertheless, we find that the major patterns predicted from theory are clear in the data, revealing the evolutionary differences between zygotic and maternal genes and between those and maternal-zygotic gene interactions.

Genes in the maternal genome play an important role in determining offspring phenotypes (Mousseau & Fox 1998). Maternal effects begin long before fertilization, during oocyte maturation, when hundreds of maternal gene products are passed to the unfertilized ovum, and they continue long after birth in mammals. By some estimates, approximately one-third of all genes acting in early development are maternal-effect genes (Wolpert et al. 2002). For offspring traits with an intermediate fitness optimum, like body size, we show why the dual genetic control of maternal provisioning and selection among maternal lineages tends to create negative genetic correlations between coadapted maternal and zygotic traits (Wolf et al. 1999; Kölliker et al. 2005). We also show how these theoretically expected correlations can be revealed at the phenotypic and genetic level with experimental cross-fostering (Kölliker et al. 2000; 2005; Agrawal et al. 2001a; Wolf et al. 2002b). Because Mendelian transmission prevents some maternal-zygotic combinations from occurring and, in addition, selection can create associations between maternal and zygotic genes, cross-fostering is an essential experimental tool for partitioning maternal from zygotic effects.

When local selection among matrilines within a population creates genetic correlations between maternal and zygotic traits, the evolution of maternal effects is particularly susceptible to population genetic subdivision (Agrawal et al. 2001b; Wolf 2003). Whenever populations are genetically structured, the maternal environment experienced by an individual tends to be more similar to that experienced by its parents than to the environment experienced by an individual from some other deme. This is especially true when populations are structured along matrilines, as is characteristic of many group-living mammals. These situations facilitate the coevolution of the maternal and zygotic genes.

For most mammalian species, an experimental approach to maternal genetic effects is difficult. Quantifying the proportion of variation in offspring trait expression that results from variations in maternal phenotypes is not feasible for the vast majority of mammalian species. For most wild populations we cannot conduct the crosses necessary to adequately control for genetic background or for variation resulting from differences in condition. For some species we can cross-foster offspring with different mothers in the laboratory or in field enclosures, which allows us to separate the effects of mothers from those of the offspring. However, in most mammal species, cross-fostering cannot be done early enough in development with a sufficiently high degree of replication to distinguish maternal from zygotic genetic effects or to detect maternal-zygotic interactions (Wolf et al. 2002b).

Molecular genetic approaches may help our understanding of how maternal-effect genes evolve in mammals. We outline molecular genetic predictions from RSC theory, emphasizing those predictions amenable to testing with comparative genomic data. RSC theory generates separate predictions for maternal-effect genes with strict maternal expression and for maternal-effect genes with maternal-zygotic interactions.

We will illustrate the theoretical predictions associated with genes with strict maternal expression, comparing levels of sequence diversity and divergence for the mouse maternal-effect gene *Mater* (also called *Nalp5*) and its zygotically expressed homolog, *Nalp2*. We show why *Mater* should have greater standing polymorphism within populations than *Nalp2* and why

Mater should have greater rates of divergence across taxa than *Nalp2*. However, our predictions for other species, including humans, depend upon assuming conservation of *Mater*'s strict maternal expression as demonstrated in the mouse; we do not know this for certain in other mammals. Microarray data suggest that the pattern of *Mater* expression is not strictly maternal in humans (see below). Furthermore, its observed level of sequence diversity relative to *Nalp2* in humans supports the inference that its pattern of expression is not strictly maternal in our species.

We illustrate the evolutionary predictions associated with maternal-zygotic interactions using GenBank sequences of the *RH* gene family (Matassi et al. 1999). We show why some members of this gene family should exhibit greater sequence variation in their coding regions than others and why they should be evolving more rapidly across mammalian taxa. Although differences in evolutionary rates have been observed before in this gene family (Matassi et al. 1999; Huang & Peng 2005), they have been interpreted as evidence for adaptive evolution by positive selection rather than as the expected result of relaxed selective constraint. Relaxed constraint explains *both* the elevated sequence diversity within species as well as the more rapid evolution among species. In contrast, rapid positive selection predicts less genetic diversity within populations at both silent and replacement sites, the opposite of the observed pattern. We show how the RSC theory makes general and testable predictions for rates of evolution within and among mammalian taxa and discuss a number of caveats in implementing tests of this theory using non-model organisms. Overall, the patterns we observe for the *RH* gene family and for *Mater* are in concordance with our earlier findings in fruit flies (*Drosophila melanogaster*) for the tandem duplicate pair of *Hox* genes, *bicoid* (with maternal expression) and *zerknüllt* (with zygotic expression; Barker et al. 2005; Demuth & Wade 2007a) and for maternal genes acting early in fly development (Cruickshank & Wade 2008).

DEFINITION OF MATERNAL EFFECTS

A maternal effect may be defined as an influence of the maternal genotype and/or phenotype on offspring phenotype that is independent of the offspring genotype (Roff 1998; Wade 1998; but see Cheverud & Wolf, this volume). Maternal effects are often detected as a difference between the mother-offspring and father-offspring regressions. However, the mother-offspring regression can also be affected by maternal inheritance, which is not a maternal effect. Maternally transmitted organelle genomes, such as the mitochondrion or chloroplast, increase the mother-offspring regression but not the father-offspring regression, just as Y-linked nuclear genes affect the latter but not the former. The mitochondrial genome is a part of the offspring genotype (as are Y-linked genes in sons) and, for cytoplasmic effects, the regression of offspring phenotype on offspring mitochondrial genotype is the same as the regression of offspring phenotype on maternal mitochondrial genotype (barring paternal leakage, maternal heteroplasmy, or new mutations). Despite the congruence of these two regressions, the direct effects of maternally inherited cytoplasmic genes acting on offspring phenotypes do not fall under the strict definition of maternal effects (see also Cheverud & Wolf, this volume). For example, human mitochondrial diseases acting late in life are caused by misexpression of the individual's own mitochondrial genes, often via the accumulation of new mutations during the individual's lifetime, and not by the expression of these genes in its mother's genome. In half-sib designs, where individual males are crossed to several mothers and offspring phenotypic variation is partitioned with respect to the paternal genomes, genetic factors, like dominance, contribute to and increase the among-mother variance over the among-father variance, but these are clearly not maternal genetic effects.

Among the most difficult effects to partition are the effects on offspring phenotype of maternal-zygotic interactions. Statistically, an interaction cannot be viewed as either a maternal or a zygotic effect. An interaction means that zygotic genes have different effects on offspring phenotypes in different maternal genetic backgrounds. And, conversely, the effect of maternal genes on offspring phenotypes differs with the genotype of the offspring. Estimates of both kinds of effects change with the frequencies of the maternal and zygotic genotypes and, because some maternal-zygotic genotypic combinations cannot be produced naturally (e.g., AABB mothers cannot have aabb offspring; Wade 1998), estimated genic main effects are likely to include some fraction of the interaction unless the "unnatural" genotypes are produced by cross-fostering. This kind of genetic interaction has been called maternal-zygotic epistasis (Wolf 2000). Below, we illustrate the evolutionary genetics of maternal-zygotic interactions using a classic example of a human maternal effect from the *RH* gene family. We show why such genes are expected to diversify among taxa even more rapidly than genes with only maternal effects and why they should tend to have higher levels of standing diversity within species.

Maternal effects are also an example of "indirect genetic effects," where the social environment of conspecifics (or heterospecifics; see Wade 2003;

2007) affects the phenotype of an individual (Wolf et al. 1998). Indirect effects complicate one of the primary goals of evolutionary genetics, which is to understand the separate contributions of nature (genes) and nurture (environment) to individual phenotypic variation. Indirect effects in general and maternal effects in particular add a third component, the nurturers, to the decomposition of individual phenotypic variation. For all mammals, mother is the central nurturer and primary conveyor of indirect effects to the phenotypes of her offspring. Just as indirect effects play a more critical role in genetically subdivided populations (Agrawal et al. 2001b), the role of maternal effects is enhanced in species in which the social and genetic structure is subdivided into groups representing different matrilines (Wade 1998; Linksvayer & Wade 2005; Bjima & Wade 2008).

Maternal genetic and environmental effects are one of the most important contexts for a developing mammalian embryo (Vandenberg, this volume). Maternal gene products, both proteins and mRNAs, in the oocyte initiate gene expression and development in offspring post-fertilization. In mammals, the maternal-zygotic transition (MZT) from maternal to zygotic genetic control of development occurs between the 2- and 8-cell stages (Jarrell et al. 1991; Shultz 2002; Vigneault et al. 2004). Because differences in gene-expression patterns affect the efficiency of natural selection, maternal and zygotic development genes can evolve differently from one another.

EXPERIMENTAL DESIGNS FOR DETECTING GENES WITH MATERNAL EFFECTS

The central statistical concept behind studies of marker-assisted detection of quantitative trait loci (QTL) is the association between individual phenotypic variation and highly variable markers, such as microsatellites, that permits detection of chromosomal regions segregating among the individuals (for discussion of QTL methods applied to complex trait analysis see Lynch & Walsh 1988; Demuth & Wade 2006; Cheverud & Wolf, this volume). These methods have proven useful for detecting genes with direct effects on individual phenotypes as well as interactions. However, when the genes causing the phenotypic variation reside in the maternal genome and not in the genome of the phenotypically varying individuals, the typical experimental procedures are inadequate for detecting genes with indirect maternal effects. For example, consider two genetically divergent mouse strains that are crossed to produce an F1 and these are crossed with one another to produce a segregating F2. None of the phenotypic variation among the F2 progeny can be attributed to genes with maternal effects, because all of the F1 mothers share the same genotype and are not variable. To detect genes with maternal effects requires not only creating an F3 generation but also cross-fostering the F3 pups among mothers. In this case, phenotypic variation among the pups can be partitioned into components of foster mother and birth mother (e.g., Wolf et al. 2002b). A maternal-effect QTL is detected when phenotypic variation in the F3 pups is associated with markers segregating among the foster maternal genomes. Without cross-fostering pups among mothers, maternal genetic effects would remain confounded with direct effects since mothers would otherwise contribute both genes and nursing environment simultaneously to their pups (see Atchley et al. 1991; Wolf et al. 2002b; Cheverud & Wolf, this volume, for further discussion). Furthermore, the early maternal-zygotic transition in the control of mammalian development (e.g., by the 4-cell stage in swine; Jarrell et al. 1991) means that even cross-fostering of embryos will not remove all maternal effects.

When Wolf et al. (2002b) implemented an F3 cross-foster design and measured growth rates in mice, they found that maternal genetic effects accounted for nearly three times more of the variation in offspring growth rate than genes with direct effects (31.5% versus 11.8%, respectively) despite the much smaller sample size and reduced statistical power. Unfortunately, for most mammalian species, especially humans and other primates, it is difficult to impose the necessary cross-fostering, early in development with sufficient replication.

GENETIC CORRELATION BETWEEN MATERNAL AND ZYGOTIC EFFECTS

Whenever maternal and zygotic genes interact, so that some gene combinations have higher and others lower fitness, then natural selection can create genetic correlations between maternal and offspring traits (Wolf & Brodie 1998). In a type of maternal effect likely to be common (Wolf et al. 2002a), the maternal genetic environment can favor certain zygotic genotypes over others, causing a component of among-family selection that can be distinguished in theory and in experiments (Wade 1998, 2001). For traits with intermediate fitness optima without interactions, like offspring body size, selection tends to favor maternal effects of opposite sign to direct effects, with maternal alleles conferring larger body size associated with zygotic genes for smaller size and vice versa. This gives rise to negative genetic maternal-zygotic correlations, but other scenarios are also possible (Kölliker et al. 2005). Regardless of the sign of the correlation, the genes conferring the maternal and zygotic effects tend to become "integrated" into adaptive trans-genomic, trans-generational combinations (Wolf & Brodie 1998), with many different maternal-zygotic gene combinations resulting in similarly sized offspring. In these cases, natural selection is relaxed relative to the case where a single gene or single gene combination is most fit and sequence diversity accumulates by mutation for those genes where selection is relaxed.

When some maternal-zygotic gene combinations have fitness effects that are greater or less than the sum of their separate additive effects, there is epistasis for fitness between the maternal and zygotic gene combinations (Wade 1998; Wolf & Brodie 1998). Differently put, from the viewpoint of zygotic genotypes, such maternal-zygotic interactions are heritable genotype-by-environment interactions (G × E), a type of genetic architecture known in ecological genetics for its ability to maintain genetic diversity (Falconer 1952; Curtsinger et al. 1994). Random mating breaks up good maternal-zygotic combinations and, like migration among environments with G × E, mixes zygotic genes across "good" and "bad" maternal environments. As a result of weakened selection, not only is gene diversity elevated for both maternal and zygotic genes, but also the prevalence of those complex genetic diseases caused by the poorer gene combinations is elevated in the population.

In mice, where it is possible to conduct embryo transplants very early in development and effect cross-fostering of zygotic genotypes, the contribution of genetic interactions between maternal and zygotic genes to offspring phenotypes can be studied. For example, Cowley (1991) found that maternal-zygotic epistasis accounted for a larger fraction of the variance in offspring size than either maternal or zygotic effects, revealing the importance of the coadaptation of maternal and zygotic genotypes. Such experiments cannot be carried out in humans or many other mammal species, but this type of natural selection is expected to leave its fingerprints on the pattern of sequence diversity as we show below in the theoretical section on maternal-zygotic selection. Later, in the empirical section, we contrast the levels of sequence diversity for maternal and zygotically acting members of the *RH* gene family, which arose through at least two duplications. First, *RH30* was duplicated to form the *RH30* and *RHAG* (also known as RH50) homologues (Kitano et al. 1998). Subsequently, *RHD* and *RHCE* arose by a second duplication of *RH30* (Matassi et al. 1999). The *RHD* gene is involved in the maternal-zygotic interaction that is a textbook case of a complex maternal effect with potentially lethal consequences for the developing embryo and its levels of sequence diversity are much higher than those of its zygotically expressed homolog, *RHAG*.

Finally, it should be noted that maternal-zygotic epistasis for fitness in a genetically subdivided metapopulation leads, in theory, to divergent evolution among different component populations (Wade 1998; Wolf & Brodie 1998). When these groups come into secondary contact post-divergence, incompatible maternal-fetal interactions may play a role in reproductive isolation and speciation (Wade & Beeman 1994; Demuth & Wade 2007b, c). These kinds of interactions between maternal and zygotic genes coupled with a high frequency of life histories consisting of small, more or less isolated social groups may be responsible for the accelerated rate of evolution observed in mammals in general and in primates in particular.

THE EVOLUTIONARY GENETICS OF ZYGOTIC, MATERNAL, AND MATERNAL-ZYGOTIC EFFECTS

Zygotic effects with selection and mutation

Loci with the same average selection coefficient, s, but different patterns of gene expression experience natural selection in different ways. Typically, we think of alleles as having effects on the phenotypes or fitnesses of the individuals that carry them. This is what we mean by a "direct zygotic effect" (Table 3-1). For this simple model, we have assumed that males and females have equal genotype frequencies and fitnesses. Note that, in Table 3-1, the frequency of the A allele in parents equals $p = (X + Y)$, and the frequency of the a allele equals $q = (1 - p) = (Y + Z)$. After selection has acted on the offspring genotypes, the frequency of the a allele changes from q in the

Table 3-1. Zygotic effects on viability. The A-locus has two alleles, *A* and *a*, which have only additive zygotic effects on fitness, so that incrementing a genotype by one *a* allele decreases viability by *s*, the selection coefficient.

Maternal genotype	Family frequency	Offspring genotype (within-family fitness)			Mean family fitness
		AA	Aa	aa	
AA	X	p (1)	q (1+s)	—	1+sq
Aa	2Y	p/2 (1)	1/2 (1+s)	q/2 (1+2s)	$1+(s/2)(1+2q)$
aa	Z	—	p (1+s)	q (1+2s)	1+(s)(1+q)
Offspring genotype frequency		p(X+Y) or p^2	(qX+Y+pZ) or 2pq	q(Y+Z) or q^2	
Mean genotypic fitnesses		1	(1+s)	(1+2s)	$W_m = 1+2qs$

parents to q' by differential deaths among the offspring genotypes. After selection, we find that q' equals $\{[pq(1 - s) + q^2(1 - 2s)]/W_z\}$, where W_z is the mean population fitness, which is calculated in the lower right-hand corner of Table 3-1. The change in allele frequency by selection, Δq_z, is $(q' - q)$ or

$$\Delta q_z = spq/W_z. \tag{1}$$

If selection acts against the a allele $(s < 0)$ but mutation continually reintroduces the allele into the population, the population will achieve "mutation-selection balance," with a stable intermediate, equilibrium frequency, q_m^*. If we assume that selection is sufficiently weak that W_z can be considered equal to 1, then it is easy to solve for q^* at mutation-selection balance. The total change in allele frequency, Δq_{total}, equals the sum of the changes owing to natural selection, Δq_z (eq. [1]), and mutation, $\Delta q_u = +up$, where u is the mutation rate from allele A to allele a and we have assumed that uq is negligible:

$$\Delta q_{total} = \Delta q_z + \Delta q_u = 0, \tag{2a}$$
$$\Delta q_{total} = spq^* + up = 0, \tag{2b}$$
$$q_z^* = -u/s. \tag{2c}$$

In the last step, we assumed that p is approximately 1. Note that when purifying selection is strong, the equilibrium value of a is lower than it is when purifying selection is weaker. At the level of the gene sequence, this means that levels of replacement site diversity are expected to be determined by the interaction of selection and mutation (u/s), while levels of silent site diversity are determined by the balance between random genetic drift and mutation $(u/[1/(2N_e)])$ or $2N_e u)$.

Maternal effects with selection and mutation

Selection on maternal effects is different from selection on zygotic effects. Here, the fitness of a given offspring genotype is determined by the genotype of its mother (Table 3-2). Note that the fitness of an offspring of genotype Aa changes from 1, to $(1 - s)$, to $(1 - 2s)$ as the mother's genotype changes from AA to Aa to aa, respectively. More generally, the marginal fitnesses of the offspring genotypes (last row of Table 3-2) are explicit functions of the genotype frequencies and allele frequencies of the parents. Note also that the mean fitness, W_m, is the same as in the zygotic model. Thus, *average viability fitness is the same in both models*. The two models differ in how the viability is distributed: in the zygotic model, viability varies with offspring genotype, while in the maternal model, it varies with family, i.e., with maternal genotype. For maternal selection, we calculate Δq_m as $(q' - q)$ or

Table 3-2. Maternal effects on viability. The A-locus has two alleles, **A** and **a**, which have only additive maternal effects on fitness, so that incrementing the mother's genotype by one **a** allele decreases viability of all her offspring, regardless of genotype, by **s**, the selection coefficient.

Maternal genotype	Family frequency	Offspring genotype (within-family fitness)			Mean family fitness
		AA	Aa	aa	
AA	X	p (1)	q (1)	—	1
Aa	2Y	$p/2$ $(1+s)$	$1/2$ $(1+s)$	$q/2$ $(1+s)$	$1+s$
Aa	Z	—	p $(1-2s)$	q $(1-2s)$	$1+2s$
Offspring genotype frequency		$p(X+Y)$ or p^2	$(qX+Y+pZ)$ or $2pq$	$q(Y+Z)$ or q^2	
Mean genotypic fitnesses		$1+(Y^s/p)$	$1+(s^Z/q)(s^Y/2pq)$	$(1+s)+(s^Z/q)$	$W_m = 1+2qs$

$$\Delta q_m = spq/2W_m. \qquad [3]$$

The primary difference between eqs. [1] and [3] is the factor of ½ in the latter. In eq. [1], the regression of offspring genotypic fitness on number of a alleles is twice the regression of maternal genotype on number of a alleles. Differently put, selection is only half as strong when acting on a maternal-effect gene as it is acting on a zygotic-effect gene. Because a alleles are not expressed in males, natural selection is relaxed by a factor of two relative to the zygotic selection model where a alleles are expressed in both sexes. Although we present an additive-effect model here, including dominance either here or in the preceding model does not change this very general conclusion (Demuth & Wade 2007a; Cruickshank & Wade 2008).

We find the equilibrium value, q_m^*, achieved at mutation-selection balance in the same way as before except that there is a factor of 2 involved:

$$q_m^* = -2u/s. \qquad [4]$$

Thus, for the same average coefficient of purifying selection, s, a maternal-effect allele is twice as common as a zygotic effect allele at mutation-selection equilibrium. This is why we say that selection on maternal-effect genes is weaker than on zygotic-effect genes. Because only half of the adult population expresses the trait (i.e., the mothers), selection is weaker on maternal-effect genes than on zygotic-effect genes. It should be noted that this prediction depends on the adult sex ratio of the population. In female-biased populations, selection on maternal-effect genes would be somewhere between 0.5 and 1 of selection on zygotic genes; conversely, in male-biased populations, selection on maternal-effect genes would be substantially less than 0.5 of selection on zygotic genes. The adjustment of q_m^* based on the sex-ratio of the population is relevant to mammal species with unusual mating systems.

Our expectations for variation at the sequence level follow from the two-fold difference between q_m and q_z at mutation-selection balance. Specifically, heterozygosity, H, is given by $H = 2pq$. Assuming p is small, we find that

$$H_m = (1/L)\Sigma(4u/s), \qquad [5a]$$
$$H_z = (1/L)\Sigma(2u/s). \qquad [5b]$$

The general implication for molecular evolutionary genetics of RSC theory is that levels of replacement site diversity determined by the interaction of selection and mutation will be twice as large for a strictly maternal-effect gene as for a zygotically expressed homologue. For silent sites or introns, however, there should be no difference in the levels of sequence diversity, because true selective neutrality is not affected by gene-expression pattern and both genes will experience the same degree of random genetic drift. This assumes that the average coefficient of selection, s, and the fraction of nonsynonymous sites which are effectively synoynmous are approximately equal for both genes, so that the quantities cancel and nucleotide diversity differs by a factor of 2 for maternal and zygotic genes. However, only 2–5% of nonsynonymous sites are neutral (in *Drosophila*), so it is unlikely that observable differences in this fraction would strongly influence the predicted pattern.

Maternal-zygotic interaction effects with selection and mutation.

We now consider the well-known maternal-fetal interactions resulting from Rh maternal-fetal incompatibilities that affect offspring viability. Haldane (1942) suggested the positive effects of placental permeability on nutritional transfer outweighed its deleterious effects involving Rh antigens. For ease of comparison with the two earlier cases, we let the dominant Rh factor be the A allele and let the recessive Rh factor be the a allele. There is selection against heterozygous Aa offspring born of aa mothers and either Aa or aa fathers. This has long been recognized as a case of selection against heterozygotes in humans (Haldane 1942) and is expected to result in *fewer* heterozygotes than expected by Hardy-Weinberg within a population and lowered heterozygosity. We will show, however, that the relaxed selective constraint owing to maternal-zygotic epistasis is expected to result in a *greater* level of replacement site diversity, i.e., increased heterozygosity, than that of a zygotically expressed homologue.

Table 3-3. Maternal-zygotic interaction effects on viability (after Wright 1969). The A-locus has two alleles, *A* and *a*, which have a maternal-zygotic interaction effects on fitness analogous to that of *RHD*.

Maternal × paternal genotype	Family frequency	Offspring genotype (within-family fitness)			Mean family fitness
		AA	Aa	aa	
AA	X	p (1)	q (1)	—	1
Aa	2Y	$p/2$ (1)	½ (1)	$q/2$ (1)	1
aa × AA	ZX	—	1 (1+s)	—	1+s
aa × Aa	2ZY	—	½ (1+s)	½	$1+(s/2)$
aa × aa	ZZ	—	—	1	1
Offspring genotype frequency		p(X+Y) or p²	(qX+Y+pZ) or 2pq	q(Y+Z) or q²	1
Mean genotypic fitnesses		1	$1+(s^Z/2q)$	1	$W_{mz} = 1+spZ$

We use the models of Haldane (1942, p. 334) and Wright (1969, pp. 148–149) to describe the somewhat simplified population genetics of these maternal-fetal interactions (Table 3-3). Here, homozygous aa mothers reduce the viability of their heterozygous Aa offspring. Whenever $s < 0$, there is marginal underdominance for fitness, that is, the average fitness of heterozygous offspring is *less than* that of either homozygote (last row of Table 3-3). In standard genotypic selection models, underdominance results in an unstable equilibrium allele frequency, where small random deviations lead to fixation or loss. However, when the selection involves maternal effects, the equilibrium can be stable (Wade & Beeman 1994; Wade 2000). Since aa mothers produce different proportions of Aa offspring with different sires, we have made the three possible one-locus, two-allele paternal genotypes explicit in Table 3-3 (rows 3–5). Selection is stronger for families of aa mothers with AA fathers than it is for aa mothers with Aa fathers, as reflected by the lowered mean family fitness (Table 3-3, last column), because all of the offspring in the former families are heterozygotes compared with only half the offspring in the latter.

After selection has acted, $q' = [q + (psZ/2)]/W_{mz}$. Note that W_{mz}, the mean population fitness, depends on the frequency of homozygous mothers, Z, and on the frequency of the A allele in sperm, p. The change in allele frequency by selection, Δq_{mz}, is $(q' - q)$ or

$$\Delta q_z = (spZ[p - q])/2W_z. \qquad [6]$$

Our eq. [5] differs from that of Wright (1969, p. 149), who used the approximation that $Z = q^2$, but agrees with that of Haldane (1942, p. 335), who did not. From eq. [6], we can see that there is an unstable equilibrium when $p = q$, since, for $s < 0$, selection acts against the a allele when it is rare, i.e., $\Delta q_z < 0$, but favors that allele when it is common. A stable frequency-dependent equilibrium requires the opposite: an allele that is favored when rare but selected against when common. Indeed, when a is common and ZX families are rare, the evolutionary dynamics of Rh are similar to those of the maternal-effect selfish gene, *Medea*, in flour beetles (Wade & Beeman 1994). Like Rh, selection on *Medea* involves a maternal-zygotic interaction: homozygous *Medea* mothers reduce the viability of their heterozygous offspring creating a selection barrier to the invasion of the population by a non-*Medea* allele.

If we assume that A is common so that $(p - q)$ can be approximated as p, then we have

$$p_{mz}^* = (-2u/sZ)^{1/2}. \qquad [7]$$

Note that sequence diversity at mutation-selection balance will be higher for this maternal-zygotic interaction than for a strictly maternal-effect gene (eq. [4]) or for a zygotic-effect gene (eq. [2d]) for two reasons: (1) Z, which is always less than 1, appears in the denominator of eq. [7]); and, (2) the exponent of ½ (square root) in eq. [7]. Both of these conditions can be seen as aspects of relaxed selection. First, heterozygous offspring are at a disadvantage only when born to aa mothers, unlike heterozygous offspring in the zygotic model, which all share the same fitness regardless of maternal genotype. Second, only aa mothers with Aa or Aa mates exhibit the maternal effect on offspring viability, whereas in the maternal effect model, all aa mothers regardless of mate's genotype have the same effect on all offspring. Such conditional expression of fitness effects is characteristic of maternal-zygotic interactions (Wade 1998; Wolf & Brodie 1998) and it leads to the expectation of higher levels of replacement site diversity within populations.

In summary, evolutionary genetic models indicate that genes with strict maternal effects will be more diverse within populations than genes expressed zygotically in both sexes, because the sex-limited expression pattern weakens selection. Furthermore, genes with maternal-zygotic epistasis for fitness will be more diverse within populations than either genes with strict maternal effects or genes with zygotic effects, because epistasis for fitness also weakens selection acting on specific genes. These predicted patterns for replacement site diversity in gene sequence assume that all genes have similar average selection pressures (s).

MACRO-EVOLUTIONARY PATTERN AS A CONSEQUENCE OF GENE EXPRESSION PATTERN

The theory of relaxed selective constraint predicts a positive relationship between gene diversity within taxa and substitution rate among taxa. In general, genes with maternal-zygotic interactions should be more divergent among taxa than strict maternal-effect genes, and they in turn should be more divergent than zygotic genes. This positive relationship between diversity within and substitution rate among taxa results when speciation, arising from the evolutionary processes of natural selection, mutation, and random genetic drift, converts genetic diversity within populations into more or less permanent genetic differences among them. While mutation and drift are random processes, the relative rate of substitution by natural selection depends upon **s**, the average selection coefficient. For positive selection ($s > 0$), the rate of nonsynonymous substitution for a maternal-effect gene relative to a zygotic-effect gene is less than 1 and the substitution ratio asymptotes at 0.50 for large s. The reason the relative rate of adaptive substitution is slower is that natural selection is weaker when acting on a strictly maternal-effect gene with a positive effect on fitness. In the limit, the substitution rate of a maternal-effect gene is only one half that of a zygotic-effect gene because half of the allelic copies of the former occur in males, where they are not expressed and, thus, hidden away from natural selection. For a maternal-effect gene to evolve faster than a zygotic-effect gene as a result of positive, adaptive natural selection, the strength of selection must be at least twice as strong as the selection acting on the zygotic gene (Barker et al. 2005; Demuth & Wade 2007a).

For purifying selection ($s < 0$), natural selection is similarly weaker when acting against deleterious maternal-effect genes, but the effect on the substitution rate is very different. For purifying natural selection, deleterious maternal alleles become fixed by random genetic drift *more often* than deleterious zygotic-effect alleles. As a result, relative rate of nonsynonymous or replacement substitution is *greater* for a maternal-effect gene under predominately purifying selection than it is for a zygotic-effect gene; so the ratio of maternal-to-zygotic substitution rates exceeds 1. When s approaches 1 as it does for lethal alleles, the substitution ratio (maternal/zygotic) approaches infinity, because genes with strong deleterious effects almost never fix when expressed zygotically (Demuth & Wade 2007a; Cruickshank & Wade 2008). Since adaptive substitution at the level of specific genes is believed to be much rarer than purifying selection, overall we expect faster substitution rates at replacement sites for genes with maternal-zygotic and maternal effects on fitness. Thus, it is expected that maternal-effect genes will be evolving rapidly, but owing to RSC and not because of adaptive natural selection.

In the following sections, we will use mammalian gene sequences taken from publicly available on-line databases to test these two theoretical predictions: (1) within species, maternal-effect genes should have higher sequence diversity than zygotic genes; and (2) across taxa, maternal-effect genes should evolve faster owing to relaxed selective constraint.

TESTING THEORETICAL PREDICTIONS USING MAMMALIAN SEQUENCES

To test our theoretical predictions in mammals, we selected genes with different patterns of expression. First, we used microarray information to determine the level of expression for each gene in female-specific, male-specific, or zygotic tissues. Our rationale was that tissue-specific expression provides an indication of the functional expression of genes. Genes with strict female-specific tissue expression should have only female-specific functional expression. Next, we obtained multiple sequences of the genes from humans and mice from publicly available databases to assess levels of coding and silent site polymorphism. Finally, we obtained sequences of these genes sampled from different mammalian species and assessed rates of divergence among mammalian taxa.

As a result of this screening, we focused on four genes: a maternal-effect gene, *Mater* (also called *Nalp5*), which is required for embryonic development beyond the two-cell stage in mice, and its zygotic homolog, *Nalp2*, and on the maternal/zygotic interaction gene, *RHD*, and its zygotic homolog, *RHAG* (*RH50*). We also analyze *RHCE* to control for relaxed constraint due to redundancy following gene duplication. *RHD*, *RHCE*, and *RHAG* are the products of two duplication events—one that first generated *RHAG* (zygotic) and *RH30* and a second that generated *RHD* (maternal) and *RHCE* (zygotic) from the ancestral *RH30*.

Sex-specific Expression of Maternal-effect Genes

To determine whether the four focal genes had maternal or zygotic expression, we downloaded tissue-specific expression profiles of the genes from

Unigene, the transcriptome database managed by National Center for Biotechnology Information (Schuler 1997; Wheeler et al. 2003). For the human transcriptome to date, 46 different tissues have been sampled for mRNA expression: 7 female-specific tissues (uterus, cervix, umbilical cord, mammary gland, embryo, ovary, and placenta), 2 male-specific tissues (testes and prostate), and 37 zygotic tissues. For the mouse transcriptome, 37 tissues had been sampled for mRNA expression: 4 female-specific tissues (female genital, mammary gland, embryo, extra-embryo), 1 male-specific tissue (male genital), and 32 zygotic tissues. In both species, there was variation in sampling intensity among the three tissue classes. To calculate the relative proportion of gene expression in maternal, paternal, and zygotic tissues, we determined the total number of focal gene transcripts sampled for each tissue class, divided that quantity by the total number of transcripts sampled in that tissue class, and for each gene scaled the values for each tissue class to 1.

Polymorphism and Divergence of Maternal-effect Genes

We combined general searches of the GenBank nucleotide database with blast searches to acquire mammalian sequences on-line for both sets of maternal/zygotic homologs. Nucleotide sequences for each gene were aligned and analyzed using Mega v3.1 (Kumar et al. 2004). Coding and noncoding variation of the focal genes was assessed for human and mouse sequences using the Kumar method (Nei & Kumar 2000), with standard errors determined by bootstrapping. To estimate and visualize distances, neighbor-joining phylogenies were constructed using the Kumar method to differentially correct for multiple substitutions at synonymous and nonsynonymous sites. Tree topology was verified using 1000 bootstrap replicates. Editing of the tree output was performed in Mesquite (Maddison & Maddison 2006). Accession numbers for each sequence used in our study are given in Table 3-4. We found published as well as unpublished sequences in the database, the latter representing in some cases ancillary or verification sequences from studies focused on other genes. We were concerned that systematic differences in methods or quality of sequencing between published and unpublished data could inflate the relative levels of sequence diversity on which our tests of theoretical prediction depend. For this reason, we analyzed the two kinds of data separately. We assume in our analysis that sequences collected for other purposes are random samples with respect to our tests. If maternal and fetal *RHD* sequences were collected together for aid in diagnosing the cause of problem pregnancies, this would clearly not be the case. Fortunately, we

Table 3-4. Accession numbers for *RH30, RHAG, Nalp2,* and *Nalp 5.*

Sex-specific expression: *RH30:* Hs.643556, Mm.195461; *RHAG:* Hs.120950, Mm.12961; nalp2: Hs.369279, Mm.246598; nalp5: Mm.333653, Hs. 356872.

Polymorphism:

Sequences from published data:

RHD: AF510070, AB049754, AF187846, S82449, NM016124, S70174, L08429, S78509.

RHCE: NM020485, DQ266400, AB018645, AB018644, AF056965, BC075081, AF510068, AB049753

RHAG: AF031549, AF031548, NM000324, X64594, BC012605, AF237382, AL121950.

Nalp2: NM017852, BC001039, BC003592, AF310106, BC039269, AC155171, AK054264, AY360473, DQ066881, NM177690

Nalp5: NM153447, AY154460, AY054986, NM153447, AC011470, AY329490, BC053384, NM011860, AF074018, AY196361, AY329487, AF143571

Sequences from unpublished data:

RHD: AB46420, AB018968, AF312679, AM396583, AM177314.

RHCE: AB030388, DQ322275, DQ266353, AY603478.

RHAG: CR621033, AF179684, AF178841, AF179685, AF031549, AF179682, CR592572, AL121950, AF237383.

Divergence:

RH30: AF057524, AF531096, BTU59270, AY831676, AF101479, L37054, S70343, L37053, AF012425, L37049, X54534.

RHAG: AF057526, AF531097, AF164575, AF177623, AF177621, AY831677, AF132980, AF177622, AB015467, AF177625, AF031548.

Nalp 2: NM017852, AY360473, XM541420, XM610622, DQ045971.

Nalp 5: NM153447, XM512922, NM001039143, NM001007814, XM533576.

found some randomly collected *RHD* sequences (see below). Finally, some *RH* sequences in the databases do not distinguish among the various duplicate copies, making them unsuitable for our tests. These concerns highlight the need for forward planning of a sampling regime to more incisively test theoretical predictions rather than the adventitious use (like ours) of sequences publicly available.

It is difficult to infer a strict pattern of maternal gene expression from microarray data. *Mater* had exclusively female-specific expression in mice (Table 3-5) as shown previously in humans, mice, and cows (Tong et al. 2002; 2004; Pennetier et al. 2004). However, only 20% of *RHD* expression in humans was sex-specific (Table 3-5). The expression of the zygotic homologs was also not strictly zygotic. *RHAG* was exclusively zygotically expressed in mice, but not in humans, where only 85% of *RHAG* expression was zygotic. For our purposes, we considered some expression in sex-specific tissue of both males and females, or dual expression in both zygotic and sex-limited tissues, as zygotic expression.

Some genes, like *gurken* in *Drosophila melanogaster,* have primarily female-

Table 3-5. Expression data for *Nalp5* (*Mater*), *Nalp2*, *RHD*, and *RHAG* in female-specific, male-specific, and zygotic tissues in humans and mice. The percentages indicate the proportion of total expression detected for the gene in the specific tissue, adjusted by sampling intensity. The number of transcripts of the gene in the sample out of the total number of transcripts is reported in parentheses.

		Female	Male	Zygotic
Nalp5	Human	NA	NA	NA
	Mouse	100% (4/90280)	0% (0/22930)	0% (0/2785334)
Nalp2	Human	50% (141/1021369)	48% (67/502745)	2% (21/4042206)
	Mouse	100% (2/90280)	0% (0/22930)	0% (0/2785334)
RHD	Human	20% (1/1073514)	0% (0/502998)	80% (15/3994791)
	Mouse	21% (10/584579)	0% (0/153763)	79% (138/2176030)
RHAG	Human	15% (2/1073514)	0% (0/502998)	85% (41/3994791)
	Mouse	0% (0/584579)	0% (0/153763)	100% (15/2176030)

limited expression in array studies as well as some limited amount of expression in other tissues. Nevertheless, knock-out studies of *gurken* show that it has a female-limited effect on fitness as theory requires for our predictions despite the detection of some expression in some other tissues. Furthermore, *gurken* displays greater sequence diversity within *D. melanogaster* and *D. simulans* as well as more rapid divergence across the genus than its homologous but zygotically expressed epidermal growth factor receptor, *spitz* (unpublished data). Analogous to knock-outs, *RHAG* nulls in human mothers have been shown *not* to produce a maternal effect or a maternal-zygotic interaction. However, experimental data for mammalian species at the level of the *Drosophila* gold standard do not yet exist for the genes in our study. Nevertheless, we will show that the levels of sequence diversity within species as well as the accelerated relative nonsynonymous substitution rates of evolution across taxa for *RHD* and *Nalp5* (*Mater*) conform to our theoretical predictions and are an independent evolutionary confirmation of our proposed categories of gene expression.

The variation among taxa in gene expression pattern is more difficult to account for in a study like this one. In insects, maternal-effect genes can evolve so rapidly owing to relaxed constraint that they are absent in some taxa that include their more slowly evolving zygotic homologs (Demuth & Wade 2007a). In humans, *Nalp2* was expressed in both female and male tissues, but in mice it was expressed only in female tissues and not expressed in zygotic tissues (Table 3-5). Thus, *Nalp2* may be maternally expressed in some parts of the mammalian clade but zygotically expressed in others.

Table 3-6. Levels of replacement and silent site variation in *Nalp5* (*Mater*), *Nalp2, RHD, RHCE,* and *RHAG.* Sequence diversity at nonsynonymous sites in the coding region is in the column headed π_n, while diversity at synonymous sites is in the column headed π_s. The number of sequences for each gene is in the column headed N, and standard errors are reported in parentheses. Accession numbers for each sequence are given in Table 3-4. Note that the replacement site diversity (π_n) of the maternal-effect gene, *Nalp5,* is greater than that of its zygotically expressed homolog, *Nalp2.* Note also that replacement diversity (π_n) of the maternal-zygotic gene, *RHD,* is much greater than that of its zygotically expressed homolog, *RHAG.* Overall sequence diversity of *RHAG* may also be reduced owing to background selection and its position near the centromere, which reduces recombination and mutation in some species.

a) Human		N	π_n (s.e.)	π_s (s.e.)	π_n/π_s
published	*RHD*	8	0.0053 (0.0014)	0.0049 (0.0018)	1.0816
	RHCE	8	0.0053 (0.0015)	0.0037 (0.0024	1.4324
	RHAG	8	0.0012 (0.0009)	0.0000 (0.0000)	—
unpublished	*RHD*	5	0.0103 (0.0022)	0.0025 (0.0018)	4.1354
	RHCE	4	0.0057 (0.0017)	0.0033 (0.0022)	1.7172
	RHAG	9	0.0016 (0.0007)	0.0007 (0.0007)	2.285
b) Human	*Nalp2*	5	0.00053 (0.0003)	0.0017 (0.00089)	0.3063
	Nalp5	6	0.00065 (0.0002)	0.0011 (0.00057)	0.5555
Mouse	*Nalp2*	5	0 (0)	0 (0)	—
	Nalp5	7	0.0063 (0.0031)	0.0047 (0.0016)	1.3389

Within-species Polymorphism

The presumptive maternal-effect genes, *RHD* and *Nalp5,* consistently showed greater levels of nonsynonymous polymorphism than their zygotic homologs (Table 3-6), although the comparison is compromised for *Nalp2* and *Nalp5* in the mouse by the lack of any sequence diversity, synonymous or nonsynonymous, in *Nalp2.* In the *RH* genes, for both published and unpublished sequences, there was greater replacement site variation relative to silent site variation in *RHD* than in *RHAG.* However, the unpublished sequences were up to 2–4 times as polymorphic as published sequences at both coding and noncoding sites. We therefore argue that only high-quality published data should be used in analyses such as ours.

Mater/Nalp5 has greater levels of replacement site polymorphism than *Nalp2* in mice, consistent with the predictions of RSC theory (Table 3-6), although neither gene shows a reduction below 1 in π_n/π_s ratio consistent with purifying selection. However, the human *Mater/Nalp5* and *Nalp2* have comparable levels of replacement site diversity and there is evidence of both genes being equivalently diverse, suggesting that *Mater* is selectively neutral in humans (Table 3-6).

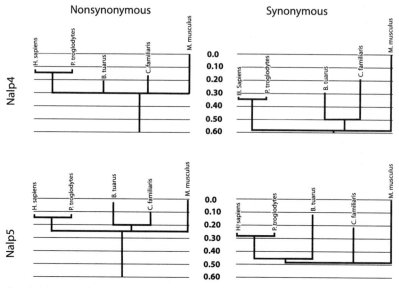

Figure 3-1. Sequence diversity at the maternal *Nalp5* locus is higher than that of the *Nalp2* locus because selection on the former is maternal, whereas selection on the latter is zygotic (see Table 3-6). As a result of this relaxed selective constraint, the coding region of *Nalp5* is evolving faster than that of *Nalp2*. Note that the branches are longer for the tree constructed using nonsynonymous or replacement substitutions for *Nalp5* than they are for the tree constructed for *Nalp2*. In contrast, the two gene trees constructed using neutral or synonymous substitutions from these same two genes are very similar. Note that both genes are experiencing net purifying selection as assumed in the theory (see text) since the trees constructed using nonsynonymous sites have smaller branch lengths for both genes than the corresponding tree constructed using synonymous substitutions.

Relative Rates of Base Pair Substitution across Mammalian Taxa

Mater and its zygotic homolog *Nalp2* have diverged across mammalian taxa (Fig. 3-1) in a different pattern than *RHD* and its zygotic homolog *RHAG* (Fig. 3-2). Recall that there is some evidence from expression data that *Nalp2* is maternal in mice but zygotic in humans. Our predicted ratio depends upon consistency of gene expression pattern as well as consistency in the average strength of selection.

The patterns of divergence across taxa of *Mater* and *Nalp2* are not those expected by RSC theory for a comparison between a strict maternal-effect gene (presumably *Mater*) and a strictly zygotic gene, *Nalp2* (Fig. 3-1). For the trees constructed using nonsynonymous substitutions, levels of divergence between *Mater* and *Nalp2* are similar and there is only slightly greater synonymous site divergence than nonsynonymous site divergence, indicating

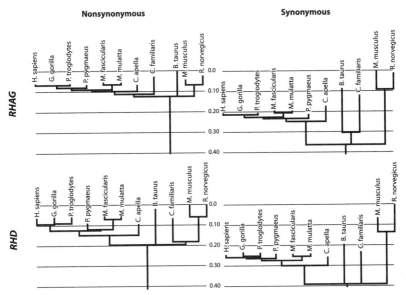

Figure 3-2. Nonsynonymous sequence diversity at the maternal-zygotic *RHD* locus is higher than that of the *RHAG* locus because selection on the former depends upon maternal-zygotic interactions, whereas selection on the latter is zygotic (see Table 3-4). As a result of the relaxed selective constraint, replacement sites of *RHD* are evolving faster than those of *RHAG*. Note the deeper branches of the tree constructed using nonsynonymous substitutions for *RHD* than for the tree constructed for *RHAG*. In contrast, the two gene trees constructed using neutral or synonymous substitutions from these same two genes are very similar. Note that both genes are experiencing net purifying selection as assumed in the theory (see text) since the trees constructed using nonsynonymous sites have smaller branch lengths for both genes than the corresponding tree constructed using synonymous substitutions.

weak overall purifying selection. Since expression pattern may be changing across taxa and the strength of selection appears to vary between these genes, it is not surprising that they do not display the expected pattern of relative divergence across these taxa. It would be better to revisit this comparison with more sequences and better information on pattern of gene expression and selection, as well as a planned sampling regime.

In contrast, the coding region of the maternal-zygotic *RHD* is evolving faster than *RHAG* (Fig. 3-2). The levels of divergence for *RHD* are nearly two times greater than those for *RHAG*. RSC theory predicts that the relative rate of divergence should be greater than 1 for purifying selection, with the relative rate increasing to very high values as average selection coefficients become increasingly negative. The greater synonymous site divergence than nonsynonymous site divergence indicates that both genes are subject to strong purifying selection.

CONCLUSIONS AND CAVEATS

Genetic models of maternal effects and maternal zygotic interactions have greatly improved our understanding of how maternal effects evolve (Cheverud & Wolf, this volume). When we can attribute a maternal effect to a single gene, we can make specific predictions about how maternal-effect genes evolve. We expect greater replacement site variation and faster rates of divergence in strictly maternal-effect genes than in their zygotic homologs. This pattern has been confirmed for one maternal-effect gene and its zygotic homolog in insects (Barker et al. 2005; Demuth & Wade 2007a; Cruickshank & Wade 2008). Here we provide the first test of this theory in mammals.

The polymorphism data provide the strongest support for the theory of relaxed selective constraint to date. Maternal-effect genes had substantially greater nonsynonymous polymorphism than their zygotic homologs in most comparisons. The divergence data for only one of the two maternal-effect gene and zygotic homolog pairs (*RHD* and *RHAG*) support the theoretical prediction that the maternal-effect gene is evolving faster than zygotic homologs throughout mammalian taxa.

Our analysis also indicates that expression-array data may have an important role in future work in the maternal-effects field. A gene with both maternal and zygotic functional expression will evolve more slowly, not faster, than a zygotically acting gene. Importantly, a gene with strict maternal expression should evolve faster than a zygotic homolog for purifying selection but not for adaptive positive selection as long as the average strength of selection is the same. Thus, we need to conduct genetic studies in conjunction with expression arrays to understand the evolutionary significance of maternal genetic effects.

Although the RSC theory is quite general, it applies only to strictly maternal-effect genes that have no functional expression in zygotic tissues or in males. For example, in fruit flies the gene *bicoid* (*bcd*) is a novel gene with maternal expression in early development that diverged from a tandem duplicate of *zerknüllt* (*zen*), a more highly conserved and zygotically expressed gene. The origin of *bcd* is relatively recent and confined to derived flies (Demuth & Wade 2007a). Array studies find low levels of zygotic expression of *bcd*, but genetic studies indicate that the zygotic expression does not contribute to the adult phenotype. As a consequence, within *Drosophilid* flies, the coding regions of *bcd* are evolving 0.59 as fast as in its zygotic homolog *zerknüllt* (*zen*), which is consistent with a maternal-effect population-genetic theory for a positively selected gene (Barker et al. 2005; Demuth & Wade 2007a).

Although the pattern found in this paper conforms to theoretical predictions as does the within-species between-gene comparison of sequence diversity, we would like to have more sequences and better information on pattern of gene expression and selection, as well as a planned, randomized sampling regime.

REFERENCES

Agrawal, A. F., Brodie, E. D., III & Brown, J. 2001a. Parent-offspring coadaptation and the dual genetic control of maternal care. *Science*, 292, 1710–1712.

Agrawal, A. F., Brodie, E. D., III & Wade, M. J. 2001b. On indirect genetic effects in structured populations. *American Naturalist*, 158, 308–323.

Atchley, W. R., Logsdon, T. E., Cowley, D. E. & Eisen, E. J. 1991. Uterine effects, epigenetics and postnatal skeletal development in the mouse. *Evolution*, 45, 891–909.

Barker, M. S., Demuth, J. P. & Wade, M. J. 2005. Maternal expression relaxes constraint on innovation of the anterior determinant, *bicoid*. *PLOS Genetics*, 1, e57.

Bjima, P. & M. J. Wade. 2008. The joint effects of kin, multilevel selection and indirect genetic effects on response to selection. *Journal of Evolutionary Biology*, in press.

Boycott, A. E. & Diver, C. 1923. On the inheritance of sinistrality in *Limnaea peregra*. *Proceedings of the Royal Society of London*, *Series B*, 98, 207–213.

Cheverud, J. M. & Wolf, J. B. 2009. The genetics and evolutionary consequences of maternal effects. In: *Maternal Effects in Mammals* (Ed. by D. Maestripieri & J. M. Mateo), pp. 11–37. Chicago: University of Chicago Press.

Cowley, D. E. 1991. Prenatal effects on mammalian growth: embryo transfer results. In: *The Unity of Evolutionary Biology*. Vol. 2. *Proceedings of the Fourth International Congress of Systematic and Evolutionary Biology* (Ed. by E. C. Dudley), pp. 762–779. Portland, OR: Dioscorides Press.

Cruickshank, T. E. & M. J. Wade. 2008. Microevolutionary support for a developmental hourglass: gene expression patterns shape sequence variation and divergence in Drosophila. *Evolution & Development*, in press.

Curtsinger, J. W., Service, P. M. & Prout, T. 1994. Antagonistic pleiotropy, reversal of dominance, and genetic polymorphism. *American Naturalist*, 144, 210–228.

Demuth, J. P. & Wade, M. J. 2006. Experimental methods for measuring gene interactions. *Annual Review of Ecology and Systematics*, 37, 289–316.

Demuth, J. P. & Wade, M. J. 2007a. Maternal expression increases the rate of bicoid evolution by relaxing selective constraint. *Genetica*, 129, 37–43.

Demuth, J. P. & Wade, M. J. 2007b. Population differentiation in the beetle *Tribolium castaneum*. I. Genetic architecture. *Evolution* 61, 494–509 Demuth, J. P. & Wade, M. J. 2007c. Population differentiation in the beetle *Tribolium castaneum*. II. Haldane's rule and incipient speciation. *Evolution* 61, 694–699.

Falconer, D. S. 1952. The problem of environment and selection. *American Naturalist*, 86, 293–298.

Haldane, J. B. S. 1942. Selection against heterozygosis in man. *Annual of Eugenics*, 11, 333–340.

Huang, C.-H. & Peng, J. 2005. Evolutionary conservation and diversification of Rh family genes and proteins. *Proceedings of the National Academy of Sciences USA*, 102, 15512–15517.

Jarrell, V. L., Day, B. N. & Prather, R. S. 1991. The transition from maternal to zygotic control of development occurs during the 4-cell stage in the domestic pig, *Sus scrofa*: quantitative aspects of protein synthesis. *Biology of Reproduction*, 44, 62–68.

Kitano, T., Sumiyama, K., Shiroishi, T. & Saitou, N. 1998. Conserved evolution of the Rh50 gene compared to its homologous Rh blood group gene. *Biochemical and Biophysical Research Communications*, 249, 78–85.

Kölliker, M., Brinkhof, M. W. G., Heeb, P., Fitze, P. S. & Richner, H. 2000. The quantitative genetic basis of offspring solicitation and parental response in a passerine bird with biparental care. *Proceedings of the Royal Society of London, Series B*, 267, 2127–2132.

Kölliker, M., Brodie, E. D., III & Moore, A. J. 2005. The coadaptation of parental supply and offspring demand. *American Naturalist*, 166, 506–516.

Kumar, S., Tamura, K. & Nei, M. 2004. MEGA 3: integrated software for molecular evolutionary genetics analysis and sequence alignment. *Briefings in Bioinformatics*, 5, 150–163.

Linksvayer, T. & Wade, M. J. 2005. The evolution of sociality in the Aculeate Hymenoptera: maternal effects, sib-social effects, and heterochrony. *Quarterly Review of Biology*, 80, 317–336.

Lynch, M. & Walsh, J. B. 1988. *Genetics and Analysis of Quantitative Traits*. Sunderland, MA: Sinauer Associates, Inc.

Maddison, W. P. & Maddison, D. R. 2006. Mesquite: a modular system for evolutionary analysis. Version 1.11. http://mesquiteproject.org

Matassi, G., Chérif-Zahar, B., Pesole, G., Raynal, V. & Cartron, J.-P. 1999. The members of the *RH* gene family (RH50 and RH30) followed different evolutionary pathways. *Journal of Molecular Evolution*, 48, 151–159.

Mousseau, T. & Fox, C. (Eds.). 1998. *Maternal Effects as Adaptations*. Oxford: Oxford University Press.

Nei, M. & Kumar, S. 2000. *Molecular Evolution and Phylogenetics*. New York: Oxford University Press.

Pennetier, S., Uzbekova, S., Perreau, C., Papillier, P., Mermillod, P. & Dalbiès-Tran, R. 2004. Spatio-temporal expression of the germ cell marker genes *MATER, ZAR1, GDF9, BMP15*, and *VASA* in adult bovine tissues, oocytes, and preimplantation embryos. *Biology of Reproduction*, 71, 1359–1366.

Roff, D. A. 1998. The detection and measurement of maternal effects. In: *Maternal Effects as Adaptations* (Ed. by T. A. Mousseau & C. W. Fox), pp. 83–96. Oxford: Oxford University Press.

Schilthuizen, M. & Davison, A. 2005. The convoluted evolution of snail chirality. *Naturwissenschaften*, 92, 504–515.

Schuler, G. D. 1997. Pieces of the puzzle: expressed sequence tags and the catalog of human genes. *Journal of Molecular Medicine*, 75, 694–698.

Shultz, R. M. 2002. The molecular foundations of the maternal to zygotic transition in the preimplantation embryo. *Human Reproduction Update*, 8, 323–331.

Sturtevant, A. H. 1923. Inheritance of direction of coiling in *Limnaea*. *Science*, 58, 269–270.

Tong, Z.-B., Bondy, C. A., Zhou, J. & Nelson, L. M. 2002. A human homologue of mouse *Mater*, a maternal effect gene essential for early embryonic development. *Human Reproduction*, 17, 903–911.

Tong, Z.-B., Gold, L., De Pol, A., Vanevski, K., Dorward, H., Sena, P., Palumbo, C., Bondy, C. A. & Nelson, L. M. 2004. Developmental expression and subcellular localization of mouse

MATER, an oocyte-specific protein essential for early development. *Endocrinology*, 145, 1427–1434.

Vigneault, C., McGraw, S., Massicotte, L. & Sirard, M.-A. 2004. Transcription factor expression patterns in bovine in vitro-derived embryos prior to maternal-zygotic transition. *Biology of Reproduction*, 70, 1701–1709.

Wade, M. J. 1998. The evolutionary genetics of maternal effects. In: *Maternal Effects as Adaptations* (Ed. by T. A. Mousseau & C. W. Fox), pp. 5–21. Oxford: Oxford University Press.

Wade, M. J. 2000. Neutrality arising from the balance between opposing levels of selection. *Evolution*, 54, 290–292.

Wade, M. J. 2001. Maternal effect genes and the evolution of sociality in haplo-diploids. *Evolution*, 55, 453–458.

Wade, M. J. 2003. Community genetics and species interactions. *Ecology*, 84, 583–585.

Wade, M. J. 2007. The coevolutionary genetics of ecological communities. *Nature Reviews Genetics*, 3, 185–195.

Wade, M. J. & Beeman, R. W. 1994. The population dynamics of maternal effect selfish genes. *Genetics*, 138, 1309–1314.

Wheeler, D. L., Church, D. M., Federhen, S., Lash, A. E., Madden, T. L., Pontius, J. U., Schuler, G. D., Schriml, L. M., Sequeira, E., Tatusova, T. A. & Wagner, L. 2003. Database resources of the National Center for Biotechnology. *Nucleic Acids Research*, 31, 28–33.

Wolf, J. B. 2000. Gene interactions from maternal effects. *Evolution*, 54, 1882–1898.

Wolf, J. B. 2003. Genetic architecture and evolutionary constraint when the environment contains genes. *Proceedings of the National Academy of Sciences USA*, 100, 4655–4660.

Wolf, J. B. & Brodie, E. D., III. 1998. Coadaptation of parental and offspring characters. *Evolution*, 52, 299–308.

Wolf, J. B., Brodie, E. D., III, Cheverud, J. M., Moore, A. J. & Wade, M. J. 1998. Evolutionary consequences of indirect genetic effects. *Trends in Ecology and Evolution*, 13, 64–69.

Wolf, J. B., Brodie, E. D., III & Moore, A. J. 1999. Interacting phenotypes and the evolutionary process. II. Selection resulting from social interactions. *American Naturalist*, 153, 254–266.

Wolf, J. B., Brodie, E. D., III & Wade, M. J. 2002a. Genotype-environment interaction and evolution when the environment contains genes. In: *Phenotypic Plasticity: Functional and Conceptual Approaches* (Ed. by T. DeWitt & S. Scheiner), pp. 173–190. Oxford: Oxford University Press.

Wolf, J. B., Vaughn, T. T., Pletscher, L. S. & Cheverud, J. M. 2002b. Contribution of maternal effect QTL to genetic architecture of early growth in mice. *Heredity*, 89, 300–310.

Wolpert, L., Beddington, R., Jessell, T., Lawrence, P., Meyerowitz, E. & Smith, J. 2002. *Principles of Development*. New York: Oxford University Press.

Wright, S. W. 1969. *Evolution and the Genetics of Population*. Vol. 2. *The Theory of Gene Frequencies*. Chicago: University of Chicago Press.

4

Maternal Effects on Evolutionary Dynamics in Wild Small Mammals

ANDREW G. MCADAM

INTRODUCTION

Maternal effects have now been documented for a variety of taxa (Roach & Wulff 1987; Bernardo 1996; Rossiter 1996; Mousseau & Fox 1998a), but they are thought to be particularly strong in mammals because of their long period of maternal dependence (Roff 1997; Reinhold 2002; Maestripieri & Mateo, this volume). Much research has focused on the adaptive nature of maternal adjustments to offspring phenotypes (Mousseau & Fox 1998b). These trans-generational effects can arise from either environmental conditions experienced by the mother or from maternal genes. When maternal contributions to offspring phenotypes are genetically based, they have the potential to contribute to the evolutionary process by providing an indirect source of genetic variation for selection to act upon (indirect genetic effects or IGEs; Moore et al. 1997; Wolf et al. 1998).

In wild mammals, maternal effects can play an important role in both adaptation to variable environmental conditions and evolutionary dynamics. Most empirical studies of maternal effects in mammals, however, have focused on environmental maternal effects on offspring traits and their fitness consequences. In contrast, the paucity of empirical studies of genetic maternal effects on evolutionary dynamics in the wild represents a potentially large gap in our understanding of the evolutionary process in wild mammals. In this chapter, I focus on genetic rather than environmental maternal effects, presenting an overview of their potential importance to evolutionary dynamics. I then review some of my work on a natural population of North American red squirrels (*Tamiasciurus hudsonicus*) in which we have quantified all of the necessary parameters for testing predictions of models of maternal-effect evolution in the wild. Finally, I discuss some of the implications of genetic maternal effects and maternal-effect evolution for internally driven population cycles in small mammals (*sensu* Chitty 1967).

MATERNAL-EFFECT EVOLUTION

In the absence of maternal effects, the change in mean value of a trait across one generation (i.e., response to selection) can be predicted using the breeders' equation, in which the response to selection (R) is predicted to equal the product of heritability (h^2) and selection (S) ($R = h^2 S$; Falconer & Mackay 1996). However, in the presence of genetically based maternal effects the potential for evolutionary change is no longer limited to direct genetic effects (i.e., h^2). Whether this maternal genetic variation accelerates or retards the response to selection, however, depends on whether the maternal effects act positively or negatively on the offspring phenotype. Here I follow Kirkpatrick and Lande's (1989) definition of the maternal-effect coefficient, m, as the effect of a maternal trait on the offspring phenotype while holding the effects of offspring genetic effects constant. In this case a negative maternal-effect coefficient indicates that larger values of the maternal trait result in smaller values of the offspring trait while controlling for offspring genetic effects. While the overall contribution of maternal effects is often presented as the proportion of offspring variance explained by maternal effects (m^2), it is important to remember that this variation results from the cumulative contributions of specific maternal traits that can each act either positively or negatively on offspring phenotype.

Predicting the response to selection is further complicated by the fact that genes responsible for maternal effects are possessed by the mother but their phenotypic and fitness consequences are experienced by her offspring. The trans-generational nature of genetic maternal effects means that the response to selection in the current generation cannot be dissociated from the strength of selection experienced in the previous generation. Maternal effects, therefore, can introduce an evolutionary time lag, in which the response to selection in the current generation also depends on the strength of selection in the previous generation (Kirkpatrick & Lande 1989; 1992; Lande & Kirkpatrick 1990; Wolf et al. 1998). As a result, populations may continue to evolve after selection has ceased and evolution may also temporarily proceed in a direction opposite to the direction of current selection (Kirkpatrick & Lande 1989).

For example, the predicted response of a hypothetical offspring trait to 5 generations of directional selection is shown in Figure 4-1 for models including positive ($m = 0.8$), negative ($m = -0.8$) and no maternal effects ($m = 0$; open dots). In the absence of maternal effects (open dots), the offspring trait is predicted to change consistently throughout the period of selection (following the breeders' equation as above) and then stop evolving and remain constant once selection is relaxed. The presence of positive or negative maternal effects, however, can result in accelerated or retarded overall responses to selection, respectively. Furthermore, in the presence of maternal effects the response to selection is not consistent throughout the period of selection, but instead is influenced by the strength of selection in the previous generation (Figure 4-1). In some cases the response of the offspring trait can even be in the opposite direction of selection (Figure 4-1, gray dots; Kirkpatrick & Lande 1989).

These dramatic and sometimes counterintuitive predictions of maternal-effect models depend on the presence of a genetic basis to maternal variation (Kirkpatrick & Lande 1989; Wolf et al. 1998). While the genetic basis to maternal effects has frequently been estimated in populations subject to generations of artificial selection or benign laboratory conditions (e.g., Cheverud 1984; Roff 1997; Wilson & Réale 2006), indirect genetic effects in nondomestic species (Agrawal et al. 2001; Hunt & Simmons 2002; Rauter & Moore 2002) and under natural field conditions have only recently been investigated (e.g., Byers et al. 1997; Thiede 1998; Table 4-1). As a result, the actual importance of maternal effects to evolutionary dynamics in natural populations remains unclear.

QUANTIFYING GENETIC MATERNAL EFFECTS

Quantifying the potential contribution of genetic maternal effects to evolutionary dynamics is a difficult task. Models of maternal-effect evolution include many more parameters than simpler models of direct evolutionary change, some of which can be difficult to measure in the wild. Populations for which large multigenerational pedigrees are available provide the opportunity to separate direct genetic contributions from maternal genetic contributions to offspring phenotypes using maximum likelihood "animal models" (see Kruuk 2004; Wilson & Festa-Bianchet, this volume). For

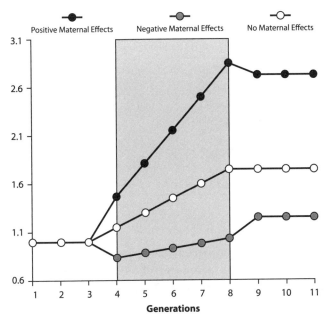

Figure 4-1. Predicted evolutionary dynamics of a hypothetical trait experiencing three generations of no selection (selection gradient on offspring trait: $\beta_0 = 0$), followed by five generations of directional selection
favoring increased values of the trait ($\beta_0 = 0.5$, area shown in gray), followed by three generations of no
selection ($\beta_0 = 0$). The predicted phenotypic response is plotted for a trait experiencing positive maternal genetic effects (maternal-effect coefficient: $m = 0.8$; solid dots), negative maternal genetic effects ($m = -0.8$;
gray dots) and no maternal genetic effects ($m = 0$; open dots). The predicted dynamics follow Kirkpatrick &
Lande (1989; 1992) for an offspring trait affected by one maternal trait (selection gradient on maternal trait:
$\beta_m = 0$; additive genetic variance in offspring trait: $G_{oo} = 0.3$; additive genetic variance in maternal trait: G_{mm}
$= 0.3$; genetic covariance between maternal and offspring trait: $G_{mo} = 0.18$; phenotypic variance in offspring
trait: $P_{oo} = 1$; phenotypic variance in maternal trait: $P_{mm} = 1$; phenotypic covariance between maternal and
offspring trait $P_{mo} = 0.7$).

example, Wilson et al. (2005a) recently used animal model approaches to
quantify sources of variation in three offspring traits for a population of
Soay sheep for which pedigree data were available for 1236 individuals born
between 1985 and 2003. All three traits experienced large maternal effects,
and genetic maternal effects accounted for a larger proportion of phenotypic
variation than direct genetic variation. This analytical approach, however,
requires substantial multigenerational pedigrees (Quinn et al. 2006) that
might not be available for many species (but see Kruuk 2004).

Alternatively, cross-fostering experiments provide a useful experimental
tool for partitioning the contribution of genetic, maternal, and environmen-

Table 4-1. Estimates of maternal genetic effects in wild mammals. The proportion of total phenotypic variation that is due to maternal genetic effects (V_{AM}/V_p) and the genetic covariance between
direct and maternal genetic ($\mathrm{cov}[A_o, A_M]$) effects are presented where available.

Species	Offspring trait	Maternal effect (V_{AM}/V_p)	$\mathrm{cov}[A_o, A_M]$	Source
Red squirrel, *Tamiasciurus hudsonicus*	Nestling growth rate	0.11	0.020	McAdam et al. 2002
Soay sheep, *Ovis aries*	Birth weight	0.12	−0.016	Wilson et al. 2005a
Bighorn sheep, *Ovis canadensis*	Body size	0.02–0.06	NA	Wilson et al. 2005b

NA: data not available.

tal variation to overall phenotypic variation within a single generation. In a
reciprocal cross-fostering experiment half the offspring from one litter are
exchanged with an equal number of offspring from a paired litter. This design, therefore, separates additive genetic variance and postnatal maternal
variance within a single statistical model (Rutledge et al. 1972). It should be
noted, however, that any maternal effects acting prior to cross-fostering will
still remain confounded with additive genetic effects. Cross-fostering experiments have been used to estimate heritabilities in laboratory populations of
mice (Rutledge et al. 1972; Atchley & Rutledge 1980; Cheverud et al. 1983; Cheverud 1984; Leamy & Cheverud 1984; Riska et al. 1984) and in some natural
populations of birds (see Roff 1997; Merilä & Sheldon 2001). In many mammals, however, offspring are difficult to manipulate (but see Humphries &
Boutin 1996; Murie et al. 1998; Mappes & Koskela 2004; Maestripieri, this
volume), which has resulted in limited use of this technique for studying
natural populations of mammals.

TESTING MODELS OF MATERNAL-EFFECT EVOLUTION USING RED SQUIRRELS

North American red squirrels have provided an excellent opportunity to examine the influence of direct and maternal genetic variation on the evolutionary dynamics of offspring traits, particularly offspring growth rates,
in a natural population of mammals. Red squirrels near Kluane Lake in
the southwest Yukon, Canada (61°N, 138°W), have been studied consistently
since 1987, so there exist both an extensive pedigree of marked individuals
with known matrilines and a comprehensive understanding of the ecology
of this system (e.g., Boutin & Larsen 1993; Stuart-Smith & Boutin 1995; Ber-

teaux & Boutin 2000; Humphries & Boutin 2000; Anderson & Boutin 2002; Réale et al. 2003b; Boutin et al. 2006; McAdam et al. 2007). In addition, this population is relatively dense (approximately 2.5 squirrels/ha) so data can be collected on a large number of animals within a single breeding season. Finally, it is possible to cross-foster offspring between litters (Humphries & Boutin 1996). This experimental manipulation of relatedness within litters provides the opportunity to disentangle genetic, environmental, and maternal effects under natural conditions.

Red squirrels in this population feed primarily on the seeds of white spruce (*Picea glauca*) cones. Spruce cones are harvested from trees in late summer and are stored for future consumption in centrally located caches (middens), within exclusively defended year-round territories (Smith 1968). The production of spruce cones has been measured in the study area since 1986 and varies dramatically from one year to the next (Humphries & Boutin 2000; McAdam & Boutin 2003b). This annual variation in the abundance of food is thought to have important implications for many aspects of red squirrel life history (Berteaux & Boutin 2000; Humphries & Boutin 2000; Réale et al. 2003b; Boutin et al. 2006) and provides an ecological context for the study of contemporary selection and evolution in this population.

My coworkers and I were interested in testing the importance of maternal effects to the evolutionary dynamics of offspring growth in body mass. Red squirrels in this area raise offspring in conspicuous grass nests or tree cavities, so offspring can be temporarily removed from their natal nest to collect mass measurements. Growth of red squirrels during the nestling period has been quantified for over 3000 juveniles in this population based on the increase in mass from soon after birth (mean ± se: 6.8 ± 0.2 days of age) to just prior to first emergence from the natal nest (27.2 ± 0.2 days of age) and is linear over this time period (McAdam et al. 2002).

During the breeding season of 1999, we reciprocally cross-fostered 33 pairs of litters (177 offspring). Variation in nestling growth rates from foster and control offspring indicated that growth rates were heritable ($h^2 = 0.10$), but offspring also experienced very strong maternal effects that accounted for 81% of the variation in nestling growth rates (McAdam et al. 2002; see McAdam & Boutin 2003a for similar results under low ambient food conditions). In addition, there was a large positive covariance between direct and maternal genetic effects, suggesting that these positive maternal effects can accelerate the overall response to selection. Although cross-fostering experiments do not directly address the genetic basis to maternal effects, we separately determined that 69% of these maternal effects resulted from variation in maternal litter size and parturition date, which are both heritable maternal traits (Réale et al. 2003a). Other maternal characteristics, such as age, reproductive experience, body size, maternal mass at parturition, and territory size were found to not be significant maternal effects on offspring growth rates (McAdam et al. 2002). We therefore calculated genetic maternal effects on offspring growth based on the heritabilities and contributions of these two maternal traits (litter size and parturition date) to maternal effects on offspring growth (see McAdam et al. 2002 for details). Overall these genetic maternal effects resulted in greater than a threefold increase in the potential for evolutionary change (total heritability including maternal genetic effects: $h_t^2 = 0.36$) in this offspring trait than would have been predicted based on direct genetic effects alone ($h^2 = 0.10$; McAdam et al. 2002). These results represented one of the first quantifications of the contribution of maternal effects to the potential for evolution in a wild vertebrate (see also Wilson et al. 2005a, b; Wilson & Festa-Bianchet, this volume) and suggested that simple heritabilities, which ignore the potential contributions of maternal effects, might severely underestimate the potential for evolutionary change.

Testing the actual importance of maternal effects to evolutionary dynamics, however, also requires measures of the strength of selection and the response to selection. Red squirrel growth rates tend to be positively associated with survival to potential breeding age overall, but the strength of selection varies widely among cohorts (McAdam & Boutin 2003b). In most years offspring with higher growth rates had a higher probability of survival, but in some years slower growth was favored by selection. Some of this variation in selection on growth rates is associated with spring temperatures and variation in the abundance of food (McAdam & Boutin 2003b) but much of the variation in selection is currently unexplained. This temporal variation in selection on growth and the high degree of overlap in red squirrel generations precludes multigenerational predictions, but a series of single-generation predictions can be made for each cohort.

Nestling growth rates are positively correlated with the abundance of spruce cones produced in the previous fall (McAdam & Boutin 2003b; Boutin et al. 2006) and experimental manipulations of food abundance have confirmed the importance of food abundance to nestling growth (e.g., McAdam & Boutin 2003a). We know that the effect of abundant food resources on offspring growth results from improved maternal provisioning because offspring are entirely dependent on maternally derived resources during this life history stage. The large temporal variation in cone production means that much of the observed interannual variation in nestling growth rate is the result of variation in environmental conditions and not an evolutionary response to selection. Red squirrels are also relatively long-lived (life expectancy given recruitment is 3.5 years and maximum life span is 8 years; McAdam et al. 2007), so an individual female is likely to raise offspring in good, poor, and moderate food conditions during her lifetime. As a result, repeated growth measurements of offspring raised by the same female but under different environmental conditions were used to quantify the environmental effect on growth for each year based on deviations from each female's lifetime mean. These environmental effects were averaged across females within each year to come up with an overall environmental effect on growth for that year. Observed growth rates were then corrected for these environmental effects prior to testing for evidence of evolutionary changes across generations. The response to selection was measured for each cohort as the difference in mean growth rate between all offspring produced by females in that cohort (i.e., across multiple years) and the mean growth rate of offspring from that cohort. For example, the mean growth rate for offspring born in 1994 (corrected for environmental effects) was 1.88 g/day and females from this cohort raised offspring over their lifetime that had a mean growth rate of 2.00 g/day (corrected for environmental effects). As a result the observed response to selection for the 1994 cohort was calculated as 0.12 g/day (see McAdam & Boutin 2004 for additional details).

Given the strong annual variation in selection on nestling growth, it is perhaps not surprising that observed changes in growth rates across generations (corrected for environmental effects) were found to vary among cohorts; responses between 1989 and 2000 ranged from 0.04 to 0.72 standard deviations (McAdam & Boutin 2004). In the absence of maternal effects significant episodes of selection are predicted to be associated with significant responses to selection. However, models of maternal-effect evolution suggest that the response to selection in the current generation also depends on the strength of selection in the previous generation (Kirkpatrick & Lande 1989). Only 1 (1992) of the 12 cohorts between 1989 and 2000 exhibited a significant response to significant selection (McAdam & Boutin 2004). Some cohorts with significant selection showed no significant response, while other cohorts experiencing weak selection often exhibited a much larger response.

Models of maternal-effect evolution make two important predictions that can be tested using data such as these. First, maternal effects are predicted to affect the overall rate of evolution (Kirkpatrick & Lande 1989). They can either accelerate or retard the response to selection depending on whether they act positively or negatively on the offspring trait. In red squirrels, there are large maternal effects that act positively on offspring growth, so we would expect a much greater response to selection than would be predicted based on heritability alone (McAdam et al. 2002). In fact, the realized heritability based on observed responses for 12 cohorts of squirrels ($h^2_r = 0.56 \pm 0.17$; McAdam & Boutin 2004) was much greater than the heritability of this trait ($h^2 = 0.10$), but was not different from the total heritability, which included the contribution of maternal genetic effects ($h_t^2 = 0.36$; McAdam et al. 2002). Second, correlations between observed responses to selection and the strength of both current and previous selection provided evidence of an evolutionary time lag in which current responses also depended on previous selection (McAdam & Boutin 2004). Therefore, the observed dynamics of red squirrel growth rates supported both of the fundamental predictions of models of maternal-effect evolution and represented one of the first empirical tests of these models in the wild (McAdam & Boutin 2004).

GENETIC MATERNAL EFFECTS IN THE WILD

Testing models of maternal-effect evolution under natural conditions is important because an organism's environment provides the context within which natural selection acts and genetic variation is expressed. Environmental effects on the expression of genetic ($G \times E$) and maternal ($M \times E$) sources of variation (e.g., Wilson et al. 2006) suggest that estimates of heritability and maternal effects expressed in captivity may not reflect levels of variation under natural conditions. While the former has received considerable attention (Weigensberg & Roff 1996; Hoffmann 2000; Charmantier & Garant 2005), we know little about the effects of captivity on estimates of maternal sources of variation (Charmantier & Garant 2005). In addition, patterns of selection shaping genetic architecture likely also differ between controlled conditions of captivity and variable natural environments. In particular, favorable combinations of direct and maternal genetic effects are expected to be driven to fixation under consistent directional selection such that negative direct-maternal genetic correlations should prevail. In fact most previous estimates of the direct-maternal genetic correlation from agricultural and laboratory species have been negative (Cheverud 1984; Roff 1997). However, temporal and spatial patterns of selection in the wild are likely to differ substantially from the consistent selection imposed directly on agricultural species through artificial selection or indirectly as a result of the consistent conditions that characterize captive populations. It is clear that selection on nestling growth rates is anything but consistent in red squirrels (McAdam & Boutin 2003b). This variation in the strength and direction of selection (together with overlapping generations) provides one explanation for the maintenance of a large positive genetic correlation between direct and maternal genetic effects and hence the large potential for evolution contributed by maternal effects. The paucity of studies examining selection across multiple temporal replicates (Kingsolver et al. 2001; but see Boyce & Perrins 1987; Przybylo et al. 2000; Sinervo et al. 2000; Kruuk et al. 2001; Svensson & Sinervo 2004; Wilson et al. 2006), however, limits our ability to assess patterns in either the temporal or spatial grain of selection in the wild. A similar positive genetic correlation between direct and maternal effects was found for emergence mass in Columbian ground squirrels (*Spermophilus columbianus*; McAdam & Murie, unpublished data). In addition, the provisioning of red squirrels with ad lib access to food reduced maternal effects by over 70% and completely eliminated the large positive direct-maternal genetic correlation (McAdam & Boutin 2003a). These results not only identify some component of maternal acquisition or provisioning of resources as the underlying mechanism responsible for maternal effects on offspring growth, but also indicate that estimates of maternal effects from animals raised on ad lib food, either in the laboratory or in captive populations, might severely underestimate the importance of maternal effects to evolutionary dynamics.

While it is clear that maternal effects can have large effects on evolutionary dynamics, it is still not yet clear how frequently they do. The results for sciurids described above contrast with estimates of the direct-maternal genetic correlation in domestic and wild ungulates, which have been found to be consistently negative (Wilson & Réale 2006; Wilson & Festa-Bianchet, this volume). In addition, the large increase in the potential for evolution of red squirrel nestling growth in body mass resulting from genetic maternal effects was not found in nestling growth in body size (McAdam et al. 2002). Maternal effects on other red squirrel traits have not yet been estimated. The contrast in the potential evolutionary importance of maternal effects between these two offspring traits in red squirrels results largely from differences in the correlation between maternal and offspring genetic effects. Additional estimates of this important maternal parameter collected from a variety of wild animals would provide extremely valuable information on the potential for maternal effects to contribute to short-term evolutionary dynamics as well as some indication of the grain of selection in the wild.

GENETIC MATERNAL EFFECTS AND POPULATION CYCLES IN SMALL MAMMALS

The indirect contributions of genetic maternal effects to the evolutionary dynamics of offspring traits have potentially important implications for adaptive density cycles of small mammal populations. Periodic fluctuations in the population densities of northern small mammals represent one of the most striking natural patterns in animal ecology. Chitty (1967) hypothesized that such cycles could be explained by natural selection favoring individuals with characteristics that are beneficial at either high or low density and that have consequences for population density. More specifically, Chitty (1967) proposed that aggressiveness or intraspecific interference is negatively related to some component of reproductive output, such that at low population density individuals with low aggression and high reproductive output are favored. The subsequent increased prevalence of individuals with high reproductive output is, therefore, predicted to increase population density resulting in selection favoring more aggressive types with a lower reproductive output. The evolution of increased aggressiveness and decreased fecundity should then result in a decline in the population density, completing the cycle. Chitty's (1967) hypothesis, therefore, is based on density-dependent selection resulting in the cyclic evolution of behavioral and reproductive types that specialize in either high- or low-density conditions. Thus, the hypothesis depends on genetically based differences among individuals that can respond to density-dependent changes in natural selection on an ecological timescale compatible with observed changes in population density.

A necessary condition for the Chitty hypothesis, therefore, is that the trait under selection has sufficient genetic variation to evolve at the same temporal scale as changes in population density. Boonstra and Boag (1987) tested this implicit assumption of the Chitty hypothesis but found insufficient heritabilities in the growth rate and body size of meadow voles (*Microtus pennsylvanicus*) to explain the observed 2–4 year cycles and, therefore, rejected the Chitty hypothesis. Specifically, female growth rates measured between 20 and 65 days of age increased by 0.138 g/day between 1982 (low-increase year) to 1983 (peak year). This observed change would require both strong directional selection (i.e., selection intensity i = 0.6; this value would fall within the top 10% of previously reported selection intensities reported from the wild; Kingsolver et al. 2001) and a very high heritability (h^2 = 0.93) to have evolved over these two generations (Boonstra & Boag 1987). The actual heritability of female growth rate (0.27) falls well below this value (Boonstra & Boag 1987) and is therefore insufficient to generate the observed phenotypic

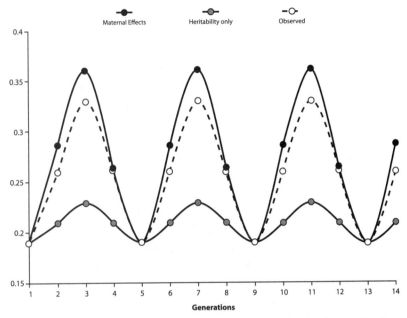

Figure 4-2. Predicted evolutionary responses to strong directional selection based on direct genetic effects alone ($h^2 = 0.27$; gray dots) suggests that there is insufficient genetic variation in the growth rate of female *Microtus pennsylvanicus* ($h^2 = 0.27$) to explain observed phenotypic changes in growth rates between peak and low phases of the cycle (open dots). A model of maternal-effect evolution, which includes maternal genetic effects (solid dots), however, is sufficient to explain the observed changes. Phenotypic changes in growth rates are extrapolated from observed changes in the growth rate of female *Microtus* between a low-increase year and a peak year (Boonstra & Boag 1987). Selection was assumed to alternate between two generations of strong directional selection for increased growth rates (standardized selection gradient on offspring trait: $\beta' = 0.6$) followed by two generations of selection for decreased growth ($\beta' = -0.6$; following Boonstra & Boag 1987). The maternal-effect model is based on a single heritable maternal trait (litter size) following Kirkpatrick & Lande (1989, 1992). Parameters for the maternal-effect model were taken from Boonstra & Boag (1987; $\beta' = 0.6$ or -0.6, $P_{oo} = 0.015$, $G_{oo} = 0.004$, $P_{mm} = 2.38$, $P_{mo} = 0.031$) or were estimated based on comparable data (McAdam et al. 2002) for the purpose of demonstration ($m = -0.15$, $G_{mm} = 0.61$, $G_{mo} = -0.04$). See Figure 4-1 legend for definitions of the model parameters.

changes (see Figure 4-2). As a result of this study (but see also Boonstra & Hochachka 1997), the literature has largely rejected the Chitty hypothesis as an explanation for small mammal population cycles (Krebs 1996).

The calculations outlined above, however, do not consider the potentially important indirect genetic contributions of maternal effects to evolutionary dynamics. The importance of environmental maternal effects has been frequently suggested as a viable mechanism by which delayed density dependence might drive population cycles (e.g., Rossiter 1994; Ginzburg 1998). These models, however, consider only trans-generational environmental effects on offspring phenotypes and do not consider the potential contribution of genetic maternal effects to adaptive evolutionary cycles. A simple maternal-effect model in which one offspring trait (e.g., juvenile growth rate) is influenced by one maternal trait (e.g., litter size), indicates that phenotypic changes in the growth rates of female *Microtus* observed by Boonstra and Boag (1987) can potentially be explained by reasonable levels of genetic maternal effects (Figure 4-2, solid dots). In the absence of maternal effects (Figure 4-2, gray dots), cyclic changes in natural selection on juvenile growth are clearly insufficient to explain the observed changes in juvenile growth rates (Figure 4-2, open dots). However, the inclusion of positive heritable maternal effects (maternal-effect coefficient: $m = -0.15$; heritability of maternal effects: $h^2_m = 0.26$; genetic correlation between direct and maternal genetic effects: $r_{g \text{ (direct-maternal)}} = -0.81$) were sufficient to generate predicted changes in growth rates across generations (Figure 4-2, solid dots) that match observed changes in growth rates in both the magnitude and frequency of the cycles.

So how realistic are these hypothetical estimates of genetic maternal effects? Maternal effects often account for a large proportion of variation in offspring traits in wild small mammals (Boonstra & Boag 1987; Boonstra & Hochachka 1997; Hansen & Boonstra 2000; McAdam et al. 2002). In particular, increased litter size typically has a negative effect on offspring growth rates (Humphries & Boutin 2000; Mappes & Koskela 2004) and negative genetic correlations between litter size and offspring growth rates have also been documented (McAdam et al. 2002; Mappes & Koskela 2004). As a result it is possible that offspring traits, such as early growth rates, might possess a much greater potential for evolution in wild mammals than would be revealed by simply measuring heritability. The goal of presenting this hypothetical model is not to determine whether maternal effects drive small mammal cycles, but rather to further highlight the potential importance of maternal effects to evolutionary dynamics and to suggest caution when inferring the potential for evolutionary change in the absence of information on genetic maternal effects. These modeling results do not provide evidence for or against the Chitty hypothesis, but instead suggest that previous estimates of genetic variation (Boonstra & Boag 1987; Boonstra & Hochachka 1997), which did not consider genetic maternal effects, should not be used as evidence against the Chitty hypothesis.

CONCLUSIONS AND FUTURE DIRECTIONS

Genetic maternal effects add an additional layer of complexity to the already formidable task of studying contemporary adaptation in the wild (e.g., Grant & Grant 2002). The prevalence of maternal effects, however, suggests that such complexity cannot be ignored if we are to understand how traits respond to episodes of natural selection. Maternal effects are widely acknowledged to be a potential source of bias in estimates of heritability (Lynch & Walsh 1998). As such, they have been implicated as one potential explanation for the prevalence of evolutionary stasis or reversed responses (Merilä et al. 2001) despite ample evidence of significant levels of both additive genetic variation (e.g., Weigensberg & Roff 1996; Hoffmann 2000) and directional selection in the wild (Endler 1986; Hoekstra et al. 2001; Kingsolver et al. 2001). However, it is also possible that unpredicted observed patterns are caused by more complex underlying evolutionary dynamics than are represented by the simple breeders' equation. For example, observed changes in red squirrel growth rates in response to episodes of strong directional selection would not have been adequately predicted by a simple model of evolution that considered only direct genetic effects. Instead, the more complex model of maternal-effect evolution was necessary to explain changes in nestling growth rates across generations.

The development of theoretical models of maternal-effect evolution over the past two decades has greatly advanced our understanding of the potential contributions of maternal effects to evolutionary change. There is much empirical work yet to be done, however, to determine the generality of these interesting and at times counterintuitive dynamics. To do this we must quantify three important parameters: the strength and direction of the maternal-effect coefficient (m), the genetic basis to maternal effects, and the genetic correlation between direct and maternal genetic effects. While the number of studies documenting the identity and magnitude of maternal effects has grown in recent years (Maestripieri & Mateo, Mateo & Maestripieri, this volume), we still know very little about genetic maternal effects in the wild (Table 4-1). Until more measurements of key maternal-effect parameters are made in natural populations, the importance of maternal-effect evolution as a general evolutionary process will remain unknown.

Wild rodent populations provide valuable opportunities to not only quantify the importance of maternal effects to offspring phenotypes but also to examine their consequences for short-term evolutionary dynamics. In particular, species for which offspring can be cross-fostered in the wild or that can be brought temporarily into captivity for the purpose of cross-fostering provide the opportunity to disentangle direct and maternal effects, which would otherwise be confounded in unmanipulated litters. Genetic maternal effects can also be estimated from multigenerational pedigrees using maximum likelihood animal models. Long-term field studies of wild mammals typically record data on mother-offspring relationships, which together with the proliferation of microsatellite paternity analysis, can provide extensive field pedigrees. These animal models have only recently been applied to natural populations of mammals (see Kruuk 2004), but there are several additional mammalian systems with sufficient long-term data to perform such analyses. In addition, the comparable ease with which a large number of individuals can be monitored and the short generation times of many rodent species provide the opportunity to test evolutionary changes over relatively short timescales that would simply not be feasible in larger mammals. Quantifying the necessary parameters of models of maternal-effect evolution is no easy task, but wild rodents appear to be promising systems in which the general importance of maternal effects to adaptation can be tested.

ACKNOWLEDGMENTS

This is contribution number 42 in the Kluane Red Squirrel Project. The investigation of maternal effects on red squirrel growth rates would not have been possible without the dedicated work of many field assistants and was funded by Natural Sciences and Engineering Research Council of Canada (NSERC) grants to Stan Boutin. Additional funds were provided by the Killam Foundation, the American Society of Mammalogists and the Arctic Institute of North America. I was supported during the writing of this chapter by the National Science Foundation and Michigan State University. Helpful comments on an earlier version of this chapter were provided by A. Wilson, J. M. Mateo, and D. Maestripieri.

REFERENCES

Agrawal, A. F., Brodie, E.0 D., III & Brown, J. 2001. Parent-offspring coadaptation and the dual genetic control of maternal care. *Science*, 292, 1710–1712.
Anderson, E. M. & Boutin, S. 2002. Edge effects on survival and behaviour of juvenile red squirrels (*Tamiasciurus hudsonicus*). *Canadian Journal of Zoology*, 80, 1038–1046.
Atchley, W. R. & Rutledge, J. J. 1980. Genetic components of size and shape. I. Dynamics of

components of phenotypic variability and covariability during ontogeny in the laboratory rat. *Evolution*, 34, 1161–1173.

Bernardo, J. 1996. Maternal effects in animal ecology. *American Zoologist*, 36, 83–105.

Berteaux, D. & Boutin, S. 2000. Breeding dispersal in female North American red squirrels. *Ecology*, 81, 1311–1326.

Boonstra, R. & Boag, P. T. 1987. A test of the Chitty Hypothesis: inheritance of life-history traits in meadow voles *Microtus pennsylvanicus*. *Evolution*, 41, 929–947.

Boonstra, R. & Hochachka, W. M. 1997. Maternal effects and additive genetic inheritance in the collared lemming *Dicrostonyx groenlandicus*. *Evolutionary Ecology*, 11, 169–182.

Boutin, S. & Larsen, K. W. 1993. Does food availability affect growth and survival of males and females differently in a promiscuous small mammal, *Tamiasciurus hudsonicus*? *Journal of Animal Ecology*, 62, 364–370.

Boutin, S., Wauters, L. A., McAdam, A. G., Humphries, M. M., Tosi, G. & Dhondt, A. A. 2006. Anticipatory reproduction and population growth in seed predators. *Science*, 314, 1928–1930.

Boyce, M. S. & Perrins, C. M. 1987. Optimizing great tit clutch size in a fluctuating environment. *Ecology*, 68, 142–153.

Byers, D. L., Platenkamp, G. A. J. & Shaw, R. G. 1997. Variation in seed characteristics in *Nemophila menziesii*: evidence of a genetic basis for maternal effects. *Evolution*, 51, 1445–1456.

Charmantier, A. & Garant, D. 2005. Environmental quality and evolutionary potential: lessons from wild populations. *Proceedings of the Royal Society of London, Series B*, 272, 1415–1425.

Cheverud, J. M. 1984. Evolution by kin selection: a quantitative genetic model illustrated by maternal performance in mice. *Evolution*, 38, 766–777.

Cheverud, J. M., Rutledge, J. J. & Atchley, W. R. 1983. Quantitative genetics of development: genetic correlations among age-specific trait values and the evolution of ontogeny. *Evolution*, 37, 895–905.

Chitty, D. 1967. The natural selection of self-regulatory behaviour in animal populations. *Proceedings of the Ecological Society of Australia*, 2, 51–78.

Endler, J. A. 1986. *Natural Selection in the Wild*. Princeton, NJ: Princeton University Press.

Falconer, D. S. & Mackay, T. F. C. 1996. *Introduction to Quantitative Genetics*. Harlow, UK: Prentice Hall.

Ginzburg, L. R. 1998. Inertial growth: population dynamics based on maternal effects. In: *Maternal Effects as Adaptations*. (Ed. by T. A. Mousseau & C. W. Fox), pp. 42–53. New York: Oxford University Press.

Grant, P. R. & Grant, B. R. 2002. Unpredictable evolution in a 30-year study of Darwin's finches. *Science*, 296, 707–711.

Hansen, T. F. & Boonstra, R. 2000. The best in all possible worlds? A quantitative genetic study of geographic variation in the meadow vole, *Microtus pennsylvanicus*. *Oikos*, 89, 81–94.

Hoekstra, H. E., Hoekstra, J. M., Berrigan, D., Vignieri, S. N., Hoang, A., Hill, C. E., Beerli, P. & Kingsolver, J. G. 2001. Strength and tempo of directional selection. *Proceedings of the National Academy of Sciences USA*, 98, 9157–9160.

Hoffmann, A. A. 2000. Laboratory and field heritabilities: some lessons from *Drosophila*. In: *Adaptive Genetic Variation in the Wild*. (Ed. by T. A. Mousseau, B. Sinervo & J. Endler). New York: Oxford University Press.

Humphries, M. M. & Boutin, S. 1996. Reproductive demands and mass gains: a paradox in female red squirrels (*Tamiasciurus hudsonicus*). *Journal of Animal Ecology*, 65, 332–338.

Humphries, M. M. & Boutin, S. 2000. The determinants of optimal litter size in free-ranging red squirrels. *Ecology*, 81, 2867–2877.

Hunt, J. & Simmons, L. W. 2002. The genetics of maternal care: direct and indirect genetic effects of phenotype in the dung beetle *Onthophagus taurus*. *Proceedings of the National Academy of Sciences USA*, 99, 6828–6832.

Kingsolver, J. G., Hoekstra, H. E., Hoekstra, J. M., Berrigan, D., Vignieri, S. N., Hill, C. E., Hoang, A., Gibert, P. & Beerli, P. 2001. The strength of phenotypic selection in natural populations. *American Naturalist*, 157, 245–261.

Kirkpatrick, M. & Lande, R. 1989. The evolution of maternal characters. *Evolution*, 43, 485–503.

Kirkpatrick, M. & Lande, R. 1992. The evolution of maternal characters: errata. *Evolution*, 46, 284.

Krebs, C. J. 1996. Population cycles revisited. *Journal of Mammalogy*, 77, 8–24.

Kruuk, L. E. B. 2004. Estimating genetic parameters in natural populations using the "animal model." *Philosophical Transactions of the Royal Society of London Series B-Biological Sciences*, 359, 873–890.

Kruuk, L. E. B., Merilä, J. & Sheldon, B. C. 2001. Phenotypic selection on a heritable size trait revisited. *American Naturalist*, 158, 557–571.

Lande, R. & Kirkpatrick, M. 1990. Selection response in traits with maternal inheritance. *Genetical Research*, 55, 189–197.

Leamy, L. & Cheverud, J. M. 1984. Quantitative genetics and the evolution of ontogeny. II. Genetic and environmental correlations among age-specific characters in randombred house mice. *Growth*, 48, 339–353.

Lynch, M. & Walsh, B. 1998. *Genetics and Analysis of Quantitative Traits*. Sunderland, MA: Sinauer Associates, Inc.

Maestripieri, D. 2009. Maternal influences on offspring growth, reproduction, and behavior in primates. In: *Maternal Effects in Mammals* (Ed. by D. Maestripieri & J. M. Mateo), pp. 256–291. Chicago: University of Chicago Press.

Maestripieri, D. & Mateo, J. M. 2009. The role of maternal effects in mammalian evolution and adaptation. In: *Maternal Effects in Mammals* (Ed. by D. Maestripieri & J. M. Mateo), pp. 1–9. Chicago: University of Chicago Press.

Mappes, T. & Koskela, E. 2004. Genetic basis of the trade-off between offspring number and quality in the bank vole. *Evolution*, 58, 645–650.

Mateo, J. M. & Maestripieri, D. 2009. Maternal effects in mammals: conclusions and future directions. In: *Maternal Effects in Mammals* (Ed. by D. Maestripieri & J. M. Mateo), pp. 322–334. Chicago: University of Chicago Press.

McAdam, A. G. & Boutin, S. 2003a. Effects of food abundance on genetic and maternal variation in the growth rate of juvenile red squirrels. *Journal of Evolutionary Biology*, 16, 1249–1256.

McAdam, A. G. & Boutin, S. 2003b. Variation in viability selection among cohorts of juvenile red squirrels (*Tamiasciurus hudsonicus*). *Evolution*, 57, 1689–1697.

McAdam, A. G. & Boutin, S. 2004. Maternal effects and the response to selection in red squirrels. *Proceedings of the Royal Society of London, Series B*, 271, 75–79.

McAdam, A. G., Boutin, S., Réale, D. & Berteaux, D. 2002. Maternal effects and the potential for evolution in a natural population of animals. *Evolution*, 56, 846–851.

McAdam, A. G., Boutin, S., Sykes, A. K., & Humphries. M. M. 2007. Life histories of female red squirrels and their contributions to population growth and lifetime fitness. *Écoscience*, 14, 362–369.

Merilä, J. & Sheldon, B. C. 2001. Avian quantitative genetics. *Current Ornithology*, 16, 179–255.

Merilä, J., Sheldon, B. C. & Kruuk, L. E. B. 2001. Explaining stasis: microevolutionary studies in natural populations. *Genetica*, 112, 199–222.

Moore, A. J., Brodie, E. D., III & Wolf, J. B. 1997. Interacting phenotypes and the evolutionary process. I. Direct and indirect genetic effects of social interactions. *Evolution*, 51, 1352–1362.

Mousseau, T. A. & Fox, C. W. (Eds.). 1998a. *Maternal Effects as Adaptations*. New York: Oxford University Press.

Mousseau, T. A. & Fox, C. W. 1998b. The adaptive significance of maternal effects. *Trends in Ecology and Evolution*, 13, 403–407.

Murie, J. O., Stevens, S. D. & Leoppky, B. 1998. Survival of captive-born cross-fostered juvenile Columbian ground squirrels in the field. *Journal of Mammalogy*, 79, 1152–1160.

Przybylo, R., Sheldon, B. C. & Merilä, J. 2000. Patterns of natural selection on morphology of male and female collared flycatchers (*Ficedula albicollis*). *Biological Journal of the Linnean Society*, 69, 213–232.

Quinn, J. L., Charmantier, A., Garant, D. & Sheldon, B. C. 2006. Data depth, data completeness, and their influence on quantitative genetic estimation in two contrasting bird populations. *Journal of Evolutionary Biology*, 19, 994–1002.

Rauter, C. M. & Moore, A. J. 2002. Evolutionary importance of parental care performance, food resources, and direct and indirect genetic effects in a burying beetle. *Journal of Evolutionary Biology*, 15, 407–417.

Réale, D., Berteaux, D., McAdam, A. G. & Boutin, S. 2003a. Lifetime selection on heritable life-history traits in a natural population of red squirrels. *Evolution*, 57, 2416–2423.

Réale, D., McAdam, A. G., Boutin, S. & Berteaux, D. 2003b. Genetic and plastic responses of a northern mammal to climate change. *Proceedings of the Royal Society of London, Series B*, 270, 591–596.

Reinhold, K. 2002. Maternal effects and the evolution of behavioral and morphological characters: a literature review indicates the importance of extended maternal care. *Journal of Heredity*, 93, 400–405.

Riska, B., Atchley, W. R. & Rutledge, J. J. 1984. A genetic analysis of targeted growth in mice. *Genetics*, 107, 79–101.

Roach, D. A. & Wulff, R. D. 1987. Maternal effects in plants. *Annual Review of Ecology and Systematics*, 18, 209–235.

Roff, D. A. 1997. *Evolutionary Quantitative Genetics*. New York: Chapman & Hall.

Rossiter, M. 1994. Maternal effects hypothesis of herbivore outbreak. *BioScience*, 44, 752–763.

Rossiter, M. 1996. Incidence and consequences of inherited environmental effects. *Annual Review of Ecology and Systematics*, 27, 451–476.

Rutledge, J. J., Robinson, O. W., Eisen, E. J. & Legates, J. E. 1972. Dynamics of genetics and maternal effects in mice. *Journal of Animal Science*, 35, 911–918.

Sinervo, B., Svensson, E. & Comendant, T. 2000. Density cycles and an offspring quantity and quality game driven by natural selection. *Nature*, 406, 985–988.

Smith, C. C. 1968. The adaptive nature of social organization in the genus of tree squirrels *Tamiasciurus*. *Ecological Monographs*, 38, 31–63.

Stuart-Smith, A. K. & Boutin, S. 1995. Predation on red squirrels during a snowshoe hare decline. *Canadian Journal of Zoology*, 73, 713–722.

Svensson, E. I. & Sinervo, B. 2004. Spatial scale and temporal component of selection in side-blotched lizards. *American Naturalist*, 163, 726–734.

Thiede, D. A. 1998. Maternal inheritance and its effects on adaptive evolution: a quantitative genetic analysis of maternal effects in a natural plant population. *Evolution*, 52, 998–1015.

Weigensberg, I. & Roff, D. A. 1996. Natural heritabilities: can they be reliably estimated in the laboratory? *Evolution*, 50, 2149–2157.

Wilson, A. J. & Festa-Bianchet, M. 2009. Maternal effects in wild ungulates. In: *Maternal Effects in Mammals* (Ed. by D. Maestripieri & J. M. Mateo), pp. 83–103. Chicago: University of Chicago Press.

Wilson, A. J. & Réale, D. 2006. Ontogeny of additive and maternal genetic effects: lessons from domestic mammals. *American Naturalist*, 167, E23–E38.

Wilson, A. J., Coltman, D. W., Pemberton, J. M., Overall, A. D. J., Byrne, K. A. & Kruuk, L. E. B. 2005a. Maternal genetic effects set the potential for evolution in a free-living vertebrate population. *Journal of Evolutionary Biology*, 18, 405–414.

Wilson, A. J., Kruuk, L. E. B. & Coltman, D. W. 2005b. Ontogenetic patterns in heritable variation for body size: using random regression models in a wild ungulate population. *American Naturalist*, 166, E177–E192.

Wilson, A. J., Pemberton, J. M., Pilkington, J. G., Coltman, D. W., Mifsud, D. V., Clutton-Brock, T. H. & Kruuk, L. E. B. 2006. Environmental coupling of selection and heritability limits evolution. *PLoS Biology*, 4, 1270–1275.

Wolf, J. B., Brodie, E. D., III, Cheverud, J. M., Moore, A. J. & Wade, M. J. 1998. Evolutionary consequences of indirect genetic effects. *Trends in Ecology and Evolution*, 13, 64–69.

5

Maternal Effects in Wild Ungulates

ALASTAIR J. WILSON AND MARCO FESTA-BIANCHET

INTRODUCTION

Ungulates, or hoofed animals, are a diverse group of mammals tradition-ally comprising the orders Perissodactyla and Artiodactyla (the "odd-toed" and "even-toed" ungulates). Together these two orders contain around 240 living species, many of which have particular significance to humans. In this chapter we review the evidence for maternal effects in ungulate popula-tions, discussing their prevalence and the mechanisms by which they are mediated. We define a maternal effect as occurring when the phenotype of a mother influences that of her offspring, beyond the direct effect of genes she transmits, and focus primarily on the implications of maternal effects for the ecology and evolution of wild ungulates. Being large, terrestrial, and long-lived, ungulates are amenable to field-based studies at the individual level (e.g., Clutton-Brock et al. 1982; Clutton-Brock & Pemberton 2004). Further-more, scrutiny of maternal effects is typically facilitated by ready identifica-tion of mother-offspring relationships through behavioral observation. This has allowed maternal effects to be tested for as an association between an offspring trait and some aspect of maternal phenotype (e.g., body condition, dominance status). The availability of several long-term studies of marked individuals with known maternal links provides a valuable opportunity to examine the role of maternal effects in ungulates.

By combining observational work with molecular pedigree analysis, extended pedigree structures have been resolved in a number of ungulate populations (e.g., Coltman et al. 1999). Because of the availability of pedigree information, ungulate studies have been at the forefront of efforts to apply quantitative genetic models (notably the animal model; Kruuk 2004) to wild populations (e.g., Réale et al. 1999; Kruuk et al. 2000). The animal model is a form of mixed model used to partition phenotypic variance into genetic and environmental components and can be readily extended to estimate maternal effects. When determined using such a mixed-model framework, the importance of maternal effects can be expressed as m^2, the proportion of variance in offspring phenotype that is explained by maternal identity. A number of studies using this approach have found significant maternal effects on aspects of offspring phenotype including body size and morphology (Milner et al. 2000), parasite resistance (Coltman et al. 2001), and lifetime fitness (Kruuk et al. 2000). In some cases maternal identity explained a substantial proportion of offspring phenotypic variance (e.g., $m^2 = 0.28$ for birth weight of male red deer, *Cervus elaphus*; Kruuk et al. 2000).

It is therefore clear that maternal effects can play an important role in determining phenotypic variation in wild ungulate populations. Here we focus on several themes that have emerged from recent studies. First, we examine the evidence for plasticity in maternal effects that may arise when long-lived iteroparous mothers reproduce in a heterogeneous environment. Second, we address the question of how long maternal influences on offspring phenotype might persist. Finally, we return to the themes discussed in previous chapters and highlight how the estimation of maternal effects, and in particular their genetic components, sheds new insight into evolutionary processes in ungulate populations. Genetic maternal effects arise from genetic differences among mothers that cause phenotypic differences among their offspring. The distinction between genetic and nongenetic (e.g., arising when different mothers experience different environmental conditions) maternal effects is important because the former represent a heritable source of phenotypic variance that may respond to selection (Wolf et al. 1998).

Although our focus is primarily on wild populations, many ungulates are also of particular importance as livestock species, most notably domesticated members of the genera *Bos*, *Ovis*, *Capra*, and *Sus*. As a result, mechanisms of phenotypic determination for commercially important traits have also been extensively studied by animal breeders, yielding important insights on the role of maternal effects. We therefore draw parallels with patterns found in domestic animals where applicable, and particularly with respect to the evidence for genetic maternal effects in ungulates.

PLASTICITY OF MATERNAL EFFECTS

Maternal effects occur because mothers differ in their performance for an offspring trait. For example, some mothers will consistently produce large offspring while others produce smaller ones. In iteroparous organisms such as ungulates, a mother's performance may also vary between reproductive episodes. For example, maternal performance for offspring size or viability may change systematically with maternal age, previous reproductive effort, or resource availability. Such changes are themselves usually related to maternal condition, experience or senescence (Green & Rothstein 1991; Mysterud et al. 2002; Stewart et al. 2005), and may be adaptive consequences of selective trade-offs on lifetime maternal reproductive strategy (Roff 2002). A mother's performance for an offspring trait may therefore show phenotypic plasticity, varying across environments and reproductive episodes. Consequently, maternal effects should cause systematic differences in the phenotypic distributions of offspring born in different environments (or to mothers of different ages; Figure 5-1).

Most studies of ungulates to date have tested for changes in the mean offspring phenotype with environmental conditions experienced by the mother (Figure 5-1a). Implicit in this approach is the assumption that maternal effects are primarily environmental in origin, and that individual mothers will respond to the environment in similar ways. However, it is worth noting that if individual mothers differ in their plastic responses, then it is also possible for the offspring phenotypic variance to change even in the absence of a shift in mean (Figure 5-1b). Although effects on the mean have been well documented, recent ungulate studies have also demonstrated changing variance in maternal performance across environments (e.g., Wilson et al. 2006). This may have important ecological and evolutionary consequences, particularly if maternal effects also include a genetic component.

Maternal Age Effects

Associations between mean offspring phenotype and maternal age have been extensively documented in free-living ungulate populations. A common observation is that maternal performance shows an initial increase with age after primiparity. For example, offspring weight and survival traits tend to increase with maternal age (Table 5-1), while the reverse may be the case for birth date (since in temperate environments earlier birth allows a longer period for offspring growth prior to winter; Rutberg 1987).

It is generally assumed that increasing maternal performance with age (at

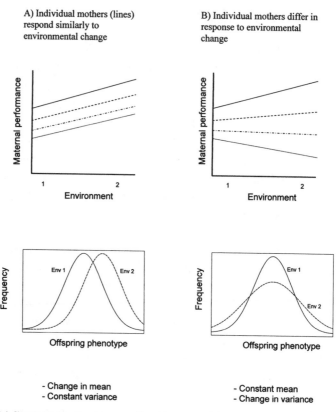

Figure 5-1. Changing maternal performance with environment conditions can induce changes in the offspring phenotypic mean (A), variance (B), or both simultaneously (not shown).

least until the onset of senescence) results from increasing maternal condition, experience, or both. Distinguishing between experience and condition as proximate mechanisms for increasing maternal performance can be difficult. In mountain goats (*Oreamnos americanus*), maternal age per se was not related to birth date or summer weight, although the latter was higher among kids born to multiparous mothers (Côté & Festa-Bianchet 2001). This is consistent with experience as a mother being important for kid mass. In contrast, in Soay sheep (*Ovis aries*), poor performance of yearling mothers is apparently due to their poor condition relative to older females. Thus, although offspring of yearling mothers are generally lighter and also have lower survival (explained in part by their reduced weight; Clutton-Brock et al. 1996), after controlling for maternal weight and size, mother's age does not significantly affect offspring survival (Jones et al. 2005).

Table 5-1. Maternal age effects on offspring traits in free-living ungulates.

Species	Offspring trait	Maternal age effect	Evidence of maternal reproductive senescence	Source(s)
Soay sheep, *Ovis aries*	Survival	Low for yearling mothers	Decrease for oldest mothers	Clutton-Brock et al. 1992
	Birth weight	Low for yearling mothers		Clutton-Brock et al. 1996
Bighorn sheep, *Ovis canadensis*	Survival	Initial increase with maternal age	Decrease with maternal age for mothers > 6 years	Festa-Bianchet 1988a; Bérubé et al. 1999
	Birth date	Decrease with maternal age		Festa-Bianchet 1988b
Mountain goat, *Oreamnos americanus*	Summer weight	Increase with maternal experience (pluriparity)		Côté & Festa-Bianchet 2001
	Male yearling weight	Increase with maternal age		Gendreau et al. 2005
	Yearling horn length	Increase with maternal age		Gendreau et al. 2005
Horse, *Equus caballus*	Survival	Increase with maternal age		Cameron et al. 2000
White tailed deer, *Odocoileus virginianus*	Survival	Increase with maternal age		Ozoga & Verme 1986
Moose, *Alces alces*	Survival (first summer)		Decrease for oldest mothers	Ericsson et al. 2001
	Birth weight	Increase with maternal age		Ericsson et al. 2001
Reindeer/caribou, *Rangifer tarandus*	Birth weight	Increase with maternal age (up to 5 years)		Adams 2005
	Calf weight		Decrease for mothers > 7 years old	Weladji et al. 2002
Red deer, *Cervus elaphus*	Birth weight	Increase with maternal age	Decrease for oldest mothers	Clutton-Brock et al. 1982
	Birth date	Initial decrease with maternal age	Increase for oldest mothers	Clutton-Brock et al. 1982

It should perhaps be noted that, in group-living ungulates, dominance hierarchies often exist among females. Dominance increases with age in red deer (Clutton-Brock et al. 1984; Thouless & Guinness 1986), bighorn sheep (*Ovis canadensis*; Festa-Bianchet 1991), and mountain goats (Côté 2000). Maternal dominance rank is also related to weaning time in Barbary sheep (*Ammotragus lervia*; Cassinello 1997), offspring rank in captive mouflon (*Ovis musimon*; Guilhem et al. 2002) and red deer (Veiberg et al. 2004), and even offspring lifetime reproductive success in male red deer (Clutton-Brock et al. 1984). Since dominance and condition will likely be positively correlated, it is not always clear what constitutes the proximate cause of these maternal effects (see Maestripieri, this volume, for primates). However, maternal rank effects on birth date and early growth traits have been demonstrated in reindeer (*Rangifer tarandus*) after controlling for maternal mass as a measure of condition (Holand et al. 2004).

Although experience, and perhaps dominance, continue to increase with age, maternal condition might actually decline later in life (Loison et al. 1999; Gaillard et al. 2000). Senescence may cause declining maternal performance for offspring traits (e.g., size, survival), reducing phenotypic values among the offspring of the oldest mothers. This effect, commonly reported in domestic animals (e.g., sheep; Mysterud et al. 2002) has also been found in wild ungulates (Table 5-1), and can be seen as part of a wider pattern of maternal reproductive senescence. Consequently, senescence-induced maternal effects on offspring will often be accompanied by age-related declines in maternal traits such as litter size, fecundity (Bérubé et al. 1999; Mysterud et al. 2002), or postreproductive survival (Ericsson et al. 2001). Senescence-based maternal effects may also be dependent on environmental conditions. For example, in Soay sheep, birth weights and neonatal survival were found to be lower for offspring of the oldest ewes, but only following winters of low adult mortality (Clutton-Brock et al. 2004). However, this may be a statistical artifact since effects due to reproductive senescence will be hard to detect if few of the oldest ewes survive to give birth (i.e., if survival senescence occurs before reproductive senescence limiting sample sizes of old mothers).

Environmental Heterogeneity

Environmental heterogeneity, whether spatial or temporal, may also cause systematic differences among maternal phenotypes, and hence the phenotypes of their offspring. In many cases this type of maternal effect might again be due to condition dependence in maternal performance. For example, in Alaskan moose (*Alces alces gigas*), females with greater rump-fat thickness gave birth to heavier offspring with greater survival (e.g., Keech et al. 2000). Similarly, after accounting for maternal age, offspring of heavy mothers tend to have greater survival in bighorn sheep (Festa-Bianchet et al. 1998), mountain goats (Côté & Festa-Bianchet 2001), and Soay sheep (Clutton-Brock et al. 1996). Condition-dependent maternal performance has also been suggested in roe deer (*Capreolus capreolus*) as maternal weight is positively correlated with offspring weight in both wild and captive animals (Andersen et al. 2000).

Extrinsic environmental variables such as climate, food abundance, or population density can all affect maternal condition (Stewart et al. 2005) and hence offspring phenotype. In red deer, offspring birth weights were found to be greater following warm springs characterized by good grazing conditions (Albon et al. 1987), suggesting condition-dependent maternal investment late in gestation. Similarly, birth weight is inversely related to levels of snowfall during gestation in Alaska caribou (Adams 2005). In other cases, postnatal environmental conditions are more important than those prior to birth in determining offspring growth (e.g., in reindeer; Weladji et al. 2003) or survival (e.g., in roe deer; Pettorelli et al. 2005). These may represent maternal effects if the environmental influence occurs via the mother as opposed to directly on the offspring (e.g., favorable environments allow mothers to increase provisioning). However, the postnatal environment may also have direct effects on offspring phenotype, for example via thermoregulation costs or food availability after weaning (Jones et al. 2005).

Variation in Plasticity of Maternal Performance

As discussed above, the effects of maternal age and environment on an offspring trait could arise from phenotypic plasticity in maternal performance. If all mothers respond similarly, then population-level analyses will be adequate for analyzing maternal effects on offspring phenotype. However, individuals may differ in their responses to environmental variables (Nussey et al. 2007), whether extrinsic (e.g., temperature, population density) or intrinsic (e.g., age, dominance). Fully understanding maternal effects may therefore require monitoring of multiple offspring from individual mothers, born at different maternal ages and in years of differing weather conditions or resource availability. This is because changes in an individual's maternal performance with environmental conditions will not necessarily mirror the average response detected at the population level.

Variation in plasticity of maternal performance has been demonstrated in red deer on the Scottish island of Rum. Although calves were generally born at higher birth weights following warm springs (Albon et al. 1987), analysis at the individual level has shown that mothers differ in their individual responses to spring temperature (Nussey et al. 2005a). Specifically, a reduced ability to increase investment in offspring weight under favorable conditions was found for mothers that had experienced high population densities early in life (Nussey et al. 2005b). This demonstrates that an environment experienced early in life can have permanent effects on a mother's future reproductive performance.

A consequence of differences in plastic responses among mothers is that the variance of an offspring trait, as well as its mean, may vary across environments as a direct consequence of maternal effects (Figure 5-1b). This phenomenon has been demonstrated for offspring birth weight in Soay sheep (Wilson et al. 2006), and when maternal effects themselves have a genetic basis, it may have considerable implications for phenotypic evolution (see below and also previous chapters).

PERSISTENCE OF MATERNAL EFFECTS

From an offspring's perspective, it is interesting to ask whether maternal effects on the phenotype decline over ontogeny, or exert a permanent influence. This question has received considerable attention in studies of plants and invertebrates, where maternal effects have commonly been shown to persist to adulthood and even into subsequent generations (e.g., grandmaternal effects; Case et al. 1996; Plaistow et al. 2006). Although rodent studies have often demonstrated a tendency towards declining maternal influence with offspring age (e.g., Riska et al. 1984; Sikes 1998), this result is not ubiquitous (e.g., Cowley et al. 1989; Jarvis et al. 2005) and may not be general to all trait types or mammalian taxa. For example, in wild baboons, Altmann & Alberts (2005) found that multiparous, high-ranking mothers produce larger offspring and that size differences remain throughout development (and also influence age of maturity in both sexes).

In most ungulates, associations between mothers and their offspring decline as offspring age. As a consequence, although maternal effects can occur both prenatally (e.g., differential placental efficiency; Dwyer et al. 2005) and postnatally (e.g., differential milk quality; Landete-Castillejos et al. 2001), the influence of the mother may decrease after weaning and offspring dispersal. Furthermore, other sources of phenotypic variance might continue to accumulate and compound as an individual ages (e.g., compounding of environmental effects and new episodes of gene expression; Houle 1998).

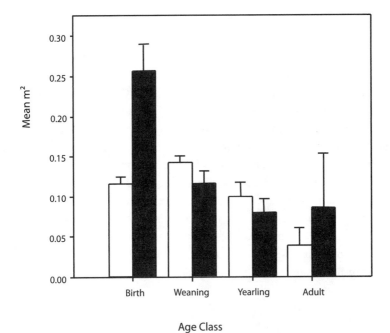

Age Class

Figure 5-2. Declining importance of maternal genetic effects with age for offspring weight in domestic sheep (white) and cattle (grey), showing mean m_G^2, the proportion of phenotypic variance explained by maternal (genetic) effects. Error bars indicate one standard error. The figure is redrawn from Wilson & Réale (2006) using data drawn from the literature.

Consequently, we might intuitively expect that, compared to other sources of phenotypic variation, maternal identity will become less important as offspring age increases.

Consistent with this expectation, weight traits in domestic cattle and sheep show declining maternal effects with offspring age (Figure 5-2). For example, m_G^2 (the proportion of phenotypic variance explained by genetic maternal effects), falls in cattle from a mean of 0.12 at birth to 0.04 in adults (Wilson & Réale 2006). Although comparisons between different offspring ages are somewhat limited in natural ungulate populations, a similar pattern is emerging. For example, in both bighorn and Soay sheep maternal identity explains a large proportion of variance in offspring weight at birth, but this proportion rapidly declines over the first few year of life (Wilson et al. 2005b; 2007). Similarly, Gendreau et al. (2005) found maternal effects on mass and horn length in yearling mountain goats but not in two-year-olds.

The rate of decline in maternal influence with offspring age will likely vary between species and populations (e.g., Wilson & Réale 2006), reflecting differences in weaning time and duration of mother-offspring associations. There may also be considerable variation within populations, for example if weaning time depends on maternal reproductive status (i.e., whether or not the mother is pregnant again; Green et al. 1993). This has been reported in bighorn sheep, in which yearling mass is positively correlated with the amount of time spent associating with mothers, and maternal–yearling associations occur only if the mother is not nursing a new lamb (L'Heureux et al. 1995). Extrinsic environmental variables (e.g., climate, food) may also influence the duration of mother-offspring associations and hence potentially influence the duration of maternal effects. For example, in red deer, the proximity of mothers and daughters declines with daughter's age as might be expected, but this decline is also more rapid at high population density (Albon et al. 1992). In bighorn sheep, mothers associate with yearlings only at high population density (L'Heureux et al. 1995).

With increasing offspring independence, maternal effects on offspring phenotype may diminish relative to other sources of phenotypic variance (e.g., additive genetic effects, environmental conditions experienced by the offspring) that may continue to accumulate throughout life. However, a meta-analysis of weight traits in domestic sheep and cattle shows that there can be absolute (as well as relative) declines in the magnitude of maternal variance with offspring age (Wilson & Réale 2006), a phenomenon also found in wild bighorn sheep (Wilson et al. 2005b). In the context of size traits, compensatory growth by offspring of poor-performing mothers might reduce the absolute variance attributable to maternal effects. However, although commonly reported in livestock systems, there is comparatively little evidence to date for compensatory growth in wild ungulates (but see Solberg et al. 2004; Gendreau et al. 2005). Selective mortality may also be an important mechanism reducing maternal variance as offspring age. For example, since small body size is often a key determinant of mortality in ungulates (e.g., Overall et al. 2005), maternal effects on traits such as birth weight will equate to maternal effects on viability. The progeny of mothers producing consistently smaller offspring will then be selectively removed from the population at an early age.

Despite a clear pattern of declining maternal influence with increasing offspring age in ungulates, not all maternal effects disappear with age. Animal breeding studies, which typically have very large sample sizes and high statistical power, commonly report weak but statistically significant maternal effects on adult livestock traits. In a few cases large effect sizes have been reported (e.g., $m^2 = 0.22$ for female weight in sheep; Näsholm & Danell 1996;

$m^2 = 0.17$ for scrotal size in cattle; Burrow 2001). In free-living ungulates, statistical power is generally lower and this may limit detection of weak maternal effects on adult offspring. Nevertheless, Tischner and Allen (2000) showed maternal effects on adult body size in feral horses, and Green & Rothstein (1993) found that bison (*Bison bison*) born later in the season were larger as adults. Although not explicitly tested in this latter case, variation in birth date often arises as a maternal effect, either through variation in maternal condition (e.g., Hogg et al. 1992) or through genetic difference at loci determining reproductive phenology (e.g., Wilson et al. 2005a). Although it is clear that maternal effects in ungulates can sometimes persist to influence adult phenotype, our understanding of when and why this occurs is limited at present. Since selection will act to remove variance, it seems plausible that maternal effects should more likely persist for traits that are not under strong selection (i.e., morphological as opposed to life history traits). There are limited data to test this hypothesis at present and more studies are required to determine the relative roles of selective mortality and postweaning sources of phenotypic variance in the persistence of maternal effects.

EVOLUTIONARY SIGNIFICANCE OF MATERNAL EFFECTS IN UNGULATES

In recent years ungulate studies have yielded great insights into the potential role maternal effects may play in shaping evolutionary trajectories. A phenotypic trait is expected to evolve if it is subject to selection and has a heritable basis of variation. Maternal effects can therefore have evolutionary significance because they may influence offspring fitness and also have a genetic, and thus potentially heritable, basis (Willham 1972). These conditions have been demonstrated in both domestic and wild ungulates.

Selection on Offspring Phenotype and Maternal Performance

Maternal effects on offspring fitness components have been extensively documented in ungulates, indicating that maternal performance for offspring phenotype can be under selection. In particular, maternal influences on neonatal and juvenile survival traits are common (e.g., Jones et al. 2005). After birth, offspring viability differences often reflect differential rates of offspring provisioning. However, it should be noted that there may also be differences among mothers in probability of fetal survival to birth. Though not typically considered in the context of maternal effects on offspring phenotype, this is one mechanism by which maternal dominance and condition can influence offspring sex ratio (i.e., sex-dependent abortion of the fetus;

Sheldon & West 2004; Hewison et al. 2005; see also Vandenbergh, Maestripieri, this volume, for rodents and primates, respectively).

Significant maternal effects on fitness components other than early offspring viability have also been reported. For example, in red deer on Rum, maternal identity accounts for large and statistically significant proportions of phenotypic variance in several fitness traits measured in adults (e.g., adult female breeding success, $m^2 = 0.171$; female longevity, $m^2 = 0.192$; male longevity, $m^2 = 0.167$; Kruuk et al. 2000). More generally, maternal effects are commonly found on birth date, birth weight, and early growth (Table 5-1; see also Kruuk et al. 2000; Wilson et al. 2005a), traits that are frequently correlated with fitness. For example, offspring weight is typically under strong positive directional selection that operates through neonatal or overwinter mortality (Festa-Bianchet et al. 1997; Singer et al. 1997; Wilson et al. 2005c).

If selection acts on an offspring trait, then by extension it may also act on maternal performance for that trait. However, it is important to note that selection on maternal performance might not always favor the same phenotypic change (or stasis) as selection on offspring phenotype. This becomes apparent when offspring provisioning (both pre- and postnatal) is seen as part of a lifetime maternal reproductive strategy. Although offspring fitness will generally increase with maternal provisioning, a mother's lifetime fitness might be optimized by producing a larger litter with reduced investment per offspring (Trivers 1972; Roff 2002). Such selective antagonism can be seen as a form of parent-offspring conflict and has been demonstrated in Soay sheep (Wilson et al. 2005c). Although offspring fitness is highest for heavy-born singletons, maternal fitness in Soay sheep is optimized by producing an average litter size greater than 1 at some cost to offspring size (and hence viability; Wilson et al. 2005c). Similarly, since reproduction in any given season can impose costs to future viability or reproductive success (e.g., in bighorn sheep; Festa-Bianchet et al. 1998) the evolution of maternal reproductive strategy may also be influenced by trade-offs between current and future reproductive success.

Genetic Maternal Effects in Ungulates

If maternal effects result, at least in part, from genetic differences among mothers, then they represent a heritable component of phenotypic variance that may respond to selection (Mousseau & Fox 1998; Wolf et al. 1998). When pedigree information is available, the magnitude of genetic maternal effects on an offspring trait can be estimated using quantitative genetic approaches for partitioning phenotypic variance (Falconer & Mackay 1996). This ap-

Table 5-2. Estimated maternal genetic effects in Soay and bighorn sheep population, showing m_G^2 the proportion of phenotypic variance explained by maternal genetic variance (and standard errors when available). Estimates are based on animal model analyses with fixed effects included to correct for known influences (e.g., sex, year of birth).

Species	Offspring trait	m_G^2	(SE)
Soay sheep,	Birth weight	0.119	(0.045)
Ovis aries[1]	Birth date	0.283	(0.051)
	Natal litter size	0.211	(0.058)
Bighorn sheep,	Lamb June weight (JW)	0.175	
Ovis canadensis[2]	Yearling JW	0.057	
	2nd year JW	0.031	
	3rd year JW	0.027	
	4th year JW	0.022	
	5th year JW	0.023	

[1] Wilson et al. (2005a), [2] Wilson et al. (2005b).

proach has been used extensively with domestic ungulates, in which maternal genetic effects have been found to play an important role in determining many commercially important traits (Näsholm 2004; Wilson & Réale 2006).

Despite the recent applications of quantitative genetic models to wild ungulate populations (Kruuk et al. 2002; Coltman et al. 2005), explicit estimates of maternal genetic variance (as opposed to variance arising from maternal effects in general) remain limited (Table 5-2). In bighorn sheep, genetic maternal effects have a large influence on the weight of very young lambs (m_G^2 = 0.18; Wilson et al. 2005b). Similarly, significant genetic maternal effects have been found for the offspring traits of birth weight, birth date, and natal litter size in Soay sheep (Wilson et al. 2005a). In all of these early-expressed traits, genetic maternal effects explained more variance than the direct additive effects (normally measured as the trait heritability, h^2). This highlights their importance for determining responses to selection acting through both maternal and offspring fitness components (Wilson et al. 2005c).

As discussed in previous chapters, covariance between maternal genetic and direct additive effects on phenotype can also occur as a result of pleiotropy or linkage. Although positive covariance will accelerate responses to selection (Wolf et al. 1998), negative covariances are commonly reported in livestock studies (Vargas et al. 2000; Burrow 2001; Wilson & Réale 2006). Negative covariances might occur through pleiotropy, for example if a gene has a positive effect on an offspring trait but a negative effect on maternal performance for that trait. In an applied context, this will reduce the efficacy of artificial selection on target traits because selection on offspring breeding value will simultaneously select for decreased maternal performance. These negative genetic covariances may also contribute to the frequent observation of phenotypic stasis in traits that appear both heritable and under selection in natural populations (Merilä et al. 2001; Kruuk et al. 2002). In Soay sheep, estimates of the direct-maternal genetic covariance for birth weight, birth date, and natal litter size provide only limited support for this hypothesis. Although estimates are negative, they do not differ significantly from 0 except for natal litter size (direct-maternal genetic correlation $r_{am} = -0.71$). Nevertheless, although nonsignificant, the effect size is also large for birth weight ($r_{am} = -0.60$; Wilson et al. 2005a), and it should be noted that the very large samples needed to obtain precise estimates of genetic covariances may limit our ability to detect these effects in natural populations (Lynch & Walsh 1998).

Natural populations are characterized by heterogeneous environments. If maternal effects have a genetic basis, plasticity in maternal performance (as discussed above) may also have important evolutionary consequences. Specifically, the expression of genetic maternal effects can be environment dependent, amounting to a maternal genotype-by-environment ($G \times E$) interaction. This is equivalent to the scenario presented in Figure 5-1b where differences among individual mothers are caused by differences at genetic loci influencing performance. Maternal $G \times E$ interactions will then result in different amounts of maternal genetic variance for an offspring trait under different environmental conditions. In one of the first empirical tests of this phenomenon (see also McAdam, this volume), maternal genetic variance, and hence the "total heritability" (Willham 1972) of birth weight in Soay sheep was shown to be higher in good environments than in harsh ones (a good environment being one in which the probability of survival is higher; Wilson et al. 2006). This phenomenon may also contribute to the phenotypic stasis observed for birth weight, since positive directional selection on birth weight (through differential neonatal viability) declines with environmental quality. Evolution of increased birth weight is therefore constrained by a lack of heritable variation in poor environments and by a reduced strength of selection in good environments (Wilson et al. 2006).

CONCLUSIONS AND FUTURE DIRECTIONS

Clearly, maternal effects are prevalent in wild ungulates and, in addition to contributing to observed phenotypic diversity, they can also affect evolutionary dynamics. Abundant evidence having been accumulated for maternal effects in wild and domestic populations, increased effort is now needed to determine what specific behavioral, physiological, and genetic mechanisms mediate maternal effects. For example, although low offspring survival is generally correlated with poor maternal condition, it is not always clear exactly why. Does maternal condition impact placental efficiency, milk quality, maternal care (e.g., restricting offspring suckling; Festa-Bianchet & Jorgenson 1998), or a combination of these (see also Holekamp & Dloniak, Maestripieri, Mateo, this volume)?

When maternal influence does have a genetic component, identifying the loci involved offers a promising avenue of future research. Quantitative trait loci (QTL) analyses have been extensively utilized for maternal performance traits (e.g., milk yield and quality) in domestic cattle, goats, and sheep (Heyen et al. 1999; Barillet et al. 2005). Extension to wild systems is possible if populations are pedigreed and linkage maps developed, although this would require significant effort. Whole genome scans have been used to detect QTL in wild red deer (Slate et al. 2002) and Soay sheep (Beraldi et al. 2006). However, in the only application of this approach to maternal performance traits to date, no significant QTL were found for either birth weight or birth date analyzed as maternal traits in Soay sheep (Beraldi et al. 2007). Given that maternal genetic variance for these traits has been demonstrated (Wilson et al. 2005a), this negative result may reflect limited statistical power for detecting QTL in the wild using linkage analysis.

Candidate gene approaches should offer a useful, and potentially more powerful, alternative for scrutinizing loci underlying genetic maternal effects. Polymorphisms in genes for prolactin, growth hormone, and a number of milk proteins (e.g., α-Lactalbumin, α_{s1}-casein) have been shown to affect maternal milk yield and quality (Pirisi et al. 1999; Dayal et al. 2006; Li et al. 2006; Zhou et al. 2006), and could underlie genetic maternal effects on early growth. In natural populations, assaying variation at candidate loci and testing for covariation with offspring traits would seem a useful strategy, particularly since the completed sequences of bovine and ovine genomes will facilitate comparative genomic studies.

Finally, if maternal effects are widespread, they present a challenge to the development of predictive models of phenotypic evolution. Quantitative genetic models will not only need to explicitly consider the genetic architecture of maternal effects, but also the nature of selection acting through maternal and offspring fitness (Wolf & Wade 2001; Wilson et al. 2005c). Furthermore, the possibility that maternal performance is also phenotypically plastic will need to be carefully considered. Data requirements for such models will inevitably be high, but might most readily be met by long-term studies of ungulate populations. Specifically, ongoing studies of red deer, Soay and big-horn sheep and mountain goat populations (e.g., Clutton-Brock et al. 2004; Gendreau et al. 2005; Wilson et al. 2005b) are providing phenotypic data and pedigree information that are difficult to obtain in many other mammalian taxa (or other vertebrates generally). Thus, while our understanding of ma-ternal effects in this group of mammals is far from complete, ungulates should continue to provide a useful model for investigating maternal effects in the wild.

ACKNOWLEDGMENTS

We thank David Coltman, Daniel Nussey, and Andrew McAdam for discus-sion and comments on an earlier version of this chapter. AJW is supported by a Leverhulme Trust research project grant and the Natural Environment Research Council (UK). MFB's research is supported by the Natural Sciences and Engineering Research Council of Canada.

REFERENCES

Adams, L. G. 2005. Effects of maternal characteristics and climatic variation on birth masses of Alaskan caribou. *Journal of Mammalogy*, 86, 506–513.

Albon, S. D., Clutton-Brock, T. H. & Guinness, F. E. 1987. Early development and popula-tion dynamics in red deer. II. Density-independent effects and cohort variation. *Journal of Animal Ecology*, 56, 69–81.

Albon, S. D., Staines, H. J., Guinness, F. E. & Clutton-Brock, T. H. 1992. Density-dependent changes in the spacing of female kin in red deer. *Journal of Animal Ecology*, 61, 131–137.

Altmann, J. & Alberts, S. C. 2005. Growth rates in a wild primate population: ecological influences and maternal effects. *Behavioral Ecology and Sociobiology*, 57, 490–501.

Andersen, R., Gaillard, J. M., Linnell, J. D. C. & Duncan, P. 2000. Factors affecting mater-nal care in an income breeder, the European roe deer. *Journal of Animal Ecology*, 69, 672–682.

Barillet, F., Arranz, J.-J. & Carta, A. 2005. Mapping quantitative trait loci for milk produc-tion and genetic polymorphisms of milk proteins in dairy sheep. *Genetics, Selection, Evolution*, 37, S109-S123.

Beraldi, D., McRae, A. F., Gratten, J., Visscher, P. M. & Pemberton, J. M. 2006. Develop-ment of a linkage map and mapping of phenotypic polymorphisms in a free-living population of Soay sheep (*Ovis aries*). *Genetics*, 173, 1521–1537.

Beraldi, D., McRae, A. F., Gratten, J., Visscher, P. M. & Pemberton, J. M. 2007. Mapping QTL underlying fitness-related traits in a free-living sheep population. *Evolution*, 61, 1403–1416.

Bérubé, C., Festa-Bianchet, M. & Jorgenson, J. T. 1999. Individual differences, longevity, and reproductive senescence in bighorn ewes. *Ecology*, 80, 2555–2565.

Burrow, H. M. 2001. Variances and covariances between productive and adaptive traits and temperament in a composite breed of tropical beef cattle. *Livestock Production Science*, 70, 213–233.

Cameron, E. Z., Linklater, W. L., Stafford, K. J. & Minot, E. O. 2000. Aging and improving reproductive success in horses: declining residual reproductive value or just older and wiser? *Behavioral Ecology and Sociobiology*, 47, 243–249.

Case, A. L., Lacey, E. P. & Hopkins, R. G. 1996. Parental effects in *Plantago lanceolata* L. II. Manipulation of grandparental temperature and parental flowering time. *Heredity*, 76, 287–295

Cassinello, J. 1997. Mother-offspring conflict in the Saharan arrui, *Ammotragus lervia saharinensis*: relation to weaning and mother's sexual activity. *Ethology*, 103, 127–137.

Cowley, D. E., Pomp, D., Atchlet, W. R., Eisen, E. J. & Hawkins, D. 1989. The impact of maternal uterine genotype on postnatal growth and adult body size in mice. *Genetics*, 122, 193–203.

Clutton-Brock, T. H. & Pemberton, J. M. (Eds.) 2004. *Soay Sheep: Dynamics and Selection in an Island Population*. Cambridge: Cambridge University Press.

Clutton-Brock, T. H., Guinness, F. E. & Albon, S. D. 1982. *Red Deer: Behavior and Ecology of Two Sexes*. Chicago: University of Chicago Press.

Clutton-Brock, T. H., Albon, S. D. & Guinness, F. E. 1984. Maternal dominance, breeding success and birth ratios in red deer. *Nature*, 308, 358–360.

Clutton-Brock, T. H., Price, O. F., Albon, S. D. & Jewell, P. A. 1992. Early development and population fluctuations in Soay sheep. *Journal of Animal Ecology*, 61, 381–396.

Clutton-Brock, T. H., Stevenson, I. R., Marrow, P., MacColl, A. D., Houston, A. I. & Mc-Namara, M. 1996. Population fluctuations, reproductive costs and life history tactics in female Soay sheep. *Journal of Animal Ecology*, 65, 675–689.

Clutton-Brock, T. H., Pemberton, J. M., Coulson, T., Stevenson, I. R. & MacColl, A. D. 2004. The sheep of St Kilda. In: *Soay Sheep: Dynamics and Selection in an Island Population* (Ed. By T. H. Clutton-Brock & J. M. Pemberton), pp. 17–52. Cambridge: Cambridge University Press.

Coltman, D. W., Bancroft, D. R., Robertson, A., Smith, J. A., Clutton-Brock, T. H. & Pemberton, J. M. 1999. Male reproductive success in a promiscuous mammal: behavioural estimates compared with genetic paternity. *Molecular Ecology*, 8, 1199–1209.

Coltman, D. W., Pilkington, J., Kruuk, L. E., Wilson, K. & Pemberton, J. M. 2001. Positive genetic correlation between parasite resistance and body size in a free-living ungulate population. *Evolution*, 55, 2116–2125.

Coltman, D. W., O'Donoghue, P., Hogg, J. T. & Festa-Bianchet, M. 2005. Selection and genetic (co)variance in bighorn sheep. *Evolution*, 59, 1372–1382.

Côté, S. D. 2000. Dominance hierarchies in female mountain goats: stability, aggressiveness and determinants of rank. *Behaviour*, 137, 1541–1566

Côté, S. D. & Festa-Bianchet, M. 2001. Birthdate, mass and survival in mountain goat kids: effects of maternal characteristics and forage quality. *Oecologia*, 127, 230–238.

Dayal, S., Bhattacharya, T. K., Vohra, V., Kumar, P. & Sharma, A. 2006. Effect of alpha-lactalbumin gene polymorphism on milk production traits in water buffalo. *Asian-Australasian Journal of Animal Sciences*, 19, 305–308.

Dwyer, C. M., Calvert, S. K., Farish, M., Donbavand, J. & Pickup, H. E. 2005. Breed, litter and parity effects on placental weight and placentome number, and consequences for the neonatal behaviour of the lamb. *Theriogenology*, 63, 1092–1110.

Ericsson, G., Wallin, K., Ball, J. & Broberg, M. 2001. Age-related reproductive effort and senescence in free-ranging moose, *Alces alces*. *Ecology*, 82, 1613–1620.

Falconer, D. S. & Mackay, T. F. C. 1996. *Introduction to Quantitative Genetics*. Harlow, UK: Prentice Hall.

Festa-Bianchet, M. 1988a. Age-specific reproduction of bighorn ewes in Alberta, Canada. *Journal of Mammalogy*, 69, 157–160.

Festa-Bianchet, M. 1988b. Birthdate and survival in bighorn lambs (*Ovis canadensis*). *Journal of Zoology*, 214, 653–661.

Festa-Bianchet, M. 1991. The social system of sheep: grouping patterns, kinship and female dominance rank. *Animal Behaviour*, 42, 71–82.

Festa-Bianchet, M. & Jorgenson, J. T. 1998. Selfish mothers: reproductive expenditure and resource availability in bighorn ewes. *Behavioral Ecology*, 9, 144–150.

Festa-Bianchet, M., Jorgenson, J. T., Bérubé, C. H., Portier, C. & Wishart, W. D. 1997. Body mass and survival of bighorn sheep. *Canadian Journal of Zoology*, 75, 1372–1379.

Festa-Bianchet, M., Gaillard, J. M. & Jorgenson, J. T. 1998. Mass- and density-dependent reproductive success and reproductive costs in a capital breeder. *American Naturalist*, 152, 367–379.

Gaillard, J. M., Festa-Bianchet, M., Yoccoz, N. G., Loison, A. & Toïgo, C. 2000. Temporal variation in fitness components and population dynamics of large herbivores. *Annual Review of Ecology and Systematics*, 31, 367–393.

Gendreau, Y., Côté, S. D. & Festa-Bianchet, M. 2005. Maternal effects on post-weaning physical and social development in juvenile mountain goats (*Oreamnos americanus*). *Behavioral Ecology and Sociobiology*, 58, 237–246.

Green, W. C. H. & Rothstein, A. 1991. Trade-offs between growth and reproduction in female bison. *Oecologia*, 86, 521–527.

Green, W. C. H. & Rothstein, A. 1993. Persistent influences of birth date on dominance, growth and reproductive success in bison. *Journal of Zoology*, 230, 177–186.

Green, W. C. H., Rothstein, A. & Griswold, J. G. 1993. Weaning and parent-offspring conflict: variation relative to interbirth interval in bison. *Ethology*, 95, 105–125.

Guilhem, C., Gerard, J. F. & Bideau, E. 2002. Rank acquisition through birth order in mouflon sheep (*Ovis gmelini*) ewes. *Ethology*, 108, 63–73.

Hewison, A. J. M., Gaillard, J. M., Kjellander, P., Toigo, C., Liberg, O. & Delorme, D. 2005. Big mothers invest more in daughters: reversed sex allocation in a weakly polygynous mammal. *Ecology Letters*, 8, 430–437.

Heyen, D. W., Weller, J. I., Ron, M., Band, M., Beever, J., Feldmesser, E., Da, Y., Wiggans, G. R., VanRaden, P. M. & Lewin, H. A. 1999. A genome scan for QTL influencing milk production and health traits in dairy cattle. *Physiological Genomics*, 1, 165–175.

Hogg, J. T., Hass, C. C. & Jenni, D. A. 1992. Sex-biased maternal expenditure in Rocky Mountain bighorn sheep. *Behavioral Ecology and Sociobiology*, 31, 243–251.

Holand, Ø., Weladji, R. B., Gjøstein, H., Kumpula, J., Smith, M. E., Nieminen, M. & Røed, K. H. 2004. Reproductive effort in relation to maternal social rank in reindeer (*Rangifer tarandus*). *Behavioral Ecology and Sociobiology*, 57, 69–76.

Holekamp, K. E. & Dloniak, S. M. 2009. Maternal effects in fissiped carnivores. In: *Maternal Effects in Mammals* (Ed. by D. Maestripieri & J. M. Mateo), pp. 227–255. Chicago: University of Chicago Press.

Houle, D. 1998. How should we explain variation in the genetic variance of traits? *Genetica*, 103, 241–253.

Jarvis, J. P., Kenney-Hunt, J., Ehrich, T. H., Pletscher, L. S., Semenkovich, C. F & Cheverud, J. M. 2005. Maternal genotype affects adult offspring lipid, obesity, and diabetes phenotypes in LGXSM recombinant inbred strains. *Journal of Lipid Research*, 46, 1692–1702.

Jones, O. R., Crawley, M. J., Pilkington, J. & Pemberton, J. M. 2005. Predictors of early survival in Soay sheep: cohort-, maternal- and individual-level variation. *Proceedings of the Royal Society of London, Series B*, 272, 2619–2625.

Keech, M. A., Bowyer, R. T., Ver Hoef, J. M., Boertje, R. D., Dale, B. W. & Stephenson, T. R. 2000. Life-history consequences of maternal condition in Alaskan moose. *Journal of Wildlife Management*, 64, 450–462.

Kruuk, L. E. B. 2004. Estimating genetic parameters in natural populations using the "animal model." *Philosophical Transactions of the Royal Society of London B*, 359, 873–890.

Kruuk, L. E. B., Clutton-Brock, T. H., Slate, J., Pemberton, J. M., Brotherstone, S. & Guinness, F. E. 2000. Heritability of fitness in a wild mammal population. *Proceedings of the National Academy of Sciences USA*, 97, 698–703.

Kruuk, L. E. B., Slate, J., Pemberton, J. M., Brotherstone, S., Guinness, F. & Clutton-Brock, T. H. 2002. Antler size in red deer: heritability and selection but no evolution. *Evolution*, 56, 1683–1695.

Landete-Castillejos, T., Garcia, A. & Gallego, L. 2001. Calf growth in captive Iberian red deer (*Cervus elaphus hispanicus*): effects of birth data and hind milk production and composition. *Journal of Animal Science*, 79, 1085–1092.

L'Heureux, N., Lucherini, M., Festa-Bianchet, M. & Jorgenson, J. T. 1995. Density-dependent mother-yearling association in bighorn sheep. *Animal Behaviour*, 49, 901–910.

Li, J. T., Wang, A. H., Chen, P., Li, H. B., Zhang, C. S. & Du, L. X. 2006. Relationship between the polymorphisms of 5′ regulation region of prolactin gene and milk traits in Chinese Holstein dairy cows. *Asian-Australasian Journal of Animal Sciences*, 19, 459–462.

Loison, A., Festa-Bianchet, M., Gaillard, J. M., Jorgenson, J. T. & Jullien, J. M. 1999. Age-specific survival in five populations of ungulates: evidence of senescence. *Ecology*, 80, 2539–2554.

Lynch, M. & Walsh, B. 1998. *Genetics and Analysis of Quantitative Traits*. Sunderland, MA: Sinauer.

Maestripieri, D. 2009. Maternal influences on offspring growth, reproduction, and behavior in primates. In: *Maternal Effects in Mammals* (Ed. by D. Maestripieri & J. M. Mateo), pp. 256–291. Chicago: University of Chicago Press.

Mateo, J. M. 2009. Maternal influences on development, social relationships and survival behaviors. In: *Maternal Effects in Mammals* (Ed. by D. Maestripieri & J. M. Mateo), pp. 133–158. Chicago: University of Chicago Press.

McAdam, A. G. 2009. Maternal effects on evolutionary dynamics in wild small mammals. In: *Maternal Effects in Mammals* (Ed. by D. Maestripieri & J. M. Mateo), pp. 64–82. Chicago: University of Chicago Press.

Merilä, J., Sheldon, B. C. & Kruuk, L. E. B. 2001. Explaining stasis: microevolutionary studies in natural populations. *Genetica*, 112, 199–222.

Milner, J. M., Pemberton, J. M., Brotherstone, S. & Albon, S. D. 2000. Estimating variance components and heritabilities in the wild: a case study using the "animal model" approach. *Journal of Evolutionary Biology*, 13, 804–813.

Mousseau, T. A. & Fox, C. W. (Eds.). 1998. *Maternal Effects as Adaptations*. Oxford: Oxford University Press.

Mysterud, A., Steinheim, G., Yoccoz, N. G., Holand, O. & Stenseth, N. C. 2002. Early onset of reproductive senescence in domestic sheep, *Ovis aries*. *Oikos*, 97, 177–183.

Näsholm, A. 2004. Direct and maternal genetic relationships of lamb live weight and carcass traits in Swedish sheep breeds. *Journal of Animal Breeding and Genetics*, 121, 66–75.

Näsholm, A. & Danell, O. 1996. Genetic relationships of lamb weight, maternal ability, and mature ewe weight in Swedish finewool sheep. *Journal of Animal Science*, 74, 329–339.

Nussey, D. H., Clutton-Brock, T. H., Albon, S. D., Pemberton, J. M. & Kruuk, L. E. B. 2005a. Constraints on plastic responses to climate variation in red deer. *Biology Letters*, 1, 457–460.

Nussey, D. H., Clutton-Brock, T. H., Elston, D. A., Albon, S. D. & Kruuk, L. E. B. 2005b. Phenotypic plasticity in a maternal trait in red deer. *Journal of Animal Ecology*, 74, 387–396.

Nussey, D. H., Wilson, A. J. & Brommer, J. E. 2007. The evolutionary ecology of individual phenotypic plasticity in naturally occurring populations. *Journal of Evolutionary Biology*, 20, 831–844.

Overall, A. D. J., Byrne, K. A., Pilkington, J. & Pemberton, J. M. 2005. Heterozygosity, inbreeding and neonatal traits in Soay sheep on St. Kilda. *Molecular Ecology*, 14, 3383–3393.

Ozoga, J. J. & Verme, L. J. 1986. Relation of maternal age to fawn-rearing success in white-tailed deer. *Journal of Wildlife Management*, 50, 480–486.

Pettorelli, N., Gaillard, J., Yoccoz, N., Duncan, P., Maillard, D., Delorme, D., Van Laere, G. & Toigo, C. 2005. The response of fawn survival to changes in habitat quality varies according to cohort quality and spatial scale. *Journal of Animal Ecology*, 74, 972–981.

Pirisi, A., Piredda, G., Papoff, C. M., Di Salvo, R., Pintus, S., Garro, G., Ferranti, P. & Chianese, L. 1999. Effects of sheep as1-casein CC, CD and DD genotypes on milk composition and cheesemaking properties. *Journal of Dairy Research*, 66, 409–419.

Plaistow, S. J., Lapsley, C. T. & Benton, T. G. 2006. Context-dependent intergenerational effects: the interaction between past and present environments and its effect on population dynamics. *American Naturalist*, 167, 206–215.

Réale, D., Festa-Bianchet, M. & Jorgenson, J. T. 1999. Heritability of body mass varies with age and season in wild bighorn sheep. *Heredity*, 83, 526–532.

Riska, B., Atchley, W. R. & Rutledge, J. J. 1984. A genetic analysis of targeted growth in mice. *Genetics*, 107, 79–101.

Roff, D. A. 2002. *Life History Evolution*. Sunderland, MA: Sinauer.

Rutberg, A. T. 1987. Adaptive hypotheses of birth synchrony in ruminants: an interspecific test. *American Naturalist*, 130, 692–710.

Sheldon, B. C. & West, S. A. 2004. Maternal dominance, maternal condition, and offspring sex ratio in ungulate mammals. *American Naturalist*, 163, 40–54.

Sikes, R. S. 1998. Tradeoffs between quality of offspring and litter size: differences do not persist into adulthood. *Journal of Mammalogy*, 79,1143–1151.

Singer, F. J., Harting, A., Symonds, K. K. & Coughenour, M. B. 1997. Density dependence, compensation, and environmental effects on elk calf mortality in Yellowstone National Park. *Journal of Wildlife Management*, 61, 12–25.

Slate, J., Visscher, P. M., MacGregor, S., Stevens, D., Tate, M. L. & Pemberton, J. M. 2002. A genome scan for quantitative trait loci in a wild population of red deer (*Cervus elaphus*). *Genetics*, 162, 1863–1873.

Solberg, E., Loison, A., Gaillard, J. & Heim, M. 2004. Lasting effects of conditions at birth on moose body mass. *Ecography*, 27, 677–687.

Stewart, K. M., Bowyer, R. T., Dick, B. L., Johnson, B. K. & Kie, J. G. 2005. Density-dependent effects on physical condition and reproduction in North American elk: an experimental test. *Oecologia*, 143, 85–93.

Thouless, C. R. & Guinness, F. E. 1986. Conflict between red deer hinds: the winner always wins. *Animal Behaviour*, 34, 1166–1171.

Tischner, M. & Allen, W. R. 2000. Maternal influence on foal development and the final size of the mature horses. *Medycyna Weterynaryjna*, 56, 283–287.

Trivers, R. L. 1972. Parental investment and sexual selection. In: *Sexual Selection and the Descent of Man* (Ed. by B. Campbell), pp. 136–179. Chicago: Aldine.

Vandenbergh, J. G. 2009. Effects of intrauterine position in litter-bearing mammals. In: *Maternal Effects in Mammals* (Ed. by D. Maestripieri & J. M. Mateo), pp. 203–226. Chicago: University of Chicago Press.

Vargas, C. A., Elzo, M. A., Chase, C. C. & Olson, T. A. 2000. Genetic parameters and relationships between hip height and weight in Brahman cattle. *Journal of Animal Science*, 78, 3045–3052.

Veiberg, V., Loe, L. E., Mysterud, A., Langvatn, R. & Stenseth, N. C. 2004. Social rank, feeding and winter weight loss in red deer: any evidence of interference competition? *Oecologia*, 138, 135–142.

Weladji, R. B., Mysterud, A., Holand, O. & Lenvik, D. 2002. Age-related reproductive effort in reindeer (*Rangifer tarandus*): evidence of senescence. *Oecologia*, 131, 79–82.

Weladji, R. B., Steinheim, G., Holand, O., Moe, S. R., Almoy, T. & Adnoy, T. 2003. Temporal patterns of juvenile body weight variability in sympatric reindeer and sheep. *Annales Zoologici Fennici*, 40, 17–26.

Willham, R. L. 1972. The role of maternal effects in animal breeding. III. Biometrical aspects of maternal effects in animals. *Journal of Animal Science*, 35, 1288–1293.

Wilson, A. J. & Réale, D. 2006. Ontogeny of additive and maternal genetic effects: lessons from domestic mammals. *American Naturalist*, 167, E23–E38.

Wilson, A. J., Coltman, D. W., Pemberton, J. M., Overall, A. D. J., Byrne, K. A. & Kruuk, L. E. B. 2005a. Maternal genetic effects set the potential for evolution in a free-living vertebrate population. *Journal of Evolutionary Biology*, 18, 405–414.

Wilson, A. J., Kruuk, L. E. B. & Coltman, D. W. 2005b. Ontogenetic patterns in heritable variation for body size: using random regression models in a wild ungulate population. *American Naturalist*, 166, E177–E192.

Wilson, A. J., Pilkington, J. G., Pemberton, J. M., Coltman, D. W., Overall, A. D. J., Byrne, K. A. & Kruuk, L. E. B. 2005c. Selection on mothers and offspring: whose phenotype is it and does it matter? *Evolution*, 59, 451–463.

Wilson, A. J., Pemberton, J. M., Pilkington, J. G., Coltman, D. W., Mifsud, D. V., Clutton-Brock, T. H. & Kruuk, L. E. B. 2006. Environmental coupling of selection and heritability limits evolution. *PLoS Biology*, 4, 1270–1275.

Wilson, A. J., Pemberton, J. M., Pilkington, J. G., Clutton-Brock, T. H., Coltman, D. W. & Kruuk, L. E. B. 2007. Quantitative genetics of growth and cryptic evolution of body weight in an island population. *Evolutionary Ecology*, 21, 1573–8477.

Wolf, J. B. & Wade, M. J. 2001. On the assignment of fitness to parents and offspring: whose fitness is it and when does it matter? *Journal of Evolutionary Biology*, 14, 347–356.

Wolf, J. B., Brodie, E. D., Cheverud, J. M., Moore, A. J. & Wade, M. J. 1998. Evolutionary consequences of indirect genetic effects. *Trends in Ecology and Evolution*, 13, 64–69.

Zhou, G. L., Zhu, Q., Jin, H. G. & Guo, S. L. 2006. Genetic variation of growth hormone gene and its relationship with milk production traits in China Holstein cows. *Asian-Australasian Journal of Animal Sciences*, 19, 315–318.

6

Maternal Effects on Offspring Size and Development in Pinnipeds

W. DON BOWEN

INTRODUCTION

The rate and direction of trait evolution depend on the mechanisms by which parents transfer their characters to progeny (Schluter & Gustafsson 1993). Although the genetic basis for such transfer has received considerable attention, maternal effects or inherited environmental effects (Rossiter 1996) may also strongly influence the response of traits to selection (Kirkpatrick & Lande 1989). Kirkpatrick & Lande (1989) considered two classes of maternal effects: those that represent non-Mendelian phenotypic transmission, called maternal inheritance, and those in which offspring fitness is determined jointly by the value of its trait and some trait in the mother, called maternal selection. Examples of maternal inheritance include the amount or quality of nutrients provided to offspring by parents, while maternal selection includes parental defense of offspring against predators or conspecifics. Both maternal inheritance and maternal selection may operate simultaneously for an individual mother and her offspring, but need not.

Pinnipeds are aquatic members of the mammalian order Carnivora and are represented by three monophyletic lineages: the Otariidae (14 fur seals and sea lions), the Phocidae (18 true or earless seals), and the Odobenidae (the walrus). They exploit diverse habitats in most aquatic environments: estuaries, continental shelves, and the deep ocean in temperate, tropical, and both Arctic and Antarctic polar seas (Bowen et al. 2002). Pinnipeds have often not been included in comparative ecological or evolutionary analyses of carnivores because they are aquatic mammals, and thus have evolved a number of morphological adaptations to exploit this medium. However, Bininda-Emonds et al. (2001) found little evidence that, as a group, pinnipeds should be considered separately from terrestrial carnivores in other functional traits such as physiology, life history, ecology, or behavior. In fact, pinnipeds share a suite of characters and many of the demographic features of other large-bodied mammals (Reiss 1989). They are large, long-lived, slow-growing mammals exhibiting delayed sexual maturity and reduced litter size, and females invest heavily in a single precocial offspring. Population abundance tends not to change dramatically from year to year, and numbers are most sensitive to changes in adult survival, followed by juvenile survival and fecundity (Eberhardt & Siniff 1977). Thus, what we learn from pinnipeds should contribute to our understanding of the ecological consequences of maternal effects in large mammals and perhaps other long-lived, large-bodied vertebrates.

As in many other mammals, pinniped females provide all of the nutrients required for dependent offspring through the provisioning of milk, while pinniped males provide no parental care. Therefore, male and female pinnipeds tend to maximize reproductive success in different ways; males by mating with many females to increase the number of offspring sired, and females by rearing a single, high-quality offspring. However, the potential for males to mate with many females depends on the competitive ability of males and the temporal and spatial distribution of receptive females, which can be influenced by resource distribution, the risk of predation, and the costs and benefits of group living (Boness et al. 2002). The diversity of mating systems exhibited by pinnipeds exposes lactating females to a wide range of densities and disturbances by conspecific adults, thus providing avenues for maternal effects.

LACTATION STRATEGIES

Parental care is an important component of reproduction in many animals (Clutton-Brock 1991). Thus, patterns of parental expenditure on offspring and the factors that influence that expenditure have long been of interest to ecologists (e.g., Trivers 1972; Gubernick & Klopfer 1981). Although parental care in pinnipeds can include offspring defense, the greatest component of parental care by far is the provisioning of milk to the offspring, which simplifies the study of parental reproductive expenditure in this group. Given the

high energetic cost and importance of lactation to maternal and offspring fitness (Millar 1977), characteristics of lactation and related reproductive traits
are presumably under strong selection. However, understanding the factors
that influence maternal reproductive effort in mammals is often complicated
by communal rearing systems (Gittleman & Oftedal 1987), allomaternal care,
multiple offspring litters, and postweaning maternal care. In pinnipeds,
however, there is no paternal care or alloparental support (Boness & Bowen
1996) and maternal care of a single pup ends at weaning (Bowen 1991). The
wide variation in lactation length, breeding habitat, and female body size,
and the fact that maternal care occurs mainly through the provisioning of
milk, make the Pinnipedia an interesting group in which to examine maternal effects on offspring.

Lactation in the pinnipeds is constrained by the spatial and temporal
separation between giving birth on land or ice and acquiring nutrients for
milk production at sea (Bartholomew 1970). Three basic lactation strategies
have been recognized (e.g., Bonner 1984; Oftedal et al. 1987). The generally
large-bodied phocid seals (family Phocidae) breed on both land and ice and
have short lactation periods (4–50 days), during which the females of the
larger species haul out onto the breeding substrate and fast until the pup
is weaned. Until recently, this fasting strategy was believed to be typical of
all phocid species (reviewed in Schulz & Bowen 2005). However, studies on
harbor seals (*Phoca vitulina*) show that females of this relatively small-bodied
species (body mass ~85 kg) cannot store sufficient energy to fuel both milk
production and maternal metabolism over the course of lactation. Thus females must feed at sea for short periods almost daily throughout the lactation period (Boness et al. 1994; Bowen et al. 2001). In contrast, the generally smaller-bodied otariids species (family Otariidae; fur seals and sea
lions) breed on land, exhibit long lactation periods (116–540 days), and a
foraging-cycle strategy, during which females alternate feeding trips to sea
(usually averaging ~7 days but up to 23 days in subantarctic fur seals; Georges
& Guinet 2000) with suckling bouts on land. Finally, the large-bodied walrus
(*Odobenus rosmarus*), the sole extant species of the family Odobenidae, exhibits a long lactation period and an aquatic nursing strategy in which offspring
accompany foraging mothers and suckle at sea.

Maternal reproductive effort in pinnipeds is most likely influenced by the
adaptive consequences of the spatio-temporal division between breeding and
feeding as well as the thermoregulatory demands of living in the ocean. A
number of factors have been advanced as influences in the evolution of lactation strategies in pinnipeds, and these factors operate as a complex adaptive

suite (Bartholomew 1970) involving multiple life-history, physiological, and behavioral traits of individuals as well as ecological factors, such as predation and breeding habitat (Schulz & Bowen 2005). Pinnipeds provide a good opportunity to examine maternal effects because they are long-lived and individual females can be studied over many reproductive events such that variation in offspring traits can be nested within females, thus largely removing the Mendelian genetic contribution to such variation.

The best-known maternal effects in pinnipeds involve maternal nutritional influences on offspring size and growth. Adult female pinnipeds are often large and store large quantities of energy in the form of blubber such that the nutrients allocated to offspring by mothers profoundly influences the growth, development, and survival of progeny. Because pinnipeds raise only single offspring, the potential trade-off between offspring number and size does not confound the interpretation of the expected response to selection. In addition to nutritional influences, other types of maternal effects are also known in pinnipeds. In this chapter, I review most of these effects beginning with the influence of timing and location of birth. With the few exceptions noted below, the data are taken from published papers. In cases where there were two studies of a particular maternal effect on the same species, I have included both in the tables if they were based on a reasonably large sample of individuals. Otherwise, I have selected the study with the largest sample size. Although studies involving < 25 mother-pup pairs may not provide good evidence of the existence or strength of a maternal effect, they were included if they represented the only data on that species.

EFFECTS OF TIMING AND LOCATION OF BIRTH

When and where offspring are born can have profound effects on offspring fitness (Emlen & Demong 1975; Rutberg 1987; Ims 1990). Within populations of most species of pinnipeds, offspring are born over a period of 3 to 12 weeks, but in several tropical or subtropical species births can occur in most months (King 1983; Boyd et al. 1999). Differences in the timing of births and the habitats used to give birth and rear offspring also occur among populations within species (Boyd 1991). For example, in the United Kingdom, grey seals (*Halichoerus grypus*) are born on land between late September and early December with births occurring progressively later as one moves clockwise from the southwest around the islands. In eastern Canada, grey seals are born on land and ice in December-January and in the Baltic Sea, they are born on ice in late February to early March. Thus both environmental and social factors may change over the course of the breeding season, favoring the offspring of females that give birth at one time vs. another—such changes are expected to affect postnatal traits such as growth rate and weaning mass (see also Mateo, this volume).

In several pinnipeds species there is evidence that both the location and timing of births are influenced by maternal traits, which in turn affect offspring phenotype and survival. For example, adult elephant seals are highly body-size dimorphic, with males often guarding harems of 20–100 females in colonies where the density of females increases over the course of the breeding season. Reiter et al. (1981) found that older northern elephant seal (*Mirounga angustirostris*) females arrived earlier in the breeding season and gave birth in higher-quality locations in the center of harems, where the risk of disturbance either by flooding of the beach during storms or by intermale aggression was significantly lower. Females giving birth on the periphery of a male's territory were younger on average and were more often separated from their pup and harassed by subordinate and subadult males. A presumed consequence of earlier birth dates was increased survivorship, as pups weaned early in the season were observed in greater numbers at both 7 months and 1 year after weaning (Reiter et al. 1978). Sydeman et al. (1991) also found that both date of arrival and date of parturition were significantly correlated with the age and experience of female northern elephant seals, but that reproductive experience and not age was responsible for the observed effects. The timing of parturition varied in a curvilinear way with arrival and parturition dates: later for inexperienced and very experienced females and earliest for females with intermediate levels of experience.

Birth location and timing of parturition also affect offspring phenotype in grey seals, another capital breeder (i.e., females that support maternal metabolism and milk production from nutrients stored prior to parturition and then fast during the lactation period). Grey seals are a body-size dimorphic species exhibiting a female-defense mating system in which males use several different tactics to secure matings (Lidgard et al. 2005). Females fast over the course of an 18-day lactation period (Iverson et al. 1993) and are mated several days before weaning their pups. Both female density and the operational adult sex ratio change markedly over the course of the breeding season (Boness et al. 1995). In the Farne Islands, Coulson & Hickling (1964) found that habitat selection by females affected pup growth rate. Pups born on the vegetated tops of islands gained mass at a rate 40% less (~2 kg/d) than those born on rocky shore sites. At North Rona, Pomeroy et al. (2001) showed that at low-lying birth locations close to sea access points, grey seal pups grew more rapidly than elsewhere in the colony. This effect remained significant after controlling for maternal size and the efficiency of nutrient transfer to pups, indicating that the location of birth had consequences for pup development. With respect to the timing of births, Boness et al. (1995) found that grey seal females that gave birth before the peak of births at Sable Island, Canada, were disturbed less by males approaching and attempting to mate. By contrast, females spent 17 times more time rejecting males after the peak of births. Females that gave birth early also spent more time suckling, with the result that their pups grew faster and weighed 16% more than pups born late in the season, even after controlling for the effects of maternal postpartum mass (MPPM) on growth rate and weaning mass (see below). Among 49 females, maternal age had no significant effect on birth date, a finding also reported by Pomeroy et al. (1999) for grey seals at North Rona. However, unlike on Sable Island (Boness et al. 1995), heavier females gave birth earlier in the season on North Rona (Anderson & Fedak 1987; Pomeroy et al. 1999).

By contrast, birth date and location appear to have little effect on offspring phenotype in harbor seals (*Phoca vitulina*). This species gives birth in relatively small colonies where female density is moderate and, because mating occurs aquatically, males do not interact with lactating females on land. Unlike the larger phocid species, harbor seal females forage during lactation and both the timing and extent of foraging are inversely related to MPPM (Bowen et al. 2001). Neither MPPM nor the age of females giving birth varied over the course of the breeding season. The birth mass and relative birth mass (pup birth mass/MPPM) of pups were not correlated with birth date, and the mean birth dates of male and female pups did not differ significantly (Bowen et al. 1994; Ellis et al. 2000). Although there was no overall effect of age on birth dates, pups with lanugo (i.e., less developed pups covered with fetal pelage) were lighter and were born earlier in the breeding season to younger and lighter females (Ellis et al. 2000). Thus, there is some evidence that the timing of birth affects offspring size and stage of development in this species, but the effects are weak compared with other species.

The effects of birth date on offspring phenotype are also evident in fur seals and sea lions. In northern fur seals (*Callorhinus ursinus*), older, heavier females gave birth earlier than other females, but maternal age was a stronger determinant of birth date than MPPM (Boltnev & York 2001). Trites (1991) also found that the oldest northern fur seal females gave birth early in the season to larger pups and that primiparous females tended to give birth in midseason to smaller pups. Thus birth mass decreases throughout the season as does maternal body size (Bigg 1986). Similar effects are found in Antarctic fur seals (*Arctocephalus gazella*), where birth mass decreases throughout the season along with a decrease in the length and mass of females (Boyd & McCann 1989; Lunn & Boyd 1993). However, maternal length accounted for almost twice as much of the variation in birth mass as did body mass. The authors speculated that larger, more experienced females were able to replenish reserves early and gave birth in prime sites when density in the colony was low. By contrast, in subantarctic fur seals (*Arctocephalus tropicalis*) larger females gave birth later in the season. Given that pup birth mass is positively correlated with maternal length in this species, birth mass increased over the course of the breeding season (Georges & Guinet 2000). In South American sea lions (*Otaria flavescens*), Campagna et al. (1992) found that mortality over the first month of life was higher in pups born early (32%) than those born during the peak of births (0.7%). Also, the mortality rate of pups born to solitary females was significantly higher than that in females that gave birth in harems. Pups born to solitary females were also lighter than similarly aged harem-born pups and were more likely to die (63%) than those born in harems (11%). Those pups died mostly from starvation and infanticide by males, which was rare in harems.

EFFECTS OF MATERNAL BEHAVIOR

Maternal behavior is likely an important source of maternal effects in pinnipeds, but has been less studied than other types of effects. In fur seals and sea lions, maternal foraging success during the lactation period largely determines the quantity of nutrients (i.e., milk) that females have available to provision their offspring. The relationships between the characteristics of maternal foraging and offspring traits are best studied in the Antarctic fur seal, where Lunn et al. (1994) showed that the average maternal foraging-trip duration in the current and previous year were negatively related to pup birth mass. Doidge & Croxall (1989) showed that the weaning mass of female pups was positively related to the time spent by the female onshore, indicating that females that spent less time foraging produced larger offspring. At Heard Island, another Antarctic fur seal colony, Goldsworthy (1995) also found that the rate of mass gain was positively related to the duration of preceding and subsequent maternal foraging trips, but the effect was observed only in male pups. In contrast, the rate of mass gain of female pups was positively related to the duration of maternal attendance. Differences in maternal condition did not explain these sex-specific effects on offspring in this or other Antarctic fur seal populations (Costa et al. 1989; Doidge & Croxall 1989).

By contrast, Georges & Guinet (2000) found that subantarctic fur seal females in better condition before beginning to nurse had longer attendance periods. Pups whose mothers were large and performed short and regular foraging trips grew faster, and were heavier at weaning. Large mothers were also more efficient when foraging as reflected by maternal mass gain during foraging trips, a pattern also found in Antarctic fur seals (Boyd et al. 1991).

Among the generally smaller phocid species, in which females feed to some extent during lactation, there is evidence that smaller, and presumably younger, females (i.e., those below the median mass) were significantly more likely to become separated from their pup during storms than larger females (Boness et al. 1992). Time spent foraging in harbor seal females varied inversely with MPPM during late lactation; similarly, the proportion of daily energy expenditure fueled by food intake during lactation varied inversely with MPPM in late lactation, but not in early lactation (Bowen et al. 2001). There is also some evidence that in Weddell seals (*Leptonychotes weddellii*) the efficiency of energy transfer to pups is negatively correlated with MPPM, indicating that lighter mothers reduce their mass loss by feeding during lactation. However, the effects of variation in maternal foraging and attendance behavior on offspring traits are presumably small in this species, as only 13% of the variation in efficiency was explained by MPPM (Testa et al. 1989). This conclusion is supported by a recent study showing that MPPM was not related to the efficiency of nutrient transfer to Weddell seal pups (Wheatley et al. 2006).

Maternal dominance and aggressive behavior toward conspecifics provide another avenue for maternal effects. For example, older, more experienced northern elephant seal females were more aggressive and won more interactions with neighboring females; however, the relative importance of successful aggression, age, and experience could not be isolated because they were highly correlated (Sydeman et al. 1991).

EFFECTS OF MATERNAL AGE AND MASS ON SIZE AND GROWTH OF OFFSPRING

Offspring size is an important life-history trait because variation in initial body size of progeny can affect their subsequent growth and survival (Mousseau & Fox 1998). Although data are available for only a few species, there is increasing evidence that offspring size also affects growth and survival in pinnipeds (see below) and is presumably positively correlated with adult size and reproductive performance. For example, in northern fur seals there is a positive relationship between body mass of males at ages 2–4 years and body mass at weaning (Baker et al. 1994). Nevertheless, the relationship between offspring size and fitness is virtually unstudied among pinnipeds.

Offspring size at weaning is largely determined by size at birth (a measure of prenatal energy investment of females), neonatal growth rate, and the duration of lactation (i.e., the quantity of nutrients and period of maternal provisioning). In pinnipeds, and particularly in members of the Phocidae, variation in offspring size and growth rate are expected to have a large non-Mendelian component as female energy stores (Pomeroy et al. 1999; Crocker et al. 2006) or access to food during lactation (e.g., Costa et al. 1989) can differ significantly from one reproductive event to another, and those differences can directly affect offspring phenotype through the quantity of milk provided. Also, pinnipeds are long-lived species (20–40 years) that continue to grow well after their first reproductive event (McLaren 1993), providing successful females with the opportunity to improve skills or enhance resources that increase the probability of successfully rearing offspring. Thus, in addition to the timing and location of birth and behavioral traits, we should expect MPPM, body condition, age, and breeding experience to affect the phenotype of pinniped offspring.

Offspring Size at Birth

The effects of maternal traits on offspring size at birth have been studied in both otariid and phocid species. The two most commonly measured maternal traits are MPPM and age. In the 10 species for which there are data, pup birth mass was positively related to MPPM, with two exceptions, the harp seal (*Phoca groenlandica*) and hooded seal (*Cystophora cristata*, Table 6-1). In these species, the relatively small sample may limit our ability to detect the effect. In most species, the effect of MPPM on pup birth mass is characterized by a linear relationship indicating that heavier females give birth to larger pups. Nevertheless, there are considerable differences among species in the amount of variation in birth mass that is explained by MPPM (Table 6-1). Explained variation averages 19.7% ± 4.8 SE in the 3 otariid species and 20.3% ± 5.9 in the 5 phocid species (24.6% if harp and hooded seals are excluded because of small sample sizes), providing no evidence of a difference between families (t-tests, $p = 0.84$ and 0.56, respectively).

In several species (northern fur seals, harbor seals, and southern elephant seals), there is evidence that pup birth mass initially increases with an

Table 6-1. Nature and strength of relationship between mean maternal postpartum body mass (MPPM) and mean pup mass in pinnipeds. R^2 and number of mother-pup pairs given in square brackets.

Species	MPPM (kg)[a]	Lactation time (d)[a]	Relationship with pup trait			
			Birth mass	Rate mass gain	Weaning mass	Sources
Antarctic fur seal	39	116	+ [25%, 47]; + [18%, 45$_m$]	+[b] [34%$_m$, 38; 58%$_f$, 41]	Nonsignificant [[d], 63]	1, 2, 3, 4
Northern fur seal	41	118	+, asymptotic [10%, 250]	N/A	N/A	5
Subantarctic fur seal	50	300	+[b] [24%, 86]	+[b]	N/A	6
Harbor seal	85	24	+, curvilinear [20%, 244]	+ [19%, 100]	+ [26%, 100]	7, 8
Harp seal	139	12	nonsignificant [13]	N/A	nonsignificant [49]	9
Grey seal	190	17	+ [23%, 251]	+ [28%, 16], + [49%[c], 97]	+ [44%, 228]	10, 11, 12
Hooded seal	236	4	nonsignificant [19]	+[c] [40%, 26]	+ [58%, 15]	13
Weddell seal	447	50	+ [7.6%, 54]	+[c] [94%, 10]	+ [[d], 47]	14, 15
Northern elephant seal	488	26	+ [38%, 27]	+ [42%, 27]	+ [44%, 27]	16
Southern elephant seal	529	24	+ asymptotic [37$_f$, 30%$_m$, 80$_f$, 71$_m$]; + [36%, 38]	+ [77%, 26]	+ [61%, 38]; + [52%$_m$, 66%$_f$, 53$_m$, 90$_f$]	17, 18, 19

[a] based on largest sample size from Schulz & Bowen (2004); [b] maternal body length; [m] male; [f] female; [c] maternal mass loss; [d] not reported; N/A = not available.

Sources: 1. Costa et al. 1988; 2. Boyd & McCann 1989; 3. Doidge & Croxall 1989; 4. Guinet et al. 2000; 5. Boltnev & York 2001; 6. Georges & Guinet 2000; 7. Ellis et al. 2000; 8. Bowen et al. 2001; 9. R. Stewart (pers comm.); 10. Bowen et al. 2006; 11. Pomeroy et al. 1999; 12. Mellish et al. 1999; 13. Kovacs & Lavigne 1992; 14. Tedman & Green 1987; 15. Wheatley et al. 2006; 16. Deutsch et al. 1994; 17. Arnbom et al. 1994; 18. Carlini 1998; 19. Arnbom et al. 1993.

increase in MPPM and then levels off despite further increases in MPPM. However, the relationships are quite different in these three species. In northern fur seals and southern elephant seals, MPPM has little effect on pup birth mass except in the lightest females, whereas in the harbor seal there is a linear relationship with MPPM over most of the range of female masses and only in the heaviest females does the effect disappear (Ellis et al. 2000). With so few species having been studied, the significance of these differing patterns among species is unclear.

The effect of maternal age on pup birth mass has been reported in only seven species (Table 6-2). In contrast to the effect of MPPM, maternal age generally has a nonlinear relationship with pup birth mass and in two species, northern fur seals and grey seals, there is evidence of a quadratic or dome-shaped relationship such that the oldest females give birth to lighter pups than intermediate-aged females (Figure 6-1) providing evidence for senescence. The effect of maternal age on pup birth mass appears to be significantly greater among phocids (41.3% ± 6.7 SE) than among otariids (mean = 5.3% ± 0.7, t-test with unequal variances, $p = 0.03$), but firm conclusions must await data on more species.

Given that maternal age and MPPM are correlated during part of the female's life, understanding the relative effects of age and size requires that these traits be studied simultaneously within the same females. Boltnev & York (2001) studied the relationships between MPPM, age, and pup birth mass in northern fur seals. They found that MPPM was positively correlated with pup birth mass in females less than the median mass, but not for females greater than the median mass. Female body mass alone explained about 17% of the observed variability in pup mass at birth. Female age also had significant linear and quadratic effects on birth mass in both male and female pups. The best fitting model, which included pup sex, female age, and mass and parturition date, explained 33% of the variation in pup birth mass, with age and MPPM accounting for 43% and 7% of the variation, respectively. When pup mass was modeled as a logistic function of MPPM and a quadratic function of maternal age (as indicated by the data), the best model again explained about a third of the variability in birth mass. However, the importance of maternal traits was quite different, with maternal age accounting for only 18% and MPPM accounting for 32% of the explained variation. This analysis underscores the difficulty in determining the relative effects of such correlated maternal traits.

The relative influence of maternal age and mass on pup size at birth has been studied in several phocid species. Northern and southern elephant

Table 6-2. Effects of maternal age and breeding experience on offspring mass in pinnipeds. R^2 and number of mother–pup pairs given in square brackets.

Species	Age range (yr)	Relationship with pup trait			Sources
		Birth mass	Rate mass gain	Weaning mass	
Antarctic fur seal	3–14	+, asymptotic [4–5%, 1037]; inconsistent	nonsignificant [100]	+ [14%, 29]	1, 2, 3
Northern fur seal	3–23	+[a] [4356]; quadratic [6%, 250]	N/A	N/A	4, 5
Harbor seal	4–16	+, asymptotic [54%, 75]	+[b] [14%, 68]; nonsignificant[c]	nonsignificant [74]	6, 7, 8
Harp seal	4–25	nonsignificant [13]	N/A	domed [9.3%, 49]	9
Grey seal	4–42	domed [39%, 251]	quadratic [24%, 228]	domed [51%, 228]	10,11
Weddell seal	[d]	nonsignificant when MPPM entered into the model	N/A	nonsignificant when MPPM entered into the model	12
Northern elephant seal	3–13+	+ [31%, 41]	+, curvilinear [17%, 39]	+ curvilinear [31%, 63]	13, 14
Southern elephant seal	4–22	N/A	nonsignificant when MPPM entered into the model	nonsignificant when MPPM entered into the model	15

[a] fetal mass; [b] through midlactation; [c] through entire lactation period; [d] not reported; N/A = not available.

Sources: 1. Lunn et al. 1994; 2. Lunn et al. 1993; 3. Lunn & Boyd 1993; 4. Trites 1991; 5. Boltnev & York 2001; 6. Ellis et al. 2000; 7 Bowen et al. 2001; 8. Bowen et al. 2003; 9. R. Stewart (pers. comm.); 10. Bowen et al. 2006; 11. Bowen et al. (unpublished); 12. Wheatley et al. 2006; 13. Sydeman et al. 1991; 14. Deutsch et al. 1994; 15. Arnbom et al. 1997.

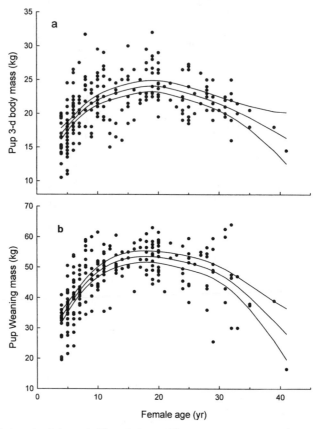

Figure 6-1. Age-related changes in (a) pup 3-day and (b) weaning mass in grey seals (after Bowen et al. 2006). Curves represent fitted values and 95% pointwise confidence limits from generalized additive models. Each symbol represents an independent mother-offspring pair.

seals are the two largest phocid species and mothers provision pups on an energy-rich milk for approximately one month. In both species, pup birth mass was positively correlated with MPPM (McCann et al. 1989; Deutsch et al. 1994) and with maternal age in the northern species, but maternal age had no significant effect on birth mass after adjusting for MPPM (Deutsch et al. 1994). Similarly, Arnbom et al. (1994) showed that heavier southern elephant seal mothers (up to ~550 kg) gave birth to heavier pups, but that there was little relationship between pup birth mass and MPPM in females > 550 kg. Maternal age was not a significant variable when MPPM was included in the model. There is also evidence that MPPM affects the sex ratio of southern elephant seal pups, in that small females (i.e., < ~350 kg) gave birth only to female pups and the sex ratio of pups increased rapidly in females up to about 425 kg, with no detectable increase thereafter (Arnbom et al. 1994). Maternal age had no significant effect on sex ratio once MPPM was entered into the analysis. By contrast, maternal age explained a significant amount of the variation in pup birth mass of grey seals after statistically controlling for the effects of MPPM (Bowen et al. 2006).

The harbor seal is a relatively small species (MPPM = 85 kg) in which most females must forage during lactation to support the energetic cost of milk production (Boness et al. 1994; Thompson et al. 1994; Bowen et al. 2001). In this species, older mothers bore heavier pups even after controlling for the effects of MPPM on birth mass (31% explained variation), but this was not true for heavier mothers after controlling for the effect of maternal age (Ellis et al. 2000). Thus age, and not body mass, appears to be the primary influence on pup birth mass in this species. Ellis et al. (2000) also used PATH analysis to simultaneously examine the effects of maternal mass and age on pup birth mass in harbor seals. They constructed an overall model using females of all ages and separate models for three age classes representing different levels of reproductive experience. In the overall model, the strongest effects were of maternal age (path coefficient = 0.79) and mass (0.15) on birth mass, but age was several times as important as mass. In the separate age class models, only the model for the youngest females (4–6 years) had interaction strengths similar to the overall model. For older females (7–10 years) only negative relationships between birth date and maternal mass on percentage lanugo (i.e., fetal pelage) were significant, and in the oldest females (11–16) none of the paths was significant. Thus, the significant effects of maternal age, but also of MPPM, on pup birth mass were limited to the youngest females indicating an asymptotic relationship between these traits and pup birth size.

Offspring Growth Rate and Weaning Mass

The rate at which females allocate nutrients to their offspring affects both the rate of offspring growth during lactation and offspring size at weaning. Therefore, factors influencing the provisioning rates of offspring by females have received considerable attention, but again only in a handful of pinniped species. Evidence for the effects of MPPM on pup growth rate and weaning mass comes largely from phocid species. To a great extent this simply reflects the greater difficulty in conducting studies on fur seals and seal lions with their long periods of lactation (often > 1 yr). However, even among the phocid species there is a bias toward research on larger species. Nevertheless, as predicted, there are positive relationships between pup growth rate and weaning mass and MPPM in all 6 phocid species studied to date, accounting for an average of 51.7% and 48.7% of the variance, respectively (Table 6-1). There are no reported studies on otariids, but two studies found a positive relationship between pup growth rate and maternal length (Table 6-1), suggesting that MPPM might have been a significant factor had it been measured. Only in the Antarctic fur seal are there data on the relationship between weaning mass and MPPM; in this species, MPPM was not a significant factor (Doidge & Croxall 1989).

The effects of maternal age on pup growth rate and weaning mass are less evident, having been found in only five (1 otariid and 4 phocids) of the 7 species studied (Table 6-2). Furthermore, the results are inconsistent among species. In the case of pup growth rate, only in grey seals and northern elephant seals was there evidence for an effect of maternal age, accounting for 24% and 17% of variance, respectively. In both species the relationship was curvilinear, but in grey seals a quadratic relationship best fit the data, again providing evidence of senescent decline. In harbor seals, there was a positive relation between pup growth rate and maternal age through mid-lactation, but this relationship was no longer significant at weaning (Bowen et al. 2001). In the only otariid studied, maternal age did not account for a significant fraction of the variability in pup growth rate (Table 6-2). Maternal age accounted for an average of 26.3% of the variability in weaning mass among the four species for which there is evidence for an age effect (Table 6-2). Age was not a significant factor after MPPM was included in the model for harbor seals, southern elephant seals, and Weddell seals (Bowen et al. 2001; Arnbom et al. 1997; Wheatley et al. 2006).

Milk Energy Intake

Studies on the energetics of lactation can provide an understanding of some of the mechanisms underlying maternal effects on growth and development. In the case of phocids, this is due to the positive correlations between MPPM and maternal age and milk energy output within species (e.g., Iverson et al. 1993; Mellish et al. 1999; Crocker et al. 2001).

Milk energy output over all or a portion of lactation has been estimated in 14 species, including 9 phocids, although most studies are based on small numbers of females or cover only a small fraction of the lactation period (reviewed in Schulz & Bowen 2004). Nevertheless, in several species the relationships among female traits, milk energy output, and offspring growth and weaning mass are well understood. One example is the grey seal (Iverson et al. 1993; Mellish et al. 1999). In this species, isotope dilution studies have shown that females at the beginning of lactation vary considerably in total body fat and protein content, but that the ratio of fat to protein is independent of MPPM. This indicates that larger females have absolutely more stored resources to allocate to offspring, but not relatively more. Females produce an energy-rich milk comprising ~60% fat through mid to late lactation. However, the fat content of the milk of individual females ranges from about 45% to 65% (Iverson et al. 1993). Furthermore, differences in milk composition appear to persist among years within females, suggesting that they have a strong genetic basis (Lang et al. 2001). Females expend almost 50% of their fat reserves over the course of lactation, resulting in a loss of 31.4% of MPPM over 17 days. MPPM accounts for about 32–53% of the variability in total milk output. Daily milk energy intake accounts for 85% of the variability in the daily rate of mass gained by pups. Lactation length is positively related to MPPM, with the heaviest females lactating for almost twice as long as the smallest females. Thus, the best predictor of pup weaning mass was total milk energy intake (i.e., daily intake × lactation length), explaining 88.4% of the variability. Pups of larger females are not only heavier at weaning but are relatively fatter than lighter pups, and this is presumably one of the factors resulting in improved survival of larger pups (see Hall et al. 2001; 2002). Similar relationships between maternal traits and pup growth rate and weaning mass are also evident in other phocid species: elephant seals (Fedak et al. 1996; Crocker et al. 2001), Weddell seals (Tedman & Green 1987; Wheatley et al. 2006), hooded seals (Bowen et al. 1987; Oftedal et al. 1993), and harp seals (Oftedal et al. 1996).

EFFECTS OF PARITY

Reproductive experience is imperfectly correlated with maternal age. Thus, it is useful to examine the effect of parity on maternal lactation performance and offspring traits. In northern fur seals, primiparous females are lighter and shorter than multiparous females and carry smaller fetuses that multiparous females (corrected for age and sampling date; Trites 1991). However, differences in fetal body mass were not entirely accounted for by differences in female body size, suggesting that physiological changes resulting from having given birth (perhaps leading to improvement in fetal nutrition) were involved. In harbor seals, primiparous females gave birth to significantly lighter pups than multiparous females, even after controlling for the effects of maternal age on birth mass (Ellis et al. 2000). In grey seals, pup birth mass was positively related to increasing reproductive experience over the first three parities, even after the effects of MPPM had been statistically removed (Bowen et al. 2006). In contrast, parity had little effect on pup weaning mass in grey seals, when MPPM was included in the analysis.

Age at primiparity had no effect on birth mass or weaning success in Antarctic fur seals (Lunn et al. 1994). Similarly, in primiparous grey seals and harbor seals, age at primiparity had no significant effect on pup birth mass (Bowen et al. 2006; Ellis et al. 2000). Pup weaning mass did increase with age at primiparity in grey seals, but the effect of age became nonsignificant once MPPM was included in the analysis (Bowen et al. 2006).

THE RELATION BETWEEN MATERNAL/OFFSPRING TRAITS AND OFFSPRING SURVIVAL

Maternal effects need not necessarily be adaptive. For example, the maternal transmission of parasites or toxins may have negative effects on offspring (Clark & Galef 1995). However, many maternal effects are presumed to be adaptive, and in some cases, there is evidence that this is indeed the case. For example, in northern elephant seals, a higher proportion of female pups born early in the season survived to reproduce and gave birth for the first time at an earlier age than female pups born during the peak or late season (Reiter et al. 1981), presumably because early-born females were larger. There is also evidence from several species linking offspring traits, particularly size and growth rate, with improved survival (see Table 6-3), suggesting that we can also expect to find that higher-quality offspring will indeed have greater reproductive success.

The most convincing evidence that maternal effects improve survival comes from studies of the larger phocid species because these species are accessible during lactation and juveniles return to relatively few haul-out sites, where they can be resighted. Effects on survival can be partitioned into those occurring prior to weaning and those occurring after weaning (Table 6-3). Reiter & Le Boeuf (1991) found that reproductive success (i.e., the proportion weaned, a measure of preweaning survival) increased with maternal age, with prime northern elephant seal females (6+ years) on Año Nuevo about twice as likely to be successful as young females. The differences between the two groups of females were greatest during years of bad weather or when the density of harems was high, suggesting that these effects are modified by environmental and social factors. Working on the Faral-

Table 6-3. Effects of maternal and offspring phenotype on offspring survival.

| Species | Effect on offspring survival | | Sources |
	Preweaning	Postweaning	
Antarctic fur seal	+, related to maternal foraging trip duration and with female pup growth rate	N/A	1
Northern fur seal	+, greater pup birth mass and length	+, greater late lactation male body mass	2, 3, 4
Subantarctic fur seal	+, greater pup birth mass	N/A	5
Harbor seal	N/A	nonlinear, asymptotic function of late autumn body mass	6
Grey seal	quadratic function of maternal age	+, greater weaning mass and body condition and lower immunoglobulin levels	7, 8, 9
Weddell seal	+, greater maternal age, multiparous > primiparous females	+, maternal age and experience	10
Northern elephant sea	+, maternal age	+, greater pup length, no effect of pup mass	11, 12, 13
Southern elephant seal	N/A	+, nonlinear function greater pup weaning mass	14

N/A = not available.

Sources: 1. Boyd et al. 1995; 2. Boltnev et al. 1998; 3. Baker & Fowler 1992; 4. Baker et al. 1994; 5. Mison-Jooste 1999; 6. Harding et al. 2005; 7. Bowen et al. 2006; 8. Hall et al. 2001; 9. Hall et al. 2002; 10. Hastings & Testa 1998; 11. Reiter et al. 1978; 12. Sydeman et al. 1991; 13. Le Boeuf et al. 1994; 14. McMahon et al. 2003.

lon Islands, Sydeman et al. (1991) showed that weaning success in northern elephant seal females increased from 3–7 years of age and then leveled off. However, after statistically controlling for previous reproductive experience, weaning success increased linearly throughout the life of a female. In Weddell seals at McMurdo Sound, Antarctica, Hastings & Testa (1998) found that pup preweaning mortality was significantly higher for primiparous females (7.1%) than multiparous females (3.2%). However, there was no effect of female parity on pup preweaning survival among multiparous females ($n =$

228 females and 627 pups) indicating that the effect was limited to first-time breeders. In grey seals breeding on Sable Island, Canada, the proportion of females successfully weaning offspring increased with maternal age and mass from ages 4–7 years, varied without trend from age ~8–33 years, and then declined sharply in the oldest females (Bowen et al. 2006). As most 6- and 7-year-old grey seal females were multiparous, the improvement in pup preweaning survival was not limited to those born to primiparous females, as in Weddell seals.

Although there is strong evidence for the effects of maternal traits on offspring size and development (Tables 6-1 & 6-2), there are few studies that directly link maternal traits with postweaning survival of offspring. Nevertheless, there is increasing indirect evidence that maternal traits can be expected to influence offspring survival. Hall et al. (2001) found that the probability of postweaning survival to age 1 year in grey seals increased with body condition at weaning, which in turn is affected by maternal condition (Pomeroy et al. 1999). Regardless of the body condition of pups or the time of year, survival of female pups was estimated to be 3.4 times as great as that of males, although the reason for this difference was unknown. Furthermore, regardless of sex, a 1-standard-deviation increase in body condition of pups was estimated to increase survival by 1.4 times. In a subsequent study at another colony, both body condition and a measure of pup immune function, serum gammaglobulin (IgG), were assessed (Hall et al. 2002). Again, increased body mass or condition of grey seal pups was positively related to survival, but higher postweaning circulating levels of IgG decreased the probability of survival. It was not known if this was because some individuals had naturally higher IgG levels or because levels were higher due to antigenic challenge. In Hawaiian monk seals (*Monachus schauinslandi*), the probability of survival from weaning to age 2 years at French Frigate Shoals was significantly related to offspring size at weaning for 1984–1987 cohorts and to measures of physical condition for the 1990–1994 cohorts (Craig & Ragen 1999). Again, the effect of offspring traits on survival was modified by environmental variation, as there was no relationship between survival of the 1988–1989 cohorts and size at weaning and the relationships differed at another breeding colony. At both study sites, the nature of the relationships between offspring traits and survival also changed over time.

In Weddell seals, although the survival of pups to 1 year of age did not vary with maternal age and parity, survival to age 6 increased with age of the mother (Hastings & Testa 1998). Annual survival of pups to all ages also increased with the body length of the mother and increased with maternal length for male offspring, but not for females. In small mothers, survival was higher for female offspring than male offspring, whereas in large mothers survival was higher for male pups than female pups during both the first and subsequent years.

In the two largest phocid species, there is conflicting evidence regarding the effect of offspring size on survival. Le Boeuf et al. (1994) found no relationship between survival to ages 1 and 2 years and weaning mass in northern elephant seals; however, there was a positive relationship between survival and length to age 1, but not to age 2. However, these estimates were made during a period of rapid population growth. Thus, the inconsistent and relatively weak effects of offspring traits on survival could reflect the presumably favorable conditions experienced by these juveniles. By contrast, in two colonies of southern elephant seals, one larger and declining (Macquarie Island) and one that had recently stabilized (Marion Island), McMahon et al. (2003) found that pups that survived were significantly heavier at weaning than those that did not survive. Survival rates differed between the two colonies, and the functional form of the relationship with weaning mass also differed. At Macquarie Island first-year survival was a polynomial function of weaning mass, whereas at Marion Island survival was best described by an exponential function of weaning mass.

Harding et al. (2005) also found that the relationship between winter survival rate of harbor seals pups and their autumn body mass was a nonlinear, asymptotic function. They calculated that increasing thermal stress with decreasing body size and low winter temperatures may result in a negative energy balance resulting in increased mortality in smaller pups. This is consistent with the finding that food intake and changes in body condition in harbor seals over the first month postweaning were a function of their body mass at weaning (Muelbert et al. 2003). Heavier pups were relatively fatter than lighter pups and therefore had significantly greater total body energy. Furthermore, the temporal pattern and composition of mass loss differed between heavier and lighter pups, with lighter pups mobilizing their blubber stores earlier than heavier pups.

Less is known about the links between maternal and offspring traits and offspring survival in otariids. In Antarctic fur seals, pup preweaning mortality was positively related to the foraging trip duration of mothers during lactation, but negatively related to growth rate in female pups only (Boyd et al. 1995). In northern fur seals, larger pups at birth survived significantly better to 40 days postpartum (i.e., ~33% of lactation) than smaller pups (Boltnev et al. 1998). High birth mass was also associated with reduced newborn mortality throughout the pupping season in Cape fur seals (*Arctocephalus tropicalis*; Mison-Jooste 1999).

The effect of offspring traits on postweaning survival in otariids is known only in the northern fur seal. On St. Paul Island, Alaska, males that survived at least two years were heavier than the mean for their cohort as pups (Baker & Fowler 1992). This effect was not evident in female pups, but the sample was too small to draw firm conclusions. In a follow-up study on cohorts born in the late 1980s and early 1990s, when pup body masses were considerably higher that during the 1960s, Baker et al. (1994) found that larger-than-average males still survived better than smaller individuals, but the difference was less evident during this period of more favorable conditions (i.e., heavier pups) compared to the 1960s.

FACTORS AFFECTING THE EXPRESSION OF MATERNAL EFFECTS

The expression of maternal effects is to some extent contingent on features of both parent and offspring environments (e.g., Riska et al. 1985; Bernardo 1996; Rossiter 1996). Environmental factors such as colony density and weather can affect the ability of mothers to maintain contact with their offspring. In northern elephant seals, the proportion of pups separated from their mothers varied from −24 to 57% at the crowded Point Harem colony over a 4-year period. Adult and subadult males caused about half of the separations with weather as a cofactor. Twenty-eight percent of separations were caused by pups wandering away from their mothers. Although separations were often short, a high proportion (48–66%) of orphaned pups (i.e., those separated for > 2 days) died, underscoring the importance of female vigilance (Reidman & Le Boeuf 1982). Density effects can also be important within season. Northern elephant seal females that arrived earlier and were more successful in female interactions were more likely to successfully wean their offspring than other females. These effects were again mediated by colony density, with females that pupped in areas of lower density being more successful (Ribic 1988). Sydeman et al. (1991) showed that weaning success in younger females declined over the breeding season even after adjusting for breeding experience, presumably because they were negatively affected by increased colony density. However, the weaning success of older females did not change. Similarly, as noted above, grey seal females that gave birth early in the breeding season, when colony density and male harassment were lower, produced heavier pups (Boness et al. 1995).

Given the high cost of lactation in mammals, interannual variability in access to food seems an obvious avenue to influence the expression of maternal effects on offspring traits. Antarctic fur seal females feed mainly on krill (*Euphausia superba*) during the breeding season and respond strongly to interannual variation in krill abundance, which is patchily distributed. During 1990–1991, a year of low krill abundance, female fur seals spent significantly greater effort, both in terms of time and activity while foraging, than usual (Boyd et al. 1994). Despite this greater foraging effort (estimated at one-third to one-half greater than normal), mothers were unable to meet the energy demands of lactation, resulting in lower pup growth rates and increased pup mortality (Lunn & Boyd 1993; Lunn et al. 1994).

There is evidence that environmental variability also affects foraging success, body condition, and offspring size and survival in several phocid species. During strong El Niño events, the daily rate of mass gained by adult female northern elephant seals was only 33% of that during weak or non–El Niño years and females returned to the breeding colony with significantly lower fat stores and reduced natality (Crocker et al. 2006). At decadal scales, less extreme variability in the foraging success and energy storage of adult northern elephant seal females had direct consequences for pup weaning mass, with reduced maternal mass gain resulting in lighter pups (Le Boeuf & Crocker 2005).

Another Pacific pinniped species, the Hawaiian monk seal, appears to benefit from strong El Niño events. Antonelis et al. (2003) found that after accounting for temporal trends from 1983 to 2001, the girths of pups were 3.7 cm (3.5%) and 2.7 cm (2.5%) greater during El Niño years at French Frigate Shoals and Laysan Island, respectively. Average pup weaning masses were significantly greater during El Niño years at French Frigate Shoals (+2.6 kg or 4.0%), but were not significantly different at Laysan Island (+1.8 kg). Pups born at French Frigate Shoals during El Niño years also survived significantly better, but this effect was not detected at Laysan Island. Given the life history of the species, the authors reasonably speculated that the increase in pup weaning mass was likely associated with increases in the abundance of available monk-seal prey that would enhance female foraging success during gestation, improve her body condition, and thereby increase the energy transfer to her offspring from milk.

CONCLUSIONS

Although relatively few species have been studied, there is evidence that offspring phenotype in pinnipeds is influenced by both maternal inheritance

and maternal selection. Pinniped offspring are affected by maternal decisions about the timing and location of birth, the foraging ability of mothers,
and the ability of mothers to defend and maintain contact with dependent
offspring. Larger, older, and in some cases more experienced females give
birth to larger pups, which then grow faster during lactation and are heavier
at weaning. There is also compelling evidence from a number of species that
offspring size at weaning increases the postweaning survival of offspring for
one or more years and to the age of first reproduction. These results have
implications for life-history theory predictions about the trade-off between
maternal growth and reproductive expenditure over the life of a female, as
a critical assumption is that the amount of resources allocated to offspring
affects fitness.

The expression of maternal effects on offspring phenotype in pinnipeds
is clearly influenced by maternal and offspring environments. These include
changes in colony density within and among breeding seasons, social structure, and variation in the availability of food either prior to, in the case of
most phocids, or both prior to and during the period of offspring provisioning, in the smaller phocids and otariids. These environmental factors modify
the expression of maternal effects to the extent that in some contexts they
may not be evident, for example, at low density or high food availability.
Thus, future studies should measure relevant environmental factors to provide an ecological context in which to interpret findings.

There are somewhat conflicting results on the relative importance of
maternal postpartum mass (MPPM), age, and reproductive experience on
offspring traits in pinnipeds. This is partly due to the positive correlation
among these maternal traits, making it more difficult to tease out the independent effects. However, other factors also potentially limit inferences,
including small sample size, the range of MPPM and age included in analyses, environmental conditions, and the demography of the study population.
Furthermore, only 30% of pinniped species have been studied, and even
within the better-studied Phocidae, there is a bias toward the larger, more accessible species. Therefore, a comparative analysis of frequency and relative
importance of maternal effects in pinnipeds is premature.

Offspring phenotype is the product of its genes, the environment in which
those genes are expressed, the gene-environment interaction, and maternal
effects. Parental environment often plays a large role in the magnitude and
nature of the maternal effects. The attributes of the parental environment
are often a permanent feature of the species' environment, suggesting their
importance as ecological and evolutionary forces (Rossiter 1996). However,

identifying the strength of maternal effects (reviewed in Rossiter 1996) requires that other influences on phenotype be accounted for. Quantitative genetics provides several approaches to estimate the relative contribution of genes and the environment to trait variation (e.g., Schluter & Gustafsson 1993). Breeding designs to estimate maternal effects may be straightforward for many taxa, but are impractical if not impossible for large, long-lived mammals such as pinnipeds and other marine mammals. Cross-fostering may be possible in some species, such as the Hawaiian monk seal in which females often suckle pups other than their own (although in practice these experiments could not be performed on such a critically endangered species); however, in most species, females do not foster. Therefore, in the absence of breeding experiments, we must be cautious in reaching conclusions about both the nature and strength of maternal effects in pinnipeds. One approach would be to make better use of long-term studies of the reproductive performance of individuals over time, thereby largely controlling for maternal nuclear gene contribution to offspring, although the contribution of sires will still partly confound interpretation.

ACKNOWLEDGMENTS

I thank S. J. Iverson and D. J. Boness for invaluable discussions about reproductive behavior and energetics in pinnipeds, and S. Lang for helpful comments on previous drafts of the paper. Many of the ideas expressed here were developed in the course of field research supported by the Department of Fisheries and Oceans, Canada, and the Natural Sciences and Engineering Research Council, Canada.

REFERENCES

Anderson, S. S. & Fedak, M. A. 1987. The energetics of sexual success of grey seals and comparison with the costs of reproduction in other pinnipeds. *Symposia of the Zoological Society of London*, 57, 319–341.

Antonelis, G. A., Baker, J. D. & Polovina, J. J. 2003. Improved body condition of weaned Hawaiian monk seal pups associated with El Niño events: potential benefits to an endangered species. *Marine Mammal Science*, 19, 590–598.

Arnbom, T., Fedak, M. A., Boyd, I. L. & McConnell, B. J. 1993. Variation in weaning mass of pups in relation to maternal mass, postweaning fast duration, and weaned pup behaviour in southern elephant seals (*Mirounga leonina*) at South Georgia. *Canadian Journal of Zoology*, 71, 1772–1781.

Arnbom, T., Fedak, M. A. & Rothery, P. 1994. Offspring sex ratio in relation to female size in southern elephant seals, *Mirounga leonina*. *Behavioral Ecology and Sociobiology*, 35, 373–375.

Arnbom, T., Fedak, M. A. & Boyd, I. L. 1997. Factors affecting maternal expenditure in southern elephant seals during lactation. *Ecology*, 78, 471–483.

Baker, J. D. & Fowler, C. W. 1992. Pup weight and survival of northern fur seals *Callorhinus ursinus*. *Journal of Zoology*, 227, 231–238.

Baker, J. D., Fowler, C. W. & Antonelis, G. A. 1994. Body weights and growth of juvenile males northern fur seals *Callorhinus ursinus*. *Marine Mammal Science*, 10, 151–162.

Bartholomew, G. A. 1970. A model for the evolution of pinniped polygyny. *Evolution*, 24, 546–559.

Bernardo, J. 1996. Maternal effects in animal ecology. *American Zoologist*, 36, 83–105.

Bigg, M. A. 1986. Arrival of northern fur seals (*Callorhinus ursinus*) off western North America. *Canadian Technical Report of Fisheries and Aquatic Science*, 1764, 1–64.

Bininda-Emonds, O. R. P., Gittleman, J. L. & Kelley, C. K. 2001. Flippers versus feet: comparative trends in aquatic and non-aquatic carnivores. *Journal of Animal Ecology*, 70, 386–400.

Boltnev, A. I. & York, A. E. 2001. Maternal investment in northern fur seals (*Callorhinus ursinus*): interrelationships among mother's age, size, parturition date, offspring size and sex ratios. *Journal of Zoology*, 254, 219–228.

Boltnev, A. I., York, A. E. & Antonelis, G. A. 1998. Northern fur seal young: interrelationships among birth size, growth, and survival. *Canadian Journal of Zoology*, 76, 843–854.

Boness, D. J. & Bowen, W. D. 1996. The evolution of maternal care in pinnipeds. *BioScience*, 46, 645–654.

Boness, D. J., Bowen, W. D., Iverson, S. J. & Oftedal, O. T. 1992. Influence of storms and maternal size on mother-pup separations and fostering in the harbour seal, *Phoca vitulina*. *Canadian Journal of Zoology*, 70, 1640–1644.

Boness, D. J., Bowen, W. D. & Oftedal, O. T. 1994. Evidence of a maternal foraging cycle resembling that of otariid seals in a small phocid, the harbor seal. *Behavioral Ecology and Sociobiology*, 34, 95–104.

Boness, D. J., Bowen, W. D. & Iverson, S. J. 1995. Does male harassment of females contribute to reproductive synchrony in the grey seal by affecting maternal performance? *Behavioral Ecology and Sociobiology*, 36, 1–10.

Boness, D. J., Clapham, P. J. & Mesnick, S. L. 2002. Life history and reproductive strategies. In: *Marine Mammals: An Evolutionary Approach* (Ed. by A. R. Hoelzel), pp. 278–324. Oxford: Blackwell Science.

Bonner, W. N. 1984. Lactation strategies in pinnipeds: problems for a marine mammalian group. *Symposia of the Zoological Society of London*, 51, 253–272.

Bowen, W. D. 1991. Behavioural ecology of pinniped neonates. In: *Behaviour of Pinnipeds* (Ed. by D. Renouf), pp. 66–127. Cambridge: Cambridge University Press.

Bowen, W. D., Boness, D. J. & Oftedal, O. T. 1987. Mass transfer from mother to pup and subsequent mass loss by the weaned pup in the hooded seal, *Cystophora cristata*. *Canadian Journal of Zoology*, 65, 1–8.

Bowen, W. D., Oftedal, O. T., Boness, D. J. & Iverson, S. J. 1994. The effect of maternal age and other factors on birth mass in the harbour seal. *Canadian Journal of Zoology*, 72, 8–14.

Bowen, W. D., Boness, D. J., Iverson, S. J. & Oftedal, O. T. 2001. Foraging effort, food intake, and lactation performance depend on maternal mass in a small phocid seal. *Functional Ecology*, 15, 325–334.

Bowen, W. D., Beck, C. A. & Austin, D. 2002. Pinniped ecology. In: *Encyclopedia of Marine Mammals* (Ed. by W. F. Perrin, B. Wursig & H. G. M. Thewissen), pp. 911–920. San Diego, CA: Academic Press.

Bowen, W. D., Ellis, S. L., Iverson, S. J. & Boness, D. J. 2003. Maternal and newborn life-history traits during periods of contrasting population trends: implications for explaining the decline of harbour seals, *Phoca vitulina*, on Sable Island. *Journal of Zoology*, 261, 155–163.

Bowen, W. D., Iverson, S. J., McMillan, J. I. & Boness, D. J. 2006. Reproductive performance in grey seals: age-related improvement and senescence in a capital breeder. *Journal of Animal Ecology*, 75, 1340–1351.

Boyd, I. L. 1991. Environmental and physiological factors controlling the reproductive cycles of pinnipeds. *Canadian Journal of Zoology*, 69, 1135–1148.

Boyd, I. L. & McCann, T. S. 1989. Pre-natal investment in reproduction by female Antarctic fur seal. *Behavioral Ecology and Sociobiology*, 24, 377–385.

Boyd, I. L., Lunn, N. J. & Barton, T. 1991. Time budgets and foraging characteristics of lactating Antarctic fur seals. *Journal of Animal Ecology*, 60, 577–592.

Boyd, I. L., Arnould, J. P. Y., Barton, T. & Croxall, J. P. 1994. Foraging behaviour of Antarctic fur seals during periods of contrasting prey abundance. *Journal of Animal Ecology*, 63, 703–713.

Boyd, I. L., Croxall, J. P., Lunn, N. J. & Reid, K. 1995. Population demography of Antarctic fur seals: the costs of reproduction and implications for life histories. *Journal of Animal Ecology*, 64, 505–518.

Boyd, I., Lockyer, C. & Marsh, H. D. 1999. Reproduction in marine mammals. In: *Biology of Marine Mammals* (Ed. by J. E. Reynolds III & S. A. Rommel), pp. 218–286. Washington, D.C.: Smithsonian Institution Press.

Campagna, C., Bisoli, C., Quintana, F., Perez, F. & Vila, A. 1992. Group breeding in sea lions: pups survive better in colonies. *Animal Behaviour*, 43, 541–548.

Carlini, A. R. 1998. Energy investment in pups of southern elephant seals and mass changes in females while at King George Island. *Berichte zur Polarforschung*, 299, 249–255.

Clark, M. M. & Galef, B. G. 1995. Prenatal influences on reproductive life history strategies. *Trends in Ecology and Evolution*, 10, 151–152.

Clutton-Brock, T. H. 1991. *The Evolution of Parental Care*. Princeton: Princeton University Press.

Costa, D. P., Trillmich, F. & Croxall, J. P. 1988. Intraspecific allometry of neonatal size in the Antarctic fur seal (*Arctocephalus galapagoensis* [sic *gazella*]). *Behavioral Ecology and Sociobiology*, 22, 361-364.

Costa, D. P., Croxall, J. P. & Duck, C. D. 1989. Foraging energetics of antarctic fur seals in relation to changes in prey availability. *Ecology*, 70, 596–606.

Coulson, J. C. & Hickling, G. 1964. The breeding biology of the grey seal, *Halichoerus grypus* (Fab.), on the Farne Islands, Northumberland. *Journal of Animal Ecology*, 33, 485–512.

Craig, M. P. & Ragen, T. J. 1999. Body size, survival, and decline of juvenile Hawaiian monk seals, *Monachus schauinslandi*. *Marine Mammal Science*, 15, 786–809.

Crocker, D. E., Williams, J. D., Costa, D. P. & Le Boeuf, B. J. 2001. Maternal traits and reproductive effort in northern elephant seals. *Ecology*, 82, 3541–3555.

Crocker, D. E., Costa, D. P., Le Boeuf, B. J., Webb, P. M. & Houser, D. S. 2006. Impact of El Niño on the foraging behaviour of female northern elephant seals. *Marine Ecology: Progress Series*, 309, 1–10.

Deutsch, C. J., Crocker, D. E., Costa, D. P. & Le Boeuf, B. J. 1994. Sex- and age-related variation in reproductive effort of northern elephant seals. In: *Elephant Seals: Population Ecology, Behavior, and Physiology* (Ed. by B. J. Le Boeuf & R. M. Laws), pp. 169–210. Berkeley: University of California Press.

Doidge, D. W. & Croxall, J. P. 1989. Factors affecting weaning weight in Antarctic fur seals *Arctocephalus gazella* at South Georgia. *Polar Biology*, 9, 155–160.

Eberhardt, L. L. & Siniff, D. B. 1977. Population dynamics and marine mammal management policies. *Journal of the Fisheries Research Board of Canada*, 34, 183–190.

Ellis, S. L., Bowen, W. D., Boness, D. J. & Iverson, S. J. 2000. Maternal effects on offspring mass and stage of development at birth in the harbor seal, *Phoca vitulina. Journal of Mammalogy*, 81, 1143–1156.

Emlen, S. T. & Demong, N. J. 1975. Adaptive significance of synchronized breeding in a colonial bird: a new hypothesis. *Science*, 188, 1029–1031.

Fedak, M. A., Arnbom, T. & Boyd, I. L. 1996. The relation between the size of southern elephant seal mothers, the growth of their pups, and the use of maternal energy, fat, and protein during lactation. *Physiological Zoology*, 69, 887–911.

Georges, J. Y. & Guinet, C. 2000. Maternal care in the subantarctic fur seals on Amsterdam Island. *Ecology*, 81, 295–308.

Gittleman, J. L. & Oftedal, O. T. 1987. Comparative growth and lactation energetics in carnivores. *Symposia of the Zoological Society of London*, 57, 41–77.

Goldsworthy, S. D. 1995. Differential expenditure of maternal resources in Antarctic fur seals, *Arctocephalus gazella*, at Heard Island, southern Indian Ocean. *Behavioral Ecology*, 6, 218–228.

Gubernick, D. J. & Klopfer, P. H. (Eds.). 1981. *Parental Care in Mammals*. New York: Plenum Press.

Guinet, C., Lea, M.-A. & Goldsworthy, S. D. 2000. Mass change in Antarctic fur seal (*Arctocephalus gazella*) pups in relation to maternal characteristics at the Kerguelen Islands. *Canadian Journal of Zoology*, 78, 476–483.

Hall, A., McConnell, B. & Barker, R. 2001. Factors affecting first-year survival in grey seals and their implications for life history strategy. *Journal of Animal Ecology*, 70, 138–149.

Hall, A. J., McConnell, B. J. & Barker, R. J. 2002. The effect of total immunoglobulin levels, mass and condition on the first-year survival of grey seal pups. *Functional Ecology*, 16, 462–474.

Harding, K. C., Fujiwara, M., Axberg, Y. & Harkonen, T. 2005. Mass-dependent energetics and survival in harbour seal pups. *Functional Ecology*, 19, 129–135.

Hastings, K. K. & Testa, J. W. 1998. Maternal and birth colony effects on survival of Weddell seal offspring from McMurdo Sound, Antarctica. *Journal of Animal Ecology*, 67, 722–740.

Ims, R. A. 1990. The ecology and evolution of reproductive synchrony. *Trends in Ecology and Evolution*, 5, 135–140.

Iverson, S. J., Bowen, W. D., Boness, D. J. & Oftedal, O. T. 1993. The effect of maternal size and milk energy output on pup growth in grey seals (*Halichoerus grypus*). *Physiological Zoology*, 66, 61–88.

King, J. E. 1983. *Seals of the World*. Ithaca, New York: Comstock Publishing Associates.

Kirkpatrick, M. & Lande, R. 1989. The evolution of maternal characters. *Evolution*, 43, 485–503.

Kovacs, K. M. & Lavigne, D. M. 1992. Mass-transfer efficiency between hooded seal (*Cystophora cristata*) mothers and their pups in the Gulf of St. Lawrence. *Canadian Journal of Zoology*, 70, 1315–1320.

Lang, S. L. C., Iverson, S. J. & Bowen, W. D. 2001. Individual female characteristics of milk composition and lactation performance. 14th Biennial Conference on the Biology of Marine Mammals. Vancouver, Canada. Nov 28–Dec 3, Abstract.

Le Boeuf, B. J. & Crocker, D. E. 2005. Ocean climate and seal condition. *BMC Biology*, 3, 9.

Le Boeuf, B. J., Morris, P. & Reiter, J. 1994. Juvenile survivorship of Northern elephant seals. In: *Elephant Seals: Population Ecology, Behavior, and Physiology* (Ed. by B. J. Le Boeuf & R. M. Laws), pp. 121–136. Berkeley: University of California Press.

Lidgard, D. C., Boness, D. J., Bowen, W. D. & McMillan, J. I. 2005. State-dependent male mating tactics in the grey seal: the importance of body size. *Behavioral Ecology*, 16, 541–549.

Lunn, N. J. & Boyd, I. L. 1993. Effects of maternal age and condition on parturition and the perinatal period of female Antarctic fur seals. *Journal of Zoology*, 229, 55–67.

Lunn, N. J., Boyd, I. L., Barton, T. & Croxall, J. P. 1993. Factors affecting the growth rate and mass at weaning of Antarctic fur seals at Bird Island, South Georgia. *Journal of Mammalogy*, 74, 908–919.

Lunn, N. J., Boyd, I. L. & Croxall, J. P. 1994. Reproductive performance of female Antarctic fur seals: the influence of age, breeding experience, environmental variation and individual quality. *Journal of Animal Ecology*, 63, 827–840.

Mateo, J. M. 2009. Maternal influences on development, social relationships, and survival behaviors. In: *Maternal Effects in Mammals* (Ed. by D. Maestripieri & J. M. Mateo), pp. 133–158. Chicago: University of Chicago Press.

McCann, T. S., Fedak, M. A. & Harwood, J. 1989. Parental investment in southern elephant seals, *Mirounga leonina. Behavioral Ecology and Sociobiology*, 25, 81–87.

McLaren, I. A. 1993. Growth in pinnipeds. *Biological Reviews*, 68, 1–79.

McMahon, C. R., Burton, H. R. & Bester, M. N. 2003. A demographic comparison of two southern elephant seal populations. *Journal of Animal Ecology*, 72, 61–74.

Mellish, J. A. E., Iverson, S. J. & Bowen, W. D. 1999. Variation in milk production and lactation performance in grey seals and consequences for pup growth and weaning characteristics. *Physiological and Biochemical Zoology*, 72, 677–690.

Millar, J. S. 1977. Adaptive features of mammalian reproduction. *Evolution*, 31, 370–386.

Mison-Jooste, V. 1999. Contribution à l'Étude de la biologie de populations de l'otarie à fourrure du Cap (*Arctocephalus pusillus pusillus*): les soins maternels different-ils en fonction au sexe du jeune? Ph.D. Thesis. Université de Lyon.

Mousseau, T. A. & Fox, C. W. 1998. The adaptive significance of maternal effects. *Trends in Ecology and Evolution*, 13, 403–407.

Muelbert, M. M. C., Bowen, W. D. & Iverson, S. J. 2003. Weaning mass affects changes in body composition and food intake in harbour seal pups during the first month of independence. *Physiological and Biochemical Zoology*, 76, 418–427.

Oftedal, O. T., Boness, D. J. & Tedman, R. A. 1987. The behavior, physiology, and anatomy of lactation in the pinnipedia. *Current Mammalogy*, 1, 175–245.

Oftedal, O. T., Bowen, W. D. & Boness, D. J. 1993. Energy transfer by lactating hooded seals and nutrient deposition in their pups during the four days of lactation. *Physiological Zoology*, 66, 412–436.

Oftedal, O. T., Bowen, W. D. & Boness, D. J. 1996. Lactation performance and nutrient deposition in pups of the harp seal, *Phoca groenlandica*, on ice floes off southeast Labrador. *Physiological Zoology*, 69, 635–657.

Pomeroy, P. P., Fedak, M. A., Rothery, P. & Anderson, S. 1999. Consequences of maternal size for reproductive expenditure and pupping success of grey seals at North Rona, Scotland. *Journal of Animal Ecology*, 68, 235–253.

Pomeroy, P. P., Worthington Wilmer, J., Amos, B. & Twiss, S. D. 2001. Reproductive performance links to fine-scale spatial patterns of female grey seal relatedness. *Proceedings of the Royal Society of London, Series B*, 268, 711–717.

Reidman, M. L. & Le Boeuf, B. J. 1982. Mother-pup separation and adoption in northern elephant seals. *Behavioral Ecology and Sociobiology*, 11, 203–215.

Reiss, M. J. 1989. *The Allometry of Growth and Reproduction*. Cambridge: Cambridge University Press.

Reiter, J. & Le Boeuf, B. J. 1991. Life history consequences of variation in age at primiparity in northern elephant seals. *Behavioral Ecology and Sociobiology*, 28, 153–160.

Reiter, J., Stinson, N. L. & Le Boeuf, B. J. 1978. Northern elephant seal development: the transition from weaning to nutritional independence. *Behavioral Ecology and Sociobiology*, 3, 337–367.

Reiter, J., Panken, K. J. & Le Boeuf, B. J. 1981. Female competition and reproductive success in northern elephant seals. *Animal Behaviour*, 29, 670–687.

Ribic, C. A. 1988. Maternal aggression in northern elephant seals: the effect of the pup. *Canadian Journal of Zoology*, 66, 1693–1698.

Riska, B., Rutledge, J. J. & Atchley, W. R. 1985. Covariance between direct and maternal genetic effects in mice, with a model of persistent environmental influences. *Genetic Research*, 45, 287–297.

Rossiter, M. 1996. Incidence and consequences of inherited environmental effects. *Annual Review of Ecology and Systematics*, 27, 451–476.

Rutberg, A. T. 1987. Adaptive hypotheses of birth synchrony in ruminants: an interspecific test. *American Naturalist*, 130, 692–710.

Schluter, D. & Gustafsson, L. 1993. Maternal inheritance of condition and clutch size in the collared flycatcher. *Evolution*, 47, 658–667.

Schulz, T. M. & Bowen, W. D. 2004. Pinniped lactation strategies: evaluation of data on maternal and offspring life-history traits. *Marine Mammal Science*, 20, 86–114.

Schulz, T. M. & Bowen, W. D. 2005. The evolution of pinniped lactation strategies: a comparative phylogenetic analysis. *Ecological Monographs*, 75, 159–177.

Sydeman, W. J., Huber, H. R., Emslie, S. D., Ribic, C. A. & Nur, N. 1991. Age-specific weaning success of northern elephant seals in relation to previous breeding experience. *Ecology*, 72, 2204–2217.

Tedman, R. & Green, B. 1987. Water and sodium fluxes and lactational energetics in suckling pups of Weddell seals (*Leptonychotes weddellii*). *Journal of Zoology*, 212, 29–42.

Testa, J. W., Hill, S. E. B. & Siniff, D. B. 1989. Diving behaviour and maternal investment in Weddell seals (*Leptonychotes weddellii*). *Marine Mammal Science*, 5, 399–405.

Thompson, P. M., Miller, D., Cooper, R. & Hammond, P. S. 1994. Changes in the distribution and activity of female harbour seals during the breeding season: implications for their lactation strategy and mating patterns. *Journal of Animal Ecology*, 63, 24–30.

Trites, A. W. 1991. Fetal growth of northern fur seals: life-history strategy and source of variation. *Canadian Journal of Zoology*, 69, 2608–2617.

Trivers, R. L. 1972. Parental investment and sexual selection. In: *Sexual Selection and the Descent of Man 1871–1971* (Ed. by B. Campbell), pp. 136–179. Chicago: Aldine.

Wheatley, K. E., Bradshaw, C. J. A., Davis, L. S., Harcourt, R. G. & Hindell, M. A. 2006. Influence of maternal mass and condition on energy transfer in Weddell seals. *Journal of Animal Ecology*, 75, 724–733.

7

Maternal Influences on Development, Social Relationships, and Survival Behaviors

JILL M. MATEO

INTRODUCTION

As a naturalist, Darwin (1859) made two observations that were particularly important for the development of his theory of natural selection—that there is significant phenotypic variation among individuals of a species, including artificially selected or domesticated species, and that offspring resemble their parents. Since the Modern Synthesis (see Fisher 1930; Wright 1931; Haldane 1932), with our growing understanding of genes and inheritance, significant advances have been made in identifying the genetic processes responsible for individual differences and family resemblance. It is now well established that phenotypes of developing young can be influenced by parents through the transfer of genetic material. However, parents can also influence the traits of their young by rearing them in particular microhabitats, providing them with food and other resources, and protecting them from predators or parasites. Until recently, scant attention has been paid to these parental effects, or the ways in which a parent's phenotype or environment influence the phenotype of its offspring. Parental effects are of special relevance to the fields of evolutionary biology and ecology for several reasons. First, they can maintain and magnify heritable individual differences across generations and thus have potent effects on the speed, strength, and direction of natural selection (see also Wade et al., McAdam, this volume). Second, parental effects lead to parent-offspring resemblance, which can be adaptive if offspring encounter similar social and environmental features as adults (assuming parents themselves were successful in that environment). Third, parental effects provide a rich source for phenotypic plasticity, including anatomical, physiological, and behavioral traits, because parents respond to dynamic cues in their environment (e.g., reduced food availability or current photoperiod) and can influence offspring accordingly. Because these intergenerational changes are plastic, parents can respond rapidly to changing environments and produce offspring whose phenotypes are well suited for current conditions, more rapidly than change based on evolution through natural selection (see also Cairns et al. 1990; Mousseau & Fox 1998). However, it should be noted that dynamic responses to environmental changes can result in negative maternal effects, such as when offspring exhibit foraging strategies similar to their mother's, yet these strategies are not longer effective in the current environment.

One of the hallmarks of mammalian behavioral development is its sensitivity to the social and physical environments provided by mothers during gestation and lactation. Therefore, opportunities for maternal effects on offspring phenotypes—morphological, physiological, and behavioral—can be particularly robust (see Reinhold 2002). A mother's set of social partners, her habitat, her physiology, her diet, and even her daily activity rhythms can have profound and lasting effects on offspring phenotypes. From a functional perspective, these effects make adaptive sense when offspring develop in similar social and physical environments as their mothers. They also contribute to individual differences within a population, with alternative phenotypes being favored by selection depending on spatial and temporal changes in social environmental conditions (reviewed in Stamps 2003; see also Mateo 2007a).

In this chapter I will present an overview of potential maternal effects on behavioral development, including pre- and postnatal effects of social experiences, stress, and seasonality on the expression of developing phenotypes. I will also discuss in detail, from a comparative approach, how maternal effects mediate the development, plasticity, and functional significance of a suite of social and survival behaviors. Examples will be drawn from studies of both captive and free-living mammals, with a focus on comparative research with ground-dwelling squirrels. Treatment of maternal effects in other contexts can be found throughout this volume, including dominance hierarchies, nutritional status, natal dispersal, intrauterine position, and population cycles. I focus on both proximate and ultimate levels of analysis of the interplay between developmental processes and their social and environmental contexts.

PROCESSES AND OUTCOMES OF MATERNAL EFFECTS

Any treatment of maternal effects should consider both the processes and the outcomes of these effects. For example, the outcome of behavioral development in species with parental investment often results in juvenile social behavior resembling parental repertoires (Bateson 1982). The process of this parental influence on the development of offspring social behavior, or any other behavioral repertoire, can range along a continuum from direct to indirect (reviewed in Mateo & Holmes 1997). Parents have a direct influence when they orient their behavior toward their young, such as by leading their young to a food source or preventing them from interacting with particular conspecifics. For a parent's influence to be considered direct, its behavior must change qualitatively or quantitatively as a function of its offspring's presence. At the other end of the continuum, parents have an indirect influence when their normal behavior inadvertently affects juvenile behavior, but is not directed specifically towards their young. Adults are thus incidental models of behavior and juveniles are inadvertent observers, and a parent's behavior is not contingent on the presence of its offspring. Examples of indirect influences include parents' own antipredator responses, their food preferences, and their reactions to territorial intrusions. Production of alarm calls (vocal signals emitted in response to predators) by adults can be an example of a direct or indirect influence depending on whether the likelihood of calling is contingent on the presence of the adults' young (e.g., Cheney & Seyfarth 1990; Hoogland 1995). Note that the effects on offspring do not necessarily differ if parental influence is direct or indirect.

Direct and indirect parental influences on the process of behavioral development share several characteristics. First, influence can have an immediate effect, such as when the response of an adult to a predator evokes almost simultaneously a response by a juvenile, or it can have a delayed effect, such as when an immature animal observes an adult's response to an alarm call but does not show its own responses until it can locomote independently. Second, neither process of influence implies complex mental states or awareness of juveniles' traits by parents (cf. Cheney & Seyfarth 1990). Third, juveniles can be passive observers or recipients of adult actions; that is, young need not seek out adults and copy their behavior. Fourth, as noted above, "direct" and "indirect" describe *processes* rather than *outcomes* of influence, and do not necessarily differ in the importance they play in juvenile behavioral development (Mateo & Holmes 1997). These processes of parental effects on juvenile ontogeny are potentially common among mammals, particularly those with extended maternal care.

MATERNAL EFFECTS ON SOCIAL DEVELOPMENT

Prenatal Environmental and Social Effects

Among mammals, development starts at conception, and because sensory and perceptual development begins in utero, the uterine environment of developing fetuses can affect their later morphology, physiology, and behavior. The onset of sensory function during development is remarkably consistent in birds and mammals, with the perceptual senses showing the following order of onset: tactile, vestibular, chemical, auditory, and visual. In altricial rodents, the latter two systems may not be functioning until after birth, and all systems continue to develop and form cortical connections during postnatal life (Gottlieb 1981; Alberts 1984; see also Alberts & Ronca 1993). Thus maternal effects can begin prenatally through a variety of sensory experiences, with fetuses moving when the mother grooms, travels, or sleeps, "smelling" what the mother eats, and experiencing changes in stress hormones when the mother experiences an agonistic social interaction, a predation attempt, or chronic food shortage. As a result of early perceptual development, therefore, experiences prior to birth can have enduring effects. For instance, fetuses can experience odors in the amniotic fluid that can later influence food preferences or social recognition of kin (Hepper 1987; Porter et al. 1991; Terry & Johanson 1996; Hudson et al. 1999; Galef, this volume). Mothers can therefore indirectly guide their offspring toward safe food items or amiable conspecifics. In summary, prenatal perceptual experiences can have long-lasting effects on offspring phenotypes (Alberts 1984; Smotherman & Robinson 1988; Grubb & Thompson 2004).

One of the best-known maternal effects in birds and mammals involves variation in the exposure of offspring to gonadal hormones (Schwabl 1999; Ryan & Vandenbergh 2002). In some mammals, exposure to gonadal hormones during gestation can have profound effects on adult morphology, physiology, and behavior, particularly among polytocous (litter size > 1) species. For example, individuals gestating between two males experience higher androgen levels than those between two females, and these intrauterine position (IUP) effects have consequences in adulthood, including variation in rates of sexual maturation, fecundity, aggressive tendencies, parental behaviors, territoriality, and sexual attractiveness. IUP can also affect sex ratios of future litters, thereby creating intergenerational transmission of IUP effects (see Vandenbergh, this volume). Observations of laboratory-born animals with known IUPs released into the wild suggest that early exposure to gonadal hormones can affect reproductive and social behaviors (Zielinski et al. 1992).

Maternally mediated gonadal-hormone exposure can have adaptive or functional effects as well. For example, females may experience higher fitness if their litters are male-biased, such as when the local breeding population is female-biased (short-term sex-ratio biasing; Fisher 1930; e.g., Creel et al. 1998; Allainé 2004) or when a female in a polygynous species is in exceptionally good condition and would be favored to overproduce sons (Trivers & Willard 1973; e.g., Clutton-Brock et al. 1984; Meikle et al. 1993; Hewison & Gaillard 1999). In such situations, daughters would more often gestate between males and be partially masculinized as adults. Those daughters may also overproduce sons themselves (as females gestating between two males tend to produce male-biased litters), and if that daughter is in exceptionally good condition herself, this would be beneficial, but if she is in moderate or poor condition, then selection would not favor her sons over her daughters, and she would experience a fitness loss. Thus the maternal effects via IUP can, over time, influence secondary sexual characteristics, sex ratios, and reproductive success depending on original maternal condition and local demographics.

Steroidal hormones can exert other effects on offspring during gestation, although some effects have not been well studied in mammals. For example, elevated maternal glucocorticoids can have negative effects on offspring morphology, development and survivability (e.g., tropical damselfish, *Pomacentrus amboinensis*: McCormick 1998; common shrew, *Sorex cinereus*: Badyaev et al. 2000; lizards, *Lacerta vivipara*: Meylan & Clobert 2004; root voles, *Microtus oeconomus*: Bian et al. 2005; barn swallows, *Hirundo rustica*: Saino et al. 2005; but see Uller & Olsson 2006 on lizards). Thus the impact of social or environmental stressors experienced by one generation can have effects on their reproductive success as well as on the phenotypes of the next generation. In addition, in quail pre-laying maternal corticosterone is correlated with female-biased clutches (Pike & Petrie 2006). Among most birds natal dispersal is female-biased, and thus if the mother experiences crowding or food shortages, for example, she would benefit from producing daughters that will leave the natal area and, perhaps, settle in less competitive environments. The costs and benefits of maternal effects can depend on the ecological and demographic patterns of a species. For instance, adult female spotted hyenas (*Crocuta crocuta*) are behaviorally dominant over males, due in part to prenatal androgenization. This dominance allows females with cubs to compete for scarce food resources (carcasses), and females with higher circulating androgens during gestation produce cubs with high levels as well. These androgens increase aggressive behaviors in cubs, which may have fitness consequences for offspring if it helps them to compete at kills (Dloniak et al. 2006; see also Holekamp & Dloniak, this volume).

Perinatal environmental and social effects

Maternal traits mediated by the external environment can also be transmitted nongenetically to offspring. Most altricial mammalian young are reared in nests, burrows, or dens and are not directly exposed to the same range of physical and social environments that the mother experiences. However, a young animal can experience the effects of food availability, climate changes (temperature, day length), or social instability before leaving the natal nest. One of the most salient perinatal maternal effects is the influence of a mother's physical condition and body weight on her offspring, with heavier females producing heavier young at birth and/or investing more in those young than females in poor condition (Boonstra & Hochachka 1997; Hansen & Boonstra 2000; see also Bowen, Cheverud & Wolf, Wilson & Festa-Bianchet, this volume). In many mammals, body weight at birth or during the juvenile period predicts adult mass (e.g., Birgersson & Ekvall 1997; Festa-Bianchet et al. 2000) or survival probability (e.g., Trombulak 1991; Neuhaus 2000), which in turn can have important consequences for adult fitness and lifetime reproductive success (Atkinson & Ramsay 1995; Festa-Bianchet et al. 2000; Wilson et al. 2005). Furthermore, positive maternal effects may be more potent for the sex that benefits most from philopatry, such as female offspring among mammals (e.g., Rieger 1996; but see Baker & Fowler 1992), and male offspring among birds (e.g., Visser & Verboven 1999), whereas poor maternal condition can have more of an effect on the sex that disperses (e.g., Labov et al. 1986; Meikle et al. 1995).

In addition, the optimal time to reproduce may be population-specific rather than species-specific, particularly when species are found along latitudinal or elevational gradients. Photoperiodic cues (day length) can be important indicators of upcoming seasonal changes and signal appropriate times for reproductive efforts. The photoperiod experienced by females during pregnancy is relayed to young during gestation and lactation through the hormone melatonin. In rodents, maternal melatonin can influence offspring growth rate, fat deposition, pelage, and sexual maturation. Young born in the spring or early summer mature quickly and can start breeding that year, but those born in the late summer remain prepubertal often until the following spring, reducing energetic demands until the reproductive season begins again. Thus perinatal melatonin from mothers adaptively primes young for somatic and reproductive growth appropriate for the time of year in which they are born (reviewed by Lee & Gorman 2000). Photoperiodic cues are not the only mechanism that can trigger breeding condition; social and dietary cues experienced by mothers can also influence development or regression of reproductive organs (Goel & Lee 1996; Demas & Nelson 1998; Lee & Gorman 2000; Ergon et al. 2001). That is, a mother can guide her offspring to partition their somatic and reproductive efforts to best suit their future fitness potential, accelerating puberty and mating if conditions are favorable, or delaying sexual maturation until breeding conditions are appropriate, in turn increasing the mother's own lifetime fitness.

If a pregnant female experiences an unstable social environment, high predation pressure, or unpredictable adverse climatic conditions, her adrenal-hormone responses can affect the hypothalamic-pituitary-adrenal (HPA) functioning of her developing offspring (e.g., Maccari et al. 1995; Barbazanges et al. 1996). For example, offspring of mothers experiencing stressors can have heightened acute stress responses compared with young of nonstressed females. In some laboratory rodents early postnatal stress decreases HPA activity and reactivity to novel objects and facilitates some forms of learning (Levine 1994; Maccari et al. 1995). However, severe exposure to stressors results in maladaptive HPA axes, with animals exhibiting inappropriate responses to novel or stressful situations, impaired learning and altered social behaviors (e.g., McEwen & Sapolsky 1995; Lupien & McEwen 1997). Thus, perinatal exposure to stress hormones, during either gestation or lactation, can fine tune sensitivity and efficacy of the HPA axis and the behaviors relating to it, producing, for example, animals with more neural receptors, lower responsiveness to stressors, and improved learning capabilities (Lupien & McEwen 1997; Catalani et al. 2002; Joëls et al. 2006; Yang et al. 2006; see also Champagne & Curley, this volume). If a mother experiences chronic stress during gestation or lactation, then this fine-tuning of her offspring's HPA axes can help them to manage similar experiences if they remain in that environment. I note, however, that there is little empirical evidence that this maternal effect operates among free-living mammals (but see below).

Postnatal Environmental and Social Effects

Young mammals experience their broader physical and social environments more directly after birth, and social experiences with parents and siblings can have significant effects on development. In laboratory rats, the development of behavioral and physiological traits can be affected by variation in maternal behaviors, such as nursing postures and rates of licking and grooming of pups. For example, offspring of high licker/groomers are less fearful and have attenuated stress responses relative to those of low licker/groomers. Cross-fostering studies (in which animals are transferred from one mother to another shortly after birth; see Mateo & Holmes 2004 for a discussion of fostering designs) have shown that these effects on offspring are not genetic, but instead are due to postnatal maternal handling. In lab rats (*Rattus norvegicus*), daughters of high-licking and grooming mothers become high-licking and grooming mothers themselves, thereby transmitting variation in parental behavior nongenetically across generations (Meaney 2001; Champagne & Curley, this volume). Note that these long-term effects of handling and grooming result from the normal range of species-specific maternal care, rather than extreme versions of neglectful or attentive mothers. However, as yet it is not clear whether these licker/groomer phenotypes exist among free-living, outbred rodent mothers, or if the effects are found in other mammals.

Experience with mothers prior to dispersal can affect the social relationships that their offspring develop as adults. For instance, in large, group-living mammals, a mother's dominance rank within a hierarchy can be inherited by her offspring (Holekamp & Dloniak, Maestripieri, this volume). Transfer of dominance rank to offspring can be mediated by physical condition, such as when offspring of high-ranking females have larger body mass or greater competitive abilities and thus become high ranking themselves, but it can also arise through social dynamics, as when mothers intervene on the offspring's behalf to support them against lower-ranking conspecifics. Although the inheritance of a low rank, regardless of an individual's actual condition, might not be adaptive for that individual, social stability within a group because of these ranks can increase the fitness of all individuals in the group through access to food resources, defense of a territory, or detection of predators. This group benefit, however, is unlikely to compensate for an individual's low rank, and such individuals may attempt to rise in the hierarchy whenever the social context permits it. Mothers can also influence their offspring's choice of social partners, if young seek other adults that exhibit phenotypes similar to hers. For example, young can learn their mother's odors, vocalizations, or pelage patterns, and preferentially interact with (or avoid) other conspecifics that exhibit similar traits. If these cues are heritable then such conspecifics would probably be kin; here learning about kin cues would likely be mediated through genetic effects, if offspring inherit both kin cues and preferences for those cues. However, preferences for kin can occur through nongenetic maternal effects if recognition cues are not heritable (e.g., Insley 2001; Sharp et al. 2005) or if learning about kin is mediated through interactions with the mother and unfamiliar kin (Holmes & Sherman 1983). Similar maternal effects can operate in mate-choice contexts, if young imprint on mother's cues and as reproductive adults avoid mating with individuals with similar cues (Dewsbury 1982; Bolhuis 1991; see also Vos 1995; Isles et al. 2002; Burley 2006; Maestripieri, this volume). Learning of maternal cues can also lead to maternal effects on sexually selected traits such as behavioral displays, as well preferences for those traits (Qvarnström & Price 2001).

MATERNAL EFFECTS ON SURVIVAL BEHAVIORS

Maternal effects can be important for the ontogeny of many adaptive behaviors, including those important for survival. This may seem counterintuitive at first, as survival behaviors might be expected to develop through a closed program (Mayr 1974). Some researchers have posited that survival behaviors, such as alarm calls or predator-avoidance tactics, are "preprogrammed" or "innate" so that the behaviors emerge independent of experiential input (Tinbergen 1953; Bolles 1970; Magurran 1990; Curio 1993). Others have proposed instead that these behaviors should be acquired or learned, through either direct or indirect experiences with predators and conspecifics (e.g., Vitale 1989; Cheney & Seyfarth 1990; Mateo 1996; Griffin 2004). Although this nature-nurture dichotomous approach is no longer formally endorsed, it still persists in the current literature despite acknowledgments by many that behavior is not innate or acquired per se, but instead develops epigenetically, through interactions between the organism and the series of environments it encounters throughout its life span (Lehrman 1970; Gottlieb 1976; Johnston 1987). For a developing mammal, one of the most salient aspects of its environment is its mother, who can have a significant influence on offspring acquisition of survival behaviors, such as antipredator strategies, foraging skills, and social acumen.

A behavioral system that required each juvenile to learn independently how to recognize and respond to predators would consume time and energy and be prone to fatal errors in learning (see Darwin 1859; Bolles 1970). Reliance on experience, whether it be practice or learning, may at first appear less than optimal, given the vulnerability of young to predators, but a flexible developmental program might be adaptive when predator contexts vary among age groups or among populations, favoring plasticity of species-typical behaviors (Johnston 1982; Shettleworth 1998; Richerson & Boyd 2001). In these instances, learning appropriate strategies from mothers would be beneficial.

A comparative study of antipredator behavior among species of ground-dwelling squirrels illustrates the potential for maternal effects on survival tactics. Ground squirrels are vulnerable to both aerial (e.g., hawks, eagles) and terrestrial (e.g., coyotes, weasels, martens, venomous snakes) predators, and most species produce vocal signals warning of danger from predators (reviewed in Owings & Hennessy 1984). Despite the obvious survival advantage of evading a hunting predator, young need to learn from which animals to flee, to which warning calls to respond, and in what manner. Belding's ground squirrels (*Spermophilus beldingi*) have two types of alarm calls that elicit two different behavioral responses by listeners. Whistles are elicited by fast-moving, typically aerial, predators and result in evasive behaviors such as running to or entering a burrow, and scanning the area only after reaching safety. Trills are elicited by slow-moving, primarily terrestrial, predators and usually cause others to post (a bipedal stance accompanied by visual scanning), with or without changing location (Mateo 1996). When juvenile *S. beldingi* first emerge aboveground at about one month of age, they do not discriminate behaviorally between these calls, or even among alarm calls and other conspecific and heterospecific vocalizations. It takes approximately one week for juveniles to learn to respond selectively to alarm calls and to show the correct response for each type of call; during this time up to 60% of juveniles disappear, many to predation (Mateo 1996).

By what processes do juveniles acquire their alarm-call responses? In *S. beldingi*, this learning is facilitated by experience hearing the calls as well as observations of adult reactions. Juveniles attend to and model adult responses, particularly those of their mother rather than those of other nearby females. Furthermore, juveniles adopt a response style similar to their mother's, remaining alert for extended periods if she does and showing more exaggerated vigilance responses if she does; these response patterns persist at later ages even when the mother is not visible at the time (Mateo 1996; Mateo & Holmes 1997). Adoption of maternal styles can be favored if mother's responses are locally adapted to the degree of predation threat in the natal area. Mothers who locate their natal burrows at the edge of meadows are more reactive to alarm calls and remain alert longer than those from the center of a meadow (Mateo 1996), which can reflect increased vulnerability to predators near the edge (Elgar 1989). Mothers' reactions, which serve as models for juvenile responses, can reflect the mothers' own vulnerabilities (indirect maternal influence), or can be a form of maternal care, becoming more vigilant if they locate their natal burrow, and thus their offspring, in a dangerous area (edge) and less vigilant if in a safer region (center; direct maternal influences). [Note that parents can even teach appropriate responses to their offspring, adjusting their behaviors as the competency of their young improves (Caro & Hauser 1992).] Because *S. beldingi* juveniles model their responses after their mother's, they are also more alert if reared on the edge of a meadow than in the center (Mateo 1996). By acquiring responses that are appropriate for a given microhabitat, *S. beldingi* can optimize both their foraging and antipredator efforts, allowing juveniles (and adults) to gain adequate body weight before hibernation without expending energy on unnecessary vigilance (Mateo & Holmes 1999). In addition, females often nest near their mothers in subsequent years (J. M. Mateo, unpublished data), so adopting location-specific responses would be favored across generations. For example, long-term studies of rock and California ground squirrels (*S. variegatus, S. beecheyi*) demonstrate that animals sympatric with predatory snakes need to develop and maintain antisnake behaviors, whereas responses to snakes are no longer evident in the antipredator repertoires of animals living in habitats without snakes (Coss & Owings 1985; Coss et al. 1993; Owings et al. 2001). Such responses, then, would not necessarily be present upon first encounter with predators, but would be acquired rapidly with additional experience, perhaps through observational learning of their mother's responses (e.g., Swaisgood et al. 1999).

In addition to serving as a model for appropriate antipredator responses, a mother's physiological phenotype can affect her offspring's physiology, which in turn can modulate their behavioral responses to predators. First emergence of young ground squirrels from natal burrows is fairly synchronous, with most litters emerging within a ten-day period. This natal emergence draws predators, and direct encounters with predators, observations of sudden, rapid responses of nearby adults, and experience with hearing loud alarm calls likely cause changes in circulating glucocorticoids in adults and offspring alike. The range of cortisol responses depends on the particular stressor as well as an individual's HPA axis. As noted above, maternal stress responses can affect the HPA functioning of their offspring, and thus a mother's hormonal patterns can have long-lasting effects on those of her young (Catalani 1997). This nongenetic transmission of adrenal functioning could have adaptive consequences for offspring. Mothers and their young that live near the edge of meadows are more vigilant and exhibit prolonged alarm-call responses, and they have lower basal cortisol levels than *S. beldingi* from the center of the meadow (J. M. Mateo, unpubl. data). This lower basal cortisol may allow animals to mount large acute responses, mobilizing energy for quick escapes from ambushing predators (Mateo 2007b).

In addition, maternal glucocorticoids can affect the rate at which young learn important survival behaviors. In laboratory rodents the influence of glucocorticoids on learning and memory has an inverted-U-shaped function. Very low or very high levels of corticoids can lead to hypo- or hyperarousal and poor selective attention to input and thus impair consolidation of new memories. Moderate levels of corticoids are optimal for attention to stimuli and consolidation of memories (reviewed in Lupien & McEwen 1997). Maternal glucocorticoids transmitted to offspring during gestation or lactation can have long-term effects on offspring hormones, and this "set point" may promote learning of antipredator responses, particularly in animals inhabiting areas with high predation pressure. Indeed, experiments in which maternal cortisol was manipulated during lactation show that juvenile *S. beldingi* with moderately elevated cortisol after emergence acquire spatial and associative-learning tasks faster than those with low or very high basal cortisol (Mateo, 2008).

Despite interacting with their offspring for just a few weeks after natal emergence, ground-squirrel mothers can also have a significant influence on the development of food preferences. Juvenile *S. beldingi* prefer foods that their mother consumed over food that she did not eat (Peacock & Jenkins 1988; see also Galef, this volume, for a review of maternal effects on food preferences). Maternal effects on food preferences might be especially evident in species that are active year round and for whom food availability changes with seasons. Maternal effects on kin preferences are well documented in ground squirrels. The most common social behavior among juveniles is play, and young prefer to play with littermates and other kin over non-kin; this discrimination among juveniles is typically mediated by odor cues (*S. columbianus*: Waterman 1986; *S. beldingi*: Holmes 1994; *S. lateralis*: Holmes 1995; Mateo 2003). In *S. beldingi*, mothers are important for the development and crystallization of these kin preferences. Juveniles reared in a seminatural enclosure without their mothers fail to develop play-partner preferences. However, when mothers are present but unable to intervene in social interactions, littermate preferences still develop, suggesting that a mother's role is indirect. Indeed, it is a mother's presence at night in the burrow system that attracts her offspring to sleep together, which results in the formation of kin preferences (Holmes & Mateo 1998). Play is thought to lay a foundation for adult kin preferences and nepotism, and therefore is expected to vary with kinship (Michener 1983; Holmes 1994). In some sciurids, however, juveniles develop kin-recognition abilities and kin biases but there is no evidence of nepotism among adults (e.g., *S. columbianus*, *S. lateralis*); note, though, that kin recognition in sciurids might also function for inbreeding avoidance (Michener 1983; Mateo 2002).

MATERNAL EFFECTS AND HABITAT SELECTION

In most mammalian species, one or both sexes of offspring disperse from the natal area sometime after the age of independence from the mother. Although considerable empirical and theoretical work has focused on the proximate and ultimate explanations for natal dispersal (e.g., Greenwood 1980; Dobson 1982; Holekamp 1983; Dobson & Jones 1985; Pusey & Wolf 1996), fewer studies have considered how animals decide *where* to settle and what role, if any, parents play in this decision. Maternal effects can operate on habitat selection when offspring "imprint" or develop preferences for features of their natal habitat and settle in an area with similar features after dispersal. Davis and Stamps (2004) term this "natal habitat preference induction," and suggest that such preferences can be adaptive for offspring because settling in the first area that resembles the natal habitat reduces the energy expended, the search time and the exposure to potential predators while searching for a place to settle. In addition, if mother and offspring have similar phenotypes and the mother was successful in the natal area, then her offspring would do well in similar areas (see also Fowler 2005; Stamps & Davis 2006). Similar phenotypes could arise because they are heritable, or because both generations were affected by their immediate environments, affecting traits such as foraging styles or antipredator strategies, and thus natal habitat preference induction can result in nongenetic transmission of traits from mothers to offspring (e.g., Dittman & Quinn 1996; Olson & Van Horne 1998; Arvedlund et al. 1999; Vogl et al. 2002).

Among territorial mammals, a mother with good foraging skills can improve the physical condition and chances of survival of her offspring not just during gestation and lactation, but after offspring dispersal as well. If mothers acquire ample food resources and defend these caches from others, then as young become independent, these caches can be transferred to them as a form of parental investment. Desert-dwelling bannertail kangaroo rats (*Dipodomys spectabilis*) cache large quantities of seeds in mounds, which are defended from conspecifics through scent marking and foot drumming (Randall 1993). These mounds are valuable resources and are maintained over many generations, with some mothers "bequeathing" the mounds to their offspring (Jones 1986). Female Columbian ground squirrels (*S. columbianus*) move their nest sites more often when their yearling daughter is present, with those daughters typically settling into the mother's old territory (Harris 1984). Prior to mating and giving birth, female red squirrels (*Tamiasciurus hudsonicus*) start building food caches that they leave for their independent offspring months later (Boutin et al. 2000). Such transfer of food resources likely benefits the inheriting offspring, although there are no long-term empirical data documenting increased lifetime reproductive success as a result of bequeathals. Related to this, some mothers, particularly older ones, will shift their territories to make room for their reproductive daughters, occasionally moving to territories that are less optimal (Sherman 1976; Berteaux & Boutin 2000; pers. obs.). However, unlike bestowal of food caches, there is little evidence that "territory bequeathal" affects offspring survival or reproductive success (Lambin 1997).

OTHER MATERNAL EFFECTS ON OFFSPRING PHENOTYPES

Timing of birth and offspring survival

In many taxonomic groups, individuals that mate early often produce offspring that are higher quality (e.g., higher body weight) and/or more likely to survive than those of parents that mate later in the breeding season (e.g., side-blotched lizards, *Uta stansburiana*: Sinervo & Doughty 1996; greater snow goose goslings, *Anser caerulescens atlantica*: Lepage et al. 1998; pink salmon, *Oncorhynchus gorbuscha*: Dickerson et al. 2002; Verreaux's sifaka, *Propithecus verreauxi*: Lewis & Kappeler 2005; tree swallows, *Tachycineta bicolor*: Nooker et al. 2005; but see Alaskan moose, *Alces alces*: Bowyer et al. 1998). Species that experience seasonal fluctuation in resources often constrain reproductive efforts so that offspring are born during peak food availability and nutritional value. This maximizes the opportunity for young to grow during times of abundance, improving their body condition and likelihood of future survival and reproductive success. Thus timing of reproduction represents a potential maternal effect. Mothers that mate earlier in the season, and thus give birth earlier in the period of resource abundance, will likely have greater reproductive success than those that mate later and give birth to young that must forage during the seasonal decline in food items and/or their nutritional value prior to independence or dispersal (Dobson & Michener 1995; Sedinger et al. 1997; Millesi et al. 1999; Réale et al. 2003). Because offspring condition can affect their future reproductive success, these offspring can also mate earlier than those born to later breeders, thus transmitting the trait over generations nongenetically (but see Sinervo & Doughty 1996 for possible genetic effects on timing of reproduction). Note, however, that there can be a trade-off between somatic investment and reproductive investment by females, particularly if they are young or are in poor condition and need access to those resources themselves before bearing the costs of gestation and lactation (e.g., Dobson et al. 1999; Millesi et al. 1999). Related, the timing of reproduction can be associated with increasing the female's future reproductive success, such as attempting to breed early so as to gain sufficient weight prior to the next breeding period (e.g., synchronous fawning among roe deer [Capreolus capreolus] because of the impending rut: Gaillard et al. 1993; see also Price & Boutin 1993).

Highly seasonal species, such as obligate hibernators, illustrate the potential importance of reproductive timing and maternal effects. Some ground-squirrel species have extremely short active seasons, being awake and socially active for just a few months each year. Females mate shortly after emerging from hibernation and rear their young for about a month in an underground natal burrow. The combined needs for young to develop in the relative safety of the natal burrow and to emerge above ground and begin foraging constrain the timing of natal emergence. Overwinter survival in some ground-dwelling squirrels is dependent upon the acquisition of adequate body fat prior to hibernation, and more than 60% of juveniles do not survive their first winter (Barash 1973; Murie & Boag 1984; Sherman & Morton 1984; J. M. Mateo, unpubl. data). Thus offspring born to females that mate earlier in the short breeding season are often heavier and have a greater chance of surviving than those that emerge later in the summer. In European ground squirrels (S. citellus), early-born litters are often larger and are nursed longer than later-born juveniles. Females that enter hibernation at a heavy weight emerge in the spring earlier and mate earlier than those that hibernate at lower weights (Millesi et al. 1999). Thus, although females can pay an energetic cost for early reproduction and prolonged lactation (weighing less at hibernation), their daughters can reap those energetic benefits and reproduce earlier the following spring themselves. In Richardson's ground squirrels (S. richardsonii), females that mate early have heavier offspring that also have more time to forage during the active season (Dobson & Michener 1995), and thus are more likely to survive hibernation. Among Uinta ground squirrels (*S. armatus*), females born early in the summer are more likely to survive their first hibernation than those born later in the season (Rieger 1996). Indeed, in *S. beldingi*, those surviving at least one year are born significantly earlier in the season than those that die or disappear (J. M. Mateo, unpubl. data). The same is true for *S. columbianus* offspring, although litter size is a greater predictor of survival than is date of birth (Dobson et al. 1999), perhaps because this species tends to have smaller litters than other species. Note, however, that very early mating can be risky if mothers experience adverse environmental conditions (e.g., Morton & Sherman 1978; Inouye et al. 2000; Farand et al. 2002) or if earliest emerging young suffer high mortality due to less dilution of predation risk. Future work could focus on similar interspecific comparisons, as well as search for sex differences in maternal effects as a function of timing of reproduction.

Length of Maternal Dependence

Among mammals, offspring are dependent on their mother at least until weaning, yet across species there is variation in the duration of care and protection afforded by mothers. This variation is, in part, correlated with the length of time young remain with their mothers. In species with extended dependency on parental care, there may be repeated opportunities for young to observe and learn from their parents' behaviors (e.g., foraging, spatial, antipredator, and social behaviors). Conversely, in species with little or no parental care after weaning, individuals often acquire behavior without assistance from adults. Thus, the length of dependence on adults provides another opportunity for a maternal effect, as it will influence the developmental process of the offspring's behavioral repertoires, and extended maternal care may allow for variable pathways of behavioral development, including social facilitation of responses (Heyes & Galef 1996; Reinhold 2002). For example, experiences with their mothers can significantly modify the responses of young vervet monkeys (*Chlorocebus aethiops*) to predators and alarm calls warning of predators, as they maintain close proximity to their mothers for at least 2 years (Cheney & Seyfarth 1990; for other species see also Yoerg & Shier 1997; White & Berger 2001; Swaisgood et al. 2003) In fish, Brown (1984) studied the development of antipredator behavior in two bass species differing in the length of time the male parent guards his young. Recently hatched young of a species with limited paternal care (*Ambloplites rupestris*) showed predator-avoidance behavior sooner than young of a species with extended care (*Micropterus salmoides*). Mexican jays (*Aphelocoma ultramarina*) with prolonged associations with experienced adults exhibit mobbing behavior at a later age than less social scrub jays (*A. coerulescens*) that fledge at an earlier age (Culley & Ligon 1976).

Comparisons among animals that differ in active-season lengths, growth rates, or periods of dependence upon mothers provide an opportunity to examine maternal effects on offspring behavioral repertoires. For instance, juveniles of some nonhibernating species, which have slower growth rates than those that hibernate (e.g., Clark 1970; Morton & Tung 1971; Pizzimenti & McClenaghan 1974; Maxwell & Morton 1975) and subsequently longer associations with their mothers, can exhibit more social facilitation of behavioral development than juveniles with limited growth periods. Depending on their latitude and elevation, *S. beecheyi* can have long active seasons and extended co-occurrence of mothers and their young (up to 12 months; Dobson & Davis 1986), and thus maternal behavior can directly or indirectly influence the behavior of *S. beecheyi* juveniles more so than responses of *S. beldingi* juveniles, which have only a 3–4 month developmental period prior to autumnal immergence into hibernacula (pers. obs.). In addition to physiological (Morton & Tung 1971; Maxwell & Morton 1975) and social (Armitage 1981) adaptations to the length of the growing period, then, selection can favor more maternal effects in slowly maturing species than in species with accelerated growth. Likewise, lengthy associations between mothers and their young might permit maternal effects to play a larger role in generating heritable trait variation, which can then be operated upon by natural selection. Interspecific differences in developmental rates as a function of ecological conditions could have significant effects on mother-offspring interactions and social development. Studies of free-living mammals therefore present a unique opportunity to examine in detail how the intertwining effects of latitude, elevation, and climate on active-season length and pre- and postweaning growth rates influence behavioral development.

SUMMARY

Mammalian behavioral development is exquisitely tuned to the physiological, environmental, and social contexts in which ontogeny occurs. The various maternal effects described in this chapter, including prenatal hormonal effects, modeling of appropriate survival behaviors, and timing of reproduction, are amenable to examination from both proximate and ultimate levels of analysis. As described above, we are beginning to understand how maternal physiology affects offspring phenotype, such as the effects of gonadal hormones, glucocorticoids, and melatonin. Much of this work has focused on laboratory rodents, but as evidenced throughout this volume, similar work is now being done with mammals in other orders and in the field. From a functional perspective, maternal effects will have a selective advantage when they increase the survival and reproductive success of offspring, but of course can have negative consequences if a mother is in poor condition or if she faces adverse conditions during reproduction. In general, maternal effects will be favored when mothers and their offspring experience similar social and physical environments and thus similar resource availability, predation pressure, and social dynamics. In concluding this chapter, I emphasize the need for more behavioral studies integrating levels of analysis as well as levels of biological organization. In addition, future work could take advantage of groups of closely related mammalian species with contrasting life-history parameters, preferred habitats, or degrees of sociality. Recent interdisciplinary research offers encouraging signs that we will soon identify and understand the links between mechanisms and functions of maternal effects.

ACKNOWLEDGMENTS

Critiques and suggestions on a previous version of this chapter by N. J. Peters and Don Owings are appreciated. Preparation of this chapter was partially supported by funds from the National Science Foundation.

REFERENCES

Alberts, J. R. 1984. Sensory-perceptual development in the Norway rat: a view toward comparative studies. In: *Comparative Perspectives on the Development of Memory* (Ed. by R. V. Kail & N. E. Spear), pp. 65–101. Hillsdale, NJ: Erlbaum.
Alberts, J. R. & Ronca, A. E. 1993. Fetal experience revealed by rats: psychobiological insights. *Early Human Development*, 35, 153–166.
Allainé, D. 2004. Sex ratio variation in the cooperatively breeding alpine marmot *Marmota marmota. Behavioral Ecology*, 15, 997–1002.
Armitage, K. B. 1981. Sociality as a life-history tactic of ground squirrels. *Oecologia*, 48, 36–49.
Arvedlund, M., McCormick, M. I., Fautin, D. G. & Bildsoe, M. 1999. Host recognition and possible imprinting in the anemonefish *Amphiprion melanopus* (Pisces: Pomacentridae). *Marine Ecology: Progress Series*, 188, 207–218.
Atkinson, S. N. & Ramsay, M. A. 1995. The effects of prolonged fasting of the body composition and reproductive success of female polar bears (*Ursus maritimus*). *Functional Ecology*, 9, 559–567.

Badyaev, A. V., Foresman, K. R. & Fernandes, M. V. 2000. Stress and developmental stability: vegetation removal causes increased fluctuating asymmetry in shrews. *Ecology*, 81, 336–345.

Baker, J. D. & Fowler, C. W. 1992. Pup weight and survival of northern fur seals *Callorhinus ursinus*. *Journal of Zoology*, 227, 231–238.

Barash, D. P. 1973. The social biology of the Olympic marmot. *Animal Behaviour Monographs*, 6, 171–245.

Barbazanges, A., Piazza, P. V., Le Moal, M. & Maccari, S. 1996. Maternal glucocorticoid secretion mediates long-term effects of prenatal stress. *Journal of Neuroscience*, 16, 3943–3949.

Bateson, P. P. G. 1982. Behavioural development and evolutionary processes. In: *Current Problems in Sociobiology* (Ed. by King's College Sociobiology Group), pp. 133–151. Cambridge: Cambridge University Press.

Berteaux, D. & Boutin, S. 2000. Breeding dispersal in female North American red squirrels. *Ecology*, 81, 1311–1326.

Bian, J. H., Wu, Y. & Liu, J. 2005. Effect of predator-induced maternal stress during gestation on growth in root voles *Microtus oeconomus*. *Acta Theriologica*, 50, 473–482.

Birgersson, B. & Ekvall, K. 1997. Early growth in male and female fallow deer fawns. *Behavioral Ecology*, 8, 493–499.

Bolhuis, J. J. 1991. Mechanisms of avian imprinting: a review. *Biological Reviews*, 66, 303–345.

Bolles, R. C. 1970. Species-specific defense reactions and avoidance learning. *Psychological Review*, 77, 32–48.

Boonstra, R. & Hochachka, W. M. 1997. Maternal effects and additive genetic inheritance in the collared lemming *Dicrostonyx groenlandicus*. *Evolutionary Ecology*, 11, 169–182.

Boutin, S., Larsen, K. W. & Berteaux, D. 2000. Anticipatory parental care: acquiring resources for offspring prior to conception. *Proceedings of the Royal Society of London, Series B*, 267, 2081–2085.

Bowen, W. D. 2009. Maternal effects on offspring size and development in pinnipeds. In: *Maternal Effects in Mammals* (Ed. by D. Maestripieri & J. M. Mateo), pp. 104–132. Chicago: University of Chicago Press.

Bowyer, R. T., Van Ballenberghe, V. & Kie, J. G. 1998. Timing and synchrony of parturition in Alaskan moose: long-term versus proximal effects of climate. *Journal of Mammalogy*, 79, 1332–1344.

Brown, J. A. 1984. Parental care and the ontogeny of predator-avoidance in two species of centrarchid fish. *Animal Behaviour*, 32, 113–119.

Burley, N. T. 2006. An eye for detail: selective sexual imprinting in zebra finches. *Evolution*, 60, 1076–1085.

Cairns, R. B., Gariepy, J. L. & Hood, K. E. 1990. Development, microevolution, and social behavior. *Psychological Review*, 97, 49–65.

Caro, T. M. & Hauser, M. D. 1992. Is there teaching in nonhuman animals? *Quarterly Review of Biology*, 67, 151–174.

Catalani, A. 1997. Neonatal exposure to glucocorticoids: long-term endocrine and behavioral effects. *Developmental Brain Dysfunction*, 10, 393–404.

Catalani, A., Casolini, P., Cigliana, G., Scaccianoce, S., Consoli, C., Cinque, C., Zuena, A. R. & Angelucci, L. 2002. Maternal corticosterone influences behavior, stress response and corticosteroid receptors in the female rat. *Pharmacology, Biochemistry and Behavior*, 73, 105–114.

Champagne, F. & Curley, J. P. 2009. The trans-generational influence of maternal care on offspring gene expression and behavior in rodents. In: *Maternal Effects in Mammals* (Ed. by D. Maestripieri & J. M. Mateo), pp. 182–202. Chicago: University of Chicago Press.

Cheney, D. L. & Seyfarth, R. M. 1990. *How Monkeys See the World*. Chicago: University of Chicago Press.

Cheverud, J. & Wolf, J. 2009. The genetics and evolutionary consequences of maternal effects. In: *Maternal Effects in Mammals* (Ed. by D. Maestripieri & J. M. Mateo), pp. 11–37. Chicago: University of Chicago Press.

Clark, T. W. 1970. Early growth, development, and behavior of the Richardson ground squirrel (*Spermophilus richardsoni elegans*). *American Midland Naturalist*, 83, 197–205.

Clutton-Brock, T. H., Albon, S. D. & Guinness, F. E. 1984. Maternal dominance, breeding success and birth sex ratios in red deer. *Nature*, 308, 358–360.

Coss, R. G. & Owings, D. H. 1985. Restraints on ground squirrel antipredator behavior: adjustments over multiple time scales. In: *Issues in the Ecological Study of Learning* (Ed. by T. D. Johnston & A. T. Pietrewicz), pp. 167–200. Hillsdale, NJ: Lawrence Erlbaum Associates.

Coss, R. G., Guse, K. L., Poran, N. S. & Smith, D. G. 1993. Development of antisnake defenses in California ground squirrels (*Spermophilus beecheyi*). II. Microevolutionary effects of relaxed selection from rattlesnakes. *Behaviour*, 124, 137–164.

Creel, S., Creel, N. M. & Monfort, S. L. 1998. Birth order, estrogens and sex-ratio adaptation in African wild dogs (*Lycaon pictus*). *Animal Reproduction Science*, 53, 315–320.

Culley, J. F., Jr. & Ligon, J. D. 1976. Comparative mobbing behavior of scrub and Mexican jays. *Auk*, 93, 116–125.

Curio, E. 1993. Proximate and developmental aspects of antipredator behavior. In: *Advances in the Study of Behavior*. Vol. 22 (Ed. by P. J. B. Slater, J. S. Rosenblatt, C. T. Snowdon & M. Milinski), pp. 135–238. New York: Academic Press.

Darwin, C. 1859. *On the Origin of Species*. London: J. Murray.

Davis, J. M. & Stamps, J. A. 2004. The effect of natal experience on habitat preferences. *Trends in Ecology and Evolution*, 19, 411–416.

Demas, G. E. & Nelson, R. J. 1998. Social, but not photoperiodic, influences on reproductive function in male *Peromyscus aztecus*. *Biology of Reproduction*, 58, 385–389.

Dewsbury, D. A. 1982. Avoidance of incestuous breeding between siblings in 2 species of *Peromyscus* mice. *Biology of Behaviour*, 7, 157–168.

Dickerson, B. R., Quinn, T. P. & Willson, M. F. 2002. Body size, arrival date, and reproductive success of pink salmon, *Oncorhynchus gorbuscha*. *Ethology, Ecology & Evolution*, 14, 29–44.

Dittman, A. H. & Quinn, T. P. 1996. Homing in Pacific salmon: mechanisms and ecological basis. *Journal of Experimental Biology*, 199, 83–91.

Dloniak, S. M., French, J. A. & Holekamp, K. E. 2006. Rank-related maternal effects of androgens on behaviour in wild spotted hyaenas. *Nature*, 440, 1190–1193.

Dobson, F. S. 1982. Competition for mates and predominant juvenile male dispersal in mammals. *Animal Behaviour*, 30, 1183–1192.

Dobson, F. S. & Davis, D. E. 1986. Hibernation and sociality in the California ground squirrel. *Journal of Mammalogy*, 67, 416–421.

Dobson, F. S. & Michener, G. R. 1995. Maternal traits and reproduction in Richardson's ground squirrels. *Ecology*, 76, 851–862.

Dobson, F. S., Risch, T. S. & Murie, J. O. 1999. Increasing returns in the life history of Columbian ground squirrels. *Journal of Animal Ecology*, 68, 73–86.

Dobson, F. S. & Jones, W. T. 1985. Multiple causes of dispersal. *American Naturalist*, 126, 855–858.

Elgar, M. A. 1989. Predator vigilance and group size in mammals and birds: a critical review of the empirical evidence. *Biological Reviews*, 64, 13–33.

Ergon, T., MacKinnon, J. L., Stenseth, N. C., Boonstra, R. & Lambin, X. 2001. Mechanisms for delayed density-dependent reproductive traits in field voles, *Microtus agrestis*: the importance of inherited environmental effects. *Oikos*, 95, 185–197.

Farand, É., Allainé, D. & Coulon, J. 2002. Variation in survival rates for the alpine marmot (*Marmota marmota*): effects of sex, age, year, and climatic factors. *Canadian Journal of Zoology*, 80, 342–349.

Festa-Bianchet, M., Jorgenson, J. T. & Réale, D. 2000. Early development, adult mass, and reproductive success in bighorn sheep. *Behavioral Ecology*, 11, 633–639.

Fisher, R. A. 1930. *The Genetical Theory of Natural Selection*. Oxford: Clarendon Press.

Fowler, M. S. 2005. Interactions between maternal effects and dispersal. *Oikos*, 110, 81–90.

Gaillard, J. M., Delorme, D., Jullien, J. M. & Tatin, D. 1993. Timing and synchrony of births in roe deer. *Journal of Mammalogy*, 74, 738–744.

Galef, B. G., Jr. 2009. Maternal influences on offspring food preferences and feeding behaviors in mammals. In: *Maternal Effects in Mammals* (Ed. by D. Maestripieri & J. M. Mateo), pp. 159–181. Chicago: University of Chicago Press.

Goel, N. & Lee, T. M. 1996. Relationship of circadian activity and social behaviors in re-entrainment rates in diurnal *Octodon degus* (Rodentia). *Physiology & Behavior*, 59, 817–826.

Gottlieb, G. 1976. The roles of experience in the development of behavior and the nervous system. In: *Studies on the Development of Behavior and the Nervous System*. Vol. 3. *Neural and Behavioral Specificity* (Ed. by G. Gottlieb), pp. 25–53. New York: Academic Press.

Gottlieb, G. 1981. Roles of early experience in species-specific perceptual development. In: *Development of Perception*. Vol. 1 (Ed. by R. N. Aslin, J. R. Alberts & M. R. Petersen), pp. 5–44. New York: Academic Press.

Greenwood, P. J. 1980. Mating systems, philopatry and dispersal in birds and mammals. *Animal Behaviour*, 28, 1140–1162.

Griffin, A. S. 2004. Social learning about predators: a review and prospectus. *Learning & Behavior*, 32, 131–140.

Grubb, M. S. & Thompson, I. D. 2004. The influence of early experience on the development of sensory systems. *Current Opinion in Neurobiology*, 14, 503–512.

Haldane, J. B. S. 1932. *The Causes of Evolution*. New York: Longmans Green.

Hansen, T. F. & Boonstra, R. 2000. The best in all possible worlds? A quantitative genetic study of geographic variation in the meadow vole, *Microtus pennsylvanicus*. *Oikos*, 89, 81–94.

Harris, M. A. 1984. Inheritance of nest sites in female Columbian ground squirrels. *Behavioral Ecology and Sociobiology*, 15, 97–102.

Hepper, P. G. 1987. The amniotic fluid: an important priming role in kin recognition. *Animal Behaviour*, 35, 1343–1346.

Hewison, A. J. M. & Gaillard, J. M. 1999. Successful sons or advantaged daughters? the Trivers–Willard model and sex-biased maternal investment in ungulates. *Trends in Ecology and Evolution*, 14, 229–234.

Heyes, C. M. & Galef, B. G., Jr. (Eds.). 1996. *Social Learning in Animals: The Roots of Culture*. San Diego: Academic Press.

Holekamp, K. E. 1983. Proximal mechanisms of natal dispersal in Belding's ground squirrels (*Spermophilus beldingi*). Ph.D. Dissertation. University of California, Berkeley.

Holekamp, K. E. & Dloniak, S. M. 2009. Maternal effects in fissiped carnivores. In: *Maternal Effects in Mammals* (Ed. by D. Maestripieri & J. M. Mateo), pp. 227–255. Chicago: University of Chicago Press.

Holmes, W. G. 1994. The development of littermate preferences in juvenile Belding's ground squirrels. *Animal Behaviour*, 48, 1071–1084.

Holmes, W. G. 1995. The ontogeny of littermate preferences in juvenile golden-mantled ground squirrels: effects of rearing and relatedness. *Animal Behaviour*, 50, 309–322.

Holmes, W. G. & Mateo, J. M. 1998. How mothers influence the development of litter-mate preferences in Belding's ground squirrels. *Animal Behaviour*, 55, 1555–1570.

Holmes, W. G. & Sherman, P. W. 1983. Kin recognition in animals. *American Scientist*, 71, 46–55.

Hoogland, J. L. 1995. *The Black-Tailed Prairie Dog: Social Life of a Burrowing Mammal*. Chicago: University of Chicago Press.

Hudson, R., Schaal, B. & Bilko, A. 1999. Transmission of olfactory information from mother to young in the European rabbit. In: *Mammalian Social Learning: Comparative and Ecological Perspectives* (Ed. by H. O. Box & K. R. Gibson), pp. 141–157. Cambridge: Cambridge University Press.

Inouye, D. W., Barr, B., Armitage, K. B. & Inouye, B. D. 2000. Climate change is affecting altitudinal migrants and hibernating species. *Proceedings of the National Academy of Sciences USA*, 97, 1630–1633.

Insley, S. J. 2001. Mother-offspring vocal recognition in northern fur seals is mutual but asymmetrical. *Animal Behaviour*, 61, 129–137.

Isles, A. R., Baum, M. J., Ma, D., Szeto, A., Keverne, E. B. & Allen, N. D. 2002. A possible role for imprinted genes in inbreeding avoidance and dispersal from the natal area in mice. *Proceedings of the Royal Society of London, Series B*, 269, 665–670.

Joëls, M., Pu, Z., Wiegert, O., Oitzl, M. S. & Krugers, H. J. 2006. Learning under stress: how does it work? *Trends in Cognitive Sciences*, 10, 152–158.

Johnston, T. D. 1982. Selective costs and benefits in the evolution of learning. In: *Advances in the Study of Behavior*. Vol. 12 (Ed. by J. S. Rosenblatt, R. A. Hinde, C. Beer & M.-C. Busnel), pp. 65–106. New York: Academic Press.

Johnston, T. D. 1987. The persistence of dichotomies in the study of behavioral development. *Developmental Review*, 7, 149–182.

Jones, W. T. 1986. Survivorship in philopatric and dispersing kangaroo rats (*Dipodomys spectabilis*). *Ecology*, 67, 202–207.

Labov, J. B., Huck, U. W., Vaswani, P. & Lisk, R. D. 1986. Sex ratio manipulation and decreased growth of male offspring of undernourished golden hamsters (*Mesocricetus auratus*). *Behavioral Ecology and Sociobiology*, 18, 241–249.

Lambin, X. 1997. Home range shifts by breeding female Townsend's voles (*Microtus townsendii*): a test of the territory bequeathal hypothesis. *Behavioral Ecology and Sociobiology*, 40, 363–372.

Lee, T. M. & Gorman, M. 2000. Environmental control of seasonal reproduction: photoperiod, maternal history and diet. In: *Reproduction in Context* (Ed. by K. Wallen & J. Schneider), pp. 191–218. Cambridge, MA: MIT Press.

Lehrman, D. S. 1970. Semantic and conceptual issues in the nature-nurture problem. In: *Development and Evolution of Behavior* (Ed. by L. R. Aronson, E. Tobach, D. S. Lehrman & J. S. Rosenblatt), pp. 17–52. San Francisco: W. H. Freeman.

Lepage, D., Gauthier, G. & Reed, A. 1998. Seasonal variation in growth of greater snow goose goslings: the role of food supply. *Oecologia*, 114, 226–235.

Levine, S. 1994. The ontogeny of the hypothalamic-pituitary-adrenal axis: the influence of maternal factors. *Annals of the New York Academy of Sciences*, 746, 275–288.

Lewis, R. J. & Kappeler, P. M. 2005. Seasonality, body condition, and timing of reproduction in *Propithecus verreauxi verreauxi* in the Kirindy Forest. *American Journal of Primatology*, 67, 347–364.

Lupien, S. J. & McEwen, B. S. 1997. The acute effects of corticosteroids on cognition: integration of animal and human model studies. *Brain Research Reviews*, 24, 1–27.

Maccari, S., Piazza, P. V., Barbazanges, A., Simon, H. & Le Moal, M. 1995. Adoption reverses the long-term impairment in glucocorticoid feedback induced by prenatal stress. *Journal of Neuroscience*, 15, 110–116.

Maestripieri, D. 2009. Maternal influences on offspring growth, reproduction, and behavior in primates. In: *Maternal Effects in Mammals* (Ed. by D. Maestripieri & J. M. Mateo), pp. 256–291. Chicago: University of Chicago Press.

Magurran, A. E. 1990. The inheritance and development of minnow anti-predator behaviour. *Animal Behaviour*, 39, 834–842.

Mateo, J. M. 1996. The development of alarm-call response behaviour in free-living juvenile Belding's ground squirrels. *Animal Behaviour*, 52, 489–505.

Mateo, J. M. 2002. Kin-recognition abilities and nepotism as a function of sociality. *Proceedings of the Royal Society of London, Series B*, 269, 721–727.

Mateo, J. M. 2003. Kin recognition in ground squirrels and other rodents. *Journal of Mammalogy*, 84, 1163–1181.

Mateo, J. M. 2007a. Ontogeny of adaptive behaviors. In: *Rodent Societies* (Ed. by J. O. Wolff & P. W. Sherman). Chicago: University of Chicago Press, pp. 195–206.

Mateo, J. M. 2007b. Ecological and physiological correlates of anti-predator behaviors of Belding's ground squirrels (*Spermophilus beldingi*). *Behavioral Ecology and Sociobiology*, 62, 37–49.

Mateo, J. M. 2008. Inverted-U shape relationship between cortisol and learning in ground squirrels. *Neurobiology of Learning and Memory*, 89, 582–590.

Mateo, J. M. & Holmes, W. G. 1997. Development of alarm-call responses in Belding's ground squirrels: the role of dams. *Animal Behaviour*, 54, 509–524.

Mateo, J. M. & Holmes, W. G. 1999. Plasticity of alarm-call response development in Belding's ground squirrels (*Spermophilus beldingi*, Sciuridae). *Ethology*, 105, 193–206.

Mateo, J. M. & Holmes, W. G. 2004. Cross-fostering as a means to study kin recognition. *Animal Behaviour*, 68, 1451–1459.

Maxwell, C. S. & Morton, M. L. 1975. Comparative thermoregulatory capabilities of neonatal ground squirrels. *Journal of Mammalogy*, 56, 821–828.

Mayr, E. 1974. Behavior programs and evolutionary strategies. *American Scientist*, 62, 650–659.

McAdam, A. G. 2009. Maternal effects on evolutionary dynamics in wild small mammals. In: *Maternal Effects in Mammals* (Ed. by D. Maestripieri & J. M. Mateo), pp. 64–82. Chicago: University of Chicago Press.

McCormick, M. I. 1998. Behaviorally induced maternal stress in a fish influences progeny quality by a hormonal mechanism. *Ecology*, 79, 1873–1883.

McEwen, B. S. & Sapolsky, R. M. 1995. Stress and cognitive function. *Current Opinion in Neurobiology*, 5, 205–216.

Meaney, M. J. 2001. Maternal care, gene expression, and the transmission of individual differences in stress reactivity across generations. *Annual Review of Neuroscience*, 24, 1161–1192.

Meikle, D. B., Drickamer, L. C., Vessey, S. H., Rosenthal, T. L. & Fitzgerald, K. S. 1993. Maternal dominance rank and secondary sex ratio in domestic swine. *Animal Behaviour*, 46, 79–85.

Meikle, D. B., Kruper, J. H. & Browning, C. R. 1995. Adult male house mice born to undernourished mothers are unattractive to oestrous females. *Animal Behaviour*, 50, 753–758.

Meylan, S. & Clobert, J. 2004. Maternal effects on offspring locomotion: influence of density and corticosterone elevation in the lizard *Lacerta vivipara*. *Physiological and Biochemical Zoology*, 77, 450–458.

Michener, G. R. 1983. Kin identification, matriarchies, and the evolution of sociality in ground-dwelling sciurids. In: *Advances in the Study of Mammalian Behavior* (Ed. by J. F. Eisenberg & D. G. Kleiman), pp. 528–572. Shippensburg, PA: American Society of Mammalogists.

Millesi, E., Huber, S., Everts, L. G. & Dittami, J. P. 1999. Reproductive decisions in female European ground squirrels: factors affecting reproductive output and maternal investment. *Ethology*, 105, 163–175.

Morton, M. L. & Sherman, P. W. 1978. Effects of a spring snowstorm on behavior, reproduction, and survival of Belding's ground squirrels. *Canadian Journal of Zoology*, 56, 2578–2590.

Morton, M. L. & Tung, H. L. 1971. Growth and development in the Belding ground squirrel (*Spermophilus beldingi beldingi*). *Journal of Mammalogy*, 52, 611–616.

Mousseau, T. A. & Fox, C. W. (Eds.). 1998. *Maternal Effects as Adaptations*. Oxford: Oxford University Press.

Murie, J. O. & Boag, D. A. 1984. The relationship of body weight to overwinter survival in Columbian ground squirrels. *Journal of Mammalogy*, 65, 688–690.

Neuhaus, P. 2000. Weight comparisons and litter size manipulation in Columbian ground squirrels (*Spermophilus columbianus*) show evidence of costs of reproduction. *Behavioral Ecology and Sociobiology*, 48, 75–83.

Nooker, J. K., Dunn, P. O. & Whittingham, L. A. 2005. Effects of food abundance, weather, and female condition on reproduction in tree swallows (*Tachycineta bicolor*). *Auk*, 122, 1225–1238.

Olson, G. S. & Van Horne, B. 1998. Dispersal patterns of juvenile Townsend's ground squirrels in southwestern Idaho. *Canadian Journal of Zoology*, 76, 2084–2089.

Owings, D. H. & Hennessy, D. F. 1984. The importance of variation in sciurid visual and vocal communication. In: *The Biology of Ground-dwelling Squirrels: Annual Cycles, Behavioral Ecology, and Sociality* (Ed. by J. O. Murie & G. R. Michener), pp. 169–200. Lincoln: University of Nebraska Press.

Owings, D. H., Coss, R. G., McKernon, D., Rowe, M. P. & Arrowood, P. C. 2001. Snake-directed antipredator behavior of rock squirrels (*Spermophilus variegatus*): population differences and snake-species discrimination. *Behaviour*, 138, 575–595.

Peacock, M. M. & Jenkins, S. H. 1988. Development of food preferences: social learning by Belding's ground squirrels *Spermophilus beldingi*. *Behavioral Ecology and Sociobiology*, 22, 393–399.

Pike, T. W. & Petrie, M. 2006. Experimental evidence that corticosterone affects offspring sex ratios in quail. *Proceedings of the Royal Society of London, Series B*, 273, 1093–1098.

Pizzimenti, J. J. & McClenaghan, L. R. 1974. Reproduction, growth and development, and behavior in the Mexican prairie dog, *Cynomys mexicanus* (Merriam). *American Midland Naturalist*, 92, 130–145.

Porter, R. H., Levy, F., Poindron, P., Litterio, M., Schaal, B. & Beyer, C. 1991. Individual olfactory signatures as major determinants of early maternal discrimination in sheep. *Developmental Psychobiology*, 24, 151–158.

Price, K. & Boutin, S. 1993. Territorial bequeathal by red squirrel mothers. *Behavioral Ecology*, 4, 144–150.

Pusey, A. & Wolf, M. 1996. Inbreeding avoidance in animals. *Trends in Ecology and Evolution*, 11, 201–206.

Qvarnström, A. & Price, T. D. 2001. Maternal effects, paternal effects and sexual selection. *Trends in Ecology and Evolution*, 16, 95–100.

Randall, J. A. 1993. Behavioural adaptations of desert rodents (Heteromyidae). *Animal Behaviour*, 45, 263–287.

Réale, D., Berteaux, D., McAdam, A. G. & Boutin, S. 2003. Lifetime selection on heritable life-history traits in a natural population of red squirrels. *Evolution*, 57, 2416–2423.

Reinhold, K. 2002. Maternal effects and the evolution of behavioral and morphological characters: a literature review indicates the importance of extended maternal care. *Journal of Heredity*, 93, 400–405.

Richerson, P. J. & Boyd, R. 2001. Built for speed, not for comfort: Darwinian theory and human culture. *History and Philosophy of the Life Sciences*, 23, 425–465.

Rieger, J. F. 1996. Body size, litter size, timing of reproduction, and juvenile survival in the Uinta ground squirrel, *Spermophilus armatus*. *Oecologia*, 107, 463–468.

Ryan, B. C. & Vandenbergh, J. G. 2002. Intrauterine position effects. *Neuroscience and Biobehavioral Reviews*, 26, 665–678.

Saino, N., Romano, M., Ferrari, R. P., Martinelli, R. & Møller, A. P. 2005. Stressed mothers lay eggs with high corticosterone levels which produce low-quality offspring. *Journal of Experimental Zoology Part A*, 303A, 998–1006.

Schwabl, H. 1999. Developmental changes and among-sibling variation of corticosterone levels in an altricial avian species. *General and Comparative Endocrinology*, 116, 403–408.

Sedinger, J. S., Lindberg, M. S., Eichholz, M. & Chelgren, N. 1997. Influence of hatch date versus maternal and genetic effects on growth of Black Brant goslings. *Auk*, 114, 129–132.

Sharp, S. P., McGowan, A., Wood, M. J. & Hatchwell, B. J. 2005. Learned kin recognition cues in a social bird. *Nature*, 434, 1127–1130.

Sherman, P. W. 1976. Natural selection among some group-living organisms. Ph.D. Dissertation. University of Michigan, Ann Arbor.

Sherman, P. W. & Morton, M. L. 1984. Demography of Belding's ground squirrels. *Ecology*, 65, 1617–1628.

Shettleworth, S. 1998. *Cognition, Evolution, and Behavior*. New York: Oxford University Press.

Sinervo, B. & Doughty, P. 1996. Interactive effects of offspring size and timing of reproduction on offspring reproduction: experimental, maternal, and quantitative genetic aspects. *Evolution*, 50, 1314–1327.

Smotherman, W. P. & Robinson, S. R. 1988. *Behavior of the Fetus*. Caldwell: Telford Press.

Stamps, J. 2003. Behavioural processes affecting development: Tinbergen's fourth question comes of age. *Animal Behaviour*, 66, 1–13.

Stamps, J. A. & Davis, J. M. 2006. Adaptive effects of natal experience on habitat selection by dispersers. *Animal Behaviour*, 72, 1279–1289.

Swaisgood, R. R., Owings, D. H. & Rowe, M. P. 1999. Conflict and assessment in a predator-prey system: ground squirrels versus rattlesnakes. *Animal Behaviour*, 57, 1033–1044.

Swaisgood, R. R., Rowe, M. P. & Owings, D. H. 2003. Antipredator responses of California ground squirrels to rattlesnakes and rattling sounds: the roles of sex, reproductive parity, and offspring age in assessment and decision-making rules. *Behavioral Ecology and Sociobiology*, 55, 22–31.

Terry, L. M. & Johanson, I. B. 1996. Effects of altered olfactory experiences on the development of infant rats' responses to odors. *Developmental Psychobiology*, 29, 353–377.

Tinbergen, N. 1953. *Social Behaviour in Animals*. New York: John Wiley & Sons.

Trivers, R. L. & Willard, D. E. 1973. Natural selection of parental ability to vary the sex ratio of offspring. *Science*, 179, 90–92.

Trombulak, S. C. 1991. Maternal influence on juvenile growth rates in Belding's ground squirrel (*Spermophilus beldingi*). *Canadian Journal of Zoology*, 69, 2140–2145.

Uller, T. & Olsson, M. 2006. Direct exposure to corticosterone during embryonic development influences behaviour in an ovoviviparous lizard. *Ethology*, 112, 390–397.

Vandenbergh, J. G. 2009. Effects of intrauterine position in litter-bearing mammals. In: *Maternal Effects in Mammals* (Ed. by D. Maestripieri & J. M. Mateo), pp. 203–226. Chicago: University of Chicago Press.

Visser, M. E. & Verboven, N. 1999. Long-term fitness effects of fledging date in great tits. *Oikos*, 85, 445–450.

Vitale, A. F. 1989. Changes in anti-predator responses of wild rabbits, *Oryctolagus cuniculus* (L.), with age and experience. *Behaviour*, 110, 47–61.

Vogl, W., Taborsky, M., Taborsky, B., Teuschl, Y. & Honza, M. 2002. Cuckoo females preferentially use specific habitats when searching for host nests. *Animal Behaviour*, 64, 843–850.

Vos, D. R. 1995. The role of sexual imprinting for sex recognition in zebra finches: a difference between males and females. *Animal Behaviour*, 50, 645–653.

Wade, M. J., Priest, N. K. & Cruickshank, T. E. 2009. A theoretical overview of genetic maternal effects: evolutionary predictions and empirical tests with mammalian data. In: *Maternal Effects in Mammals* (Ed. by D. Maestripieri & J. M. Mateo), pp. 38–63. Chicago: University of Chicago Press.

Waterman, J. M. 1986. Behaviour and use of space by juvenile Columbian ground squirrels. *Canadian Journal of Zoology*, 64, 1121–1127.

White, K. S. & Berger, J. 2001. Antipredator strategies of Alaskan moose: are maternal trade-offs influenced by offspring activity? *Canadian Journal of Zoology*, 79, 2055–2062.

Wilson, A. J. & Festa-Bianchet, M. 2009. Maternal effects in wild ungulates. In: *Maternal Effects in Mammals* (Ed. by D. Maestripieri & J. M. Mateo), pp. 83–103. Chicago: University of Chicago Press.

Wilson, A. J., Pilkington, J. G., Pemberton, J. M., Coltman, D. W., Overall, A. D. J., Byrne, K. A. & Kruuka, L. E. B. 2005. Selection on mothers and offspring: whose phenotype is it and does it matter? *Evolution*, 59, 451–463.

Wright, S. 1931. Evolution in Mendelian populations. *Genetics*, 16, 97–159.

Yang, J. L., Han, H. L., Cao, J., Li, L. J. & Xu, L. 2006. Prenatal stress modifies hippocampal synaptic plasticity and spatial learning in young rat offspring. *Hippocampus*, 16, 431–436.

Yoerg, S. I. & Shier, D. M. 1997. Maternal presence and rearing condition affect responses to a live predator in kangaroo rats (*Dipodomys heermanni arenae*). *Journal of Comparative Psychology*, 111, 362–369.

Zielinski, W. J., vom Saal, F. S. & Vandenbergh, J. G. 1992. The effect of intrauterine position on the survival, reproduction and home range size of female house mice (*Mus musculus*). *Behavioral Ecology and Sociobiology*, 30, 185–191.

8

Maternal Influences on Offspring Food Preferences and Feeding Behaviors In Mammals

BENNETT G. GALEF, JR.

INTRODUCTION: THE CHALLENGE OF WEANING

For some time after birth, every mammal has access to its mother's milk, a single food that provides a complete and balanced diet. However, for most mammals, such reliance on a single food is restricted to a relatively brief period early in life.

As the willingness of a mother to invest further resources in her young wanes (Trivers 1974), the young must undertake a potentially perilous transition from obligate monophages ingesting only mother's milk to independent foragers composing a balanced diet by ingesting a variety of foods, no one of which contains all the nutrients needed for growth and survival (Galef 1981). The transition from monophage to dietary generalist is particularly challenging because weanlings must construct a balanced diet before they exhaust their relatively limited internal reserves of any critical nutrient and without eating debilitating amounts of indigestible or toxic substances that they may encounter as they search for nutritionally valuable foods (Rozin & Schulkin 1990; Galef 1996a).

Success in making the transition from mother's milk to a diet of solid food depends on three quite different processes. First, each animal exhibits sensory-affective biases reflecting the inherent organization of its nervous system that cause it to find some flavors attractive and others aversive. To the extent that such unlearned responses to chemicals in potential foods bias weanlings to ingest nutritive substances and reject toxic ones, they predispose weanlings to choose appropriate substances to ingest (Garcia & Hankins 1975; Scott 1990).

Because discovery of taste-related genes is relatively recent, study of genetic contributions to variance in taste reactivity is still in its infancy (for review, see Reed et al. 2006). Still, convincing demonstrations of genetic effects on responses to flavors are already in hand. For example, mice not sensitive to the taste of the bitter substance PTC can be made to taste PTC by introducing the human taster receptor gene into their taste receptor cells (Mueller et al. 2005).

Second, individual experience of the consequences of ingesting potential foods, i.e., simple trial-and-error learning, can strongly influence which substances an animal subsequently eats and which it avoids eating. Indeed, with but a single exception, the common vampire bat (*Desmodus rotundus*), an obligate monophage on mammalian blood (Ratcliffe et al. 2003), members of all mammalian species examined to date, from spotted hyenas (*Crocuta crocuta*: Yoerg 1991) to Antillean fruit-eating bats (*Brachyphylla cavernarum*: Ratcliffe et al. 2003), learn to avoid repeated ingestion of a potential food when ingestion of that food is followed by aversive gastrointestinal events (Garcia & Koelling 1966). Conversely, mammals also learn to associate food flavors with positive consequences resulting from ingestion of calories or protein (Booth 1985; Sclafani 1995).

However, even taken together, inherent hedonic responses to flavors and the ability to learn by trial and error what foods to eat may be only marginally effective in protecting naïve young mammals from making inappropriate choices of substances to ingest (Beck & Galef 1989; Galef & Beck 1990). Some toxins elicit positive hedonic responses and some nutritious foods elicit negative ones. Consequently, congenital responses to flavors provide only a general guide to identifying either toxins or valuable foods (Galef 1996a). Similarly, although an ability to learn from experience whether an ingested substance is associated with deleterious or favorable consequences can be useful, in demanding environments, trial-and-error learning often fails to lead to selection of appropriate substances to ingest (Beck & Galef 1989; Galef 1991).

Given the inadequacy of congenital flavor preferences and personal experience of consequences of ingesting various substances as guides in diet construction, weanlings might be expected to attend to any additional information they can acquire as to the potential costs and benefits of ingesting each of the myriad potentially ingestible substances that they encounter. In particular, naïve weanlings struggling to compose a nutritionally adequate diet might be able to benefit from opportunities provided by interactions with adults of their species.

Adults, almost by definition, have learned to select appropriate substances to eat and, most important, any adult with whom a juvenile interacts is likely to have identified valuable foods in the environment where the juvenile is struggling to achieve nutritional independence. Consequently, juveniles that use the feeding behavior of adults to guide development of their own dietary repertoires should be able to compose adequate diets without relying on either unreliable sensory-affective responses to flavors or the inherently risky process of learning by trial and error.

It is with this third source of information as to which substances to eat, information that juveniles acquire from adult conspecifics, that this chapter is concerned. Such social learning about foods is normative in mammals because all spend their prenatal life within their mothers and all take milk from her in the period immediately following their birth. Consequently, early exposure of all mammals to gustatory and olfactory stimuli is affected by the chemical environment that their mother provides.

IS MOTHER SPECIAL?

Some of the behavioral processes that support social learning about foods involve information that a juvenile can acquire only from its mother. For example, the food choices of a pregnant female, but not those of other adults, can affect the gustatory or olfactory experiences of the young she gestates, and such prenatal experiences have the potential to influence the food choices of young when they wean. In other cases, juveniles can extract useful information about food from any adult conspecific with whom they come in contact. However, even then, a mother's social bond with her young may markedly increase the probability that she, rather than other adults, will serve as a source of information for her offspring. For example, juveniles in many ungulate species tend both to remain near their mother as she forages and to eat foods that they see adults eat (for review, see Provenza 1994). The physical proximity of mother and young should increase the likelihood that socially learned food preferences more frequently reflect the feeding patterns of mothers than of other adults. Consequently, vertical transmission (transmission from mother to young) of food preferences should be more frequent than their oblique transmission (transmission from unrelated adults to young; Boyd & Richerson 1985).

More generally, as Coussi-Korbel & Fragaszy (1995) have suggested, the social structure of a population should influence the paths along which socially learned behaviors travel. Consequently, diffusion of socially learned food choices through a population will, at least in theory, be "directed" in the sense that the identity or characteristics of potential demonstrators will affect the probability that they will serve as models for naïve individuals. For example, various formal models predict that inexperienced individuals should be more likely to adopt behaviors of successful, older, or familiar individuals than of unsuccessful, young, or unfamiliar ones (for review see Laland 2004). However, to date, relatively little experimental evidence of the importance of directed social learning in mammals is available.

EARLY STUDIES OF SOCIAL INFLUENCES ON FOOD CHOICE OF MAMMALS

Experimental studies of social influences on food selection in mammals were initiated in response to the work of an applied ecologist, Fritz Steiniger, whose professional interest lay in improving the efficiency of methods for poisoning rodent pests. In the course of his work, Steiniger (1950) attempted to reduce the need for expensive return visits to sites where rat control was under way by placing relatively permanent stations containing poison bait in problem areas. Although Steiniger's bait stations were inexpensive to service, they failed to eliminate target populations of rats. As Steiniger soon discovered, when a bait station was first placed in a rat-infested area, members of the population ate significant amounts of poison and died in large numbers. However, later acceptance of the bait was extremely poor, and baited colonies soon regained their initial size.

Steiniger's observations indicated that the failure of his permanent baiting stations resulted from information acquired by young born to the few adult colony members that ate only a small amount of poison bait upon initial contact with it, and although becoming ill, survived their encounter and subsequently avoided ingesting the bait. Offspring of these survivors refused even to taste the bait that the surviving adults in their colony had learned to avoid.

Such avoidance by juvenile rats of foods that adults of a colony have learned to avoid is a robust phenomenon, easily captured in the laboratory. Field-caught adult wild Norway rats (*Rattus norvegicus*) placed in mixed-sex groups in laboratory enclosures can be taught easily to ingest only one of two distinctively flavored foods offered them. All that need be done is to adulterate one of the two foods with a sublethal concentration of a relatively mild toxin (Galef & Clark 1971b). When unadulterated samples of both foods are subsequently offered to members of such colonies, they continue to avoid the previously adulterated alternative for weeks.

As Steiniger had discovered, when young born to female colony members first start to eat solid foods, and for at least 2 weeks thereafter, the young eat only the food that the adults of their colony are eating and totally avoid the alternative, even when both available foods are safe to eat (Galef & Clark 1971b). Much work during the past 35 years has been dedicated to analyzing the behavioral mechanisms that underlie such social influence on the food choices of Norway rats.

PRENATAL MATERNAL EFFECTS ON OFFSPRING FOOD PREFERENCES

Experience by mammalian fetuses of gustatory and olfactory stimuli can be modified by flavorants introduced into the amniotic fluid of their mother. Further, fetal exposure to a flavorant, as a result either of direct infusion into the amnion (e.g., Smotherman 1982) or ingestion by a pregnant female (Hepper 1988), can affect the postnatal response of young to flavors that they experienced in utero (e.g., Schaal et al. 1995). However, almost all studies examining effects of fetal exposure to flavors have been conducted by researchers interested primarily in development of the olfactory and gustatory systems, rather than in development of food preferences. Consequently, the dependent variables most frequently used in studies of effects of prenatal exposure to flavors are frequently not directly relevant to understanding maternal effects on food choice. For example, in a pioneering study, Hepper (1988) either fed or did not feed garlic to pregnant rats late in gestation, and within an hour of their giving birth, gave their litters to mothers that had never eaten garlic to rear. When these foster-reared pups were 12 days of age, Hepper offered each a choice between two dishes: one containing garlic and the other onion. Hepper found that pups born to mothers fed garlic during pregnancy tended to stay close to the dish containing garlic, whereas pups born to mothers that had not eaten garlic during pregnancy responded equally to garlic and onion. Similarly, human infants gestated by mothers that consumed anise-flavored food during pregnancy showed an enhanced preference for the odor of anise during the immediate postnatal period (Schaal et al. 2000). Wells & Hepper (2006) have recently provided similar evidence of prenatal olfactory learning in domestic dogs.

Using a different method for introducing flavors into the uterine environment, Smotherman (1982) found that when offered a choice between apple juice and water as adults, rats that as fetuses were exposed to apple juice injected directly into their mothers' amniotic fluid showed an enhanced preference for apple juice. Taken together, the results of the studies by Hepper (1988) and Smotherman (1982) suggest that flavors ingested by pregnant rat mothers should be preferred by their offspring at weaning. There is, however, no evidence of effects of flavors *ingested* by pregnant females affecting the subsequent *food choices* of their offspring (but see below, Bilko et al. 1994).

Because of the importance of ethanol ingestion in human affairs, effects of prenatal exposure to ethanol on subsequent ethanol ingestion have been examined relatively frequently. However, effects, if any, of prenatal exposure to ethanol ingested by a mother on subsequent active ingestion of ethanol by her young remain unclear (for review, see Spear & Molina 2005). For example, young rats exposed in utero to ethanol intubated directly into their mother's stomach ingested more ethanol delivered into their oral cavities by catheter on day 14 postpartum than did control subjects not exposed to ethanol before birth (Dominguez et al. 1998). However, the relevance of such findings to effects of mother rats' ingestion of ethanol on spontaneous ingestion of ethanol by their offspring remains to be demonstrated. For example, when Honey & Galef (2003) looked for effects of a rat mother's ingestion of ethanol throughout pregnancy on ingestion of ethanol by her young when adolescent, they found none. Thus, although exposure to alcohol in utero can affect subsequent response to ethanol by rats, it is not yet clear that *ingestion* of ethanol by a rat mother leads to increased *ingestion* of ethanol by her offspring.

In humans, evidence that prenatal exposure to flavors increases postnatal ingestion of foods containing those flavors is similarly ambiguous. For example, Mennella et al. (2001) report reduced negative affective facial expressions in response to carrot-flavored cereal in newborn infants whose mothers drank carrot juice while pregnant. However, infants whose mothers drank carrot juice while pregnant did not eat significantly more carrot-flavored cereal than did control infants not exposed in utero to carrot flavor.

Indeed, given the importance of social learning to almost every aspect of behavioral development in *Homo sapiens*, maternal effects on food choices in humans are generally surprisingly, almost shockingly, weak. For example, although all research on familial resemblances in food preferences reveals significant family correlations of youngsters' food choice with those of their parents, paternal effects are as strong as maternal ones, and sibling effects are stronger than those of either parent. Some, but not all, researchers find increased concordance in the food choices of mothers and their children as children increase in age and their dietary repertoires broaden (e.g., Birch 1980; Rozin et al. 1984; Pliner & Pelchat 1986).

In the only study of which I know providing evidence of robust effects of prenatal exposure to a mother eating a food on postnatal preference for that food in her offspring, Bilko et al. (1994), found that young domesticated rabbits (*Oryctolagus cuniculus*) born to mothers fed juniper-berry-flavored food during pregnancy showed a preference for juniper-berry-flavored food at weaning. The robust influence of prenatal exposure to a flavor on ingestion of foods of that flavor at weaning observed by Bilko et al. (1994) may reflect the unusual maternal behavior of rabbits. The rabbit doe contacts her young for less than 5 minutes each day. The resulting absence of extensive postpartum contact between mothers and young seen in most mammalian species restricts opportunities for postpartum social learning in rabbits and may have led to evolution of specialized mechanisms for prenatal learning of food preferences.

MATERNAL EFFECTS ON OFFSPRING FOOD PREFERENCES MEDIATED BY MOTHER'S MILK OR FECES

Mother's Milk

During the period following birth, when most young mammals are dependent on their mothers to meet their nutritional needs, the young have access to several potential sources of information as to what foods to ingest. Among the more obvious, flavor cues contained in maternal milk could allow juveniles to learn the flavors of foods that their mother is eating. Indeed, there is evidence from several species (e.g., rats: Galef & Sherry 1973; mice: Mainardi et al. 1989; rabbits: Bilko et al. 1994; cattle: Morrill & Dayton 1978; swine: Campbell 1976; and sheep: Nolte & Provenza 1992) suggesting that flavor cues present in mother's milk influence the food choices of young when they wean to solid food. For example, Bilko et al. (1994) offered juniper berries and laboratory chow to rabbit pups whose mothers had been fed only chow but that nursed each day from a foster mother fed either juniper berries or laboratory chow. At weaning, the young were offered a choice between juniper berries and chow. Pups that had been nursed by a foster mother fed juniper berries ate significantly more juniper berries than did pups nursed by a foster mother fed chow.

Of course, in the Bilko et al. (1994) study, the influence of foster mothers on pups' food choices may not have been mediated by maternal milk. For example, mothers eating juniper may have had the odor of juniper on their breath, in their feces, or in their fur, and foster-nursed young may have been responding to these cues rather than to cues present in maternal milk. In general, the same caveat applies to all studies where young nurse directly from a mother other than their mother and an influence of the diet eaten by the foster mother on the food choices of the young has been found.

To control for such potential confounds, Galef & Sherry (1973) manually expressed milk from lactating rats that had been fed flavored diets and then fed the expressed milk to infant rats. They found that young rats could iden-tify the flavor of a mother's diet by ingesting her milk. Galef & Henderson (1972) used a different method to control for dietary cues outside mother's milk. They compared food preferences of young rats exposed daily to either lactating foster mothers or nonlactating females in whom maternal behavior had been induced, when all females had eaten the same distinctively flavored diet. Those pups that had contact with a lactating rat showed an enhanced preference for her diet at weaning, whereas those interacting with a nonlac-tating female did not (Galef & Henderson 1972).

Occasionally, effects of flavors in mother's milk have paradoxical effects on the subsequent food preferences of their offspring. For example, Men-nella & Beauchamp (1999) have reported that human babies nursed by moth-ers who drank carrot juice ate less (not more) carrot-flavored cereal than did babies lacking experience with carrot-flavored mother's milk. Although such a change in preference is clearly a maternal effect, why babies should have behaved in such a fashion is not clear. Perhaps the presence of carrot in mother's milk causes gastrointestinal upset in babies and they learn to avoid carrot flavor. Perhaps human infants seek variety in their food (DiBattista 2002; Galef & Whiskin 2003a; 2005).

Feces

Rabbit does, like many other mammalian mothers, deposit fecal pellets in the nest with their young. Bilko et al. (1994) put fecal pellets from a mother fed juniper berries into the nest of a mother fed laboratory chow. They found that, when offered a choice between juniper berries and laboratory chow at weaning, pups that had been exposed to fecal pellets of mothers fed juniper ate significantly more juniper berries than did pups exposed to feces taken from a female eating chow.

The finding of Bilko et al. (1994) in rabbits is quite different from those of Galef (1979) and Galef & Henderson (1972), who found that, although rat pups consume appreciable amounts of maternal feces before weaning, they do not show an enhanced intake of a distinctively flavored diet after eating feces taken from a female eating that diet. Perhaps the flavor of juniper berries is particularly salient in maternal feces. Perhaps young rats and rabbits differ in their response to flavors or odors experienced in maternal feces.

MATERNAL EFFECTS ON FOOD PREFERENCES MEDIATED BY BEHAVIORAL INTERACTIONS

Mouth Licking

In a number of rodent species, for example spiny mice (*Acomys*), hamsters (*Cricetus*), and gerbils (*Meriones*), young actively lick the mouths of adults (for references, see Ewer 1968). In captivity, young of all three species direct mouth licking towards their mothers. However, in captivity, mothers are often the only adults with whom young are allowed to come in contact.

Little is known of the consequences of mouth licking. Results of a pair of suggestive studies showed that: (1) young spiny mice (*Acomys cahirinus*) lick the mouths of their mother more frequently after she has eaten a novel than a familiar food, and (2) young *Acomys* show an enhanced preference for an unpalatable diet eaten by their mother after licking her mouth (McFadyen-Ketchum & Porter 1989). However, whether mouth licking per se was responsible for the observed shift in the food preferences of young *Acomys* remains to be determined. Possibly, simple proximity of the nose of the young to the mouth of the mother resulted in enhanced intake by young of the food that their mother had eaten (See "Muzzle-muzzle" below).

Food Snatching and Food Sharing

In many species of mammals, infants obtain their first solid food by taking scraps from feeding adults. Often, close proximity of mother and young ensures that the mother is the source of most food scraps eaten by young early in weaning. Of course, as an infant matures and spends increasing time away from its mother, solicitation of foods from other adults may increase in frequency.

Galef et al. (2001) report that young Norway rats exhibit an enhanced preference for an unfamiliar food that they have snatched from an adult and eaten that they do not show after picking the same food off the ground and eating it. Whether mother rats are more tolerant of their own young taking food than are other adults, and young are therefore more likely to snatch food from their mother than from other adults of their colony, is not known.

Food snatching similar to that seen in Norway rats has been observed in numerous species (for a recent review of food transfer in nonhuman primates, see Brown et al. 2004). However, whether snatched food in species other than rats has an effect on subsequent food preference that food taken from the substrate does not has rarely been investigated.

In primates, the frequency of food transfer from adults to infants varies tremendously from species to species. For example, Boinski & Fragaszy (1989) report no instances of food transfer from adults to infants in free-ranging squirrel monkeys (*Saimiri oerstedi*), whereas Hiraiwa-Hasegawa (1990) reported that infant male chimpanzees (*Pan troglodytes*) during their first year of life solicited food from their feeding mother approximately once every 3 minutes.

When food sharing between adults and young does occur, whether mothers are a disproportionate source of transferred food also varies from one mammalian species to another depending on species-typical social organization. For example, young captive chimpanzees were more successful in soliciting food from their mothers (80% of attempts) than from other adult females (20–30% of attempts; Silk 1979), whereas infant tufted capuchin monkeys (*Cebus apella*) in captive groups were no more likely to take food from their mothers than from other adult group members (Fragaszy et al. 1997).

Meerkat (*Suricata suricatta*) females (both mothers and helpers) tending young of weaning age refrain from consuming some of the prey that they capture and run with it in their jaws to their young. The provisioning meerkat then runs to and fro in front of the young, allowing them to snatch the prey. If the young fail to take the food, it is put down in front of one of them (Ewer 1969).

In many mammalian carnivores, parental behavior includes some way of introducing young to prey by, for example, carrying it back to the young (e.g., domestic cat), or as is often the case in Canidae, eating and then regurgitating food for young (e.g., coyotes and domestic dogs). In some carnivores, mothers with weaning young refrain from killing prey they have captured, and bring live prey to their young, releasing it for the young to capture and kill (e.g., cheetah, domestic cat, and Canadian otter). In yet other species (e.g., grasshopper mice) prey are killed, but only partially eaten and are left for the young, giving them the opportunity to become familiar with foods that their mother is ingesting (for references, see Ewer 1968).

Although all investigators of food transfer from mother to young speculate that infants learn about prey by sharing the foods taken from their mother (e.g., King 1994; Brown et al. 2004), until very recently (see below, Thornton & McAuliffe 2006) strong evidence of such maternal effects facilitating weanlings' food choices has been lacking, and there has been no unambiguous evidence that the experience gained by young mammals facilitates either their identification of prey or prey capture (for review, see Galef et al. 2004). Indeed, some findings suggest that food sharing may not serve similar functions in all species. For example, young rats are more likely to snatch unfamiliar than familiar foods from adults (Galef et al. 2001). However, when an unfamiliar food was offered to a group of adult golden-lion tamarins (*Leontopithecus rosalia*), they were less likely to share food with young than when the food was familiar, suggesting that in this species, food sharing may not function to introduce young to new foods (Price & Feistner 1993).

Co-feeding

In many species, young tend to prefer to feed near adults of their species. For example, weaning Norway rat pups use visual cues to locate feeding adults at a distance from the nest site and take their first meals of solid food while an adult is eating, first crawling from under the belly of an adult to its head and then feeding under its chin (Galef & Clark 1971a, b).

In rats, there is no evidence from either laboratory or field that young approach their mothers at feeding sites in preference to other adults. However, in other mammalian species, where mother and young remain in close proximity during the period when young are weaning to solid food, feeding with the mother, resulting in enhanced preferences for the foods that she eats, has been reported. For example, Lynch et al. (1983) examined wheat acceptance by Merino lambs exposed to wheat at different ages and for different periods either with or without their mothers present. Only suckling lambs exposed to wheat with their mothers showed acceptance of wheat when tested alone after weaning. The effect on lambs of co-feeding with their mother on the food choices of the young were still observable 3 years after mothers and young co-fed (Green et al. 1984).

More recent studies by Provenza and colleagues have found that, in both sheep and goats, species where young imprint on their mother and tend to follow her about, the opportunity to feed in the presence of the mother increases future intake of foods eaten while co-feeding (e.g., Nolte et al. 1990; Distel & Provenza 1991; Nolte et al. 1992). However, the behavioral processes involved in such social effects on food preference have not been thoroughly explored in either sheep or goats.

In an innovative laboratory study, and one unusual in that members of a carnivorous species served as subjects, Wyrwicka (1974; 1978; 1981) and Wyrwicka & Long (1980) used rewarding electrical brain stimulation to induce mother cats to eat foods, such as banana slices, that they would normally avoid. When kittens were placed with their mother during her feeding periods, they too began to eat bananas, and when subsequently tested alone, continued for some weeks to prefer banana to meat pellets (Wyrwicka 1978). Although the method of transfer of a preference for banana from mother to young was not explored, the observation that kittens initially ate from the exact site where their mother had eaten suggests either close attention to visual cues the mother provided while eating or odors with which she marked food she was eating were responsible for the observed social induction of food preference.

In an observational study of the interactions of infant and adult mantled howler monkeys (*Alouatta palliata*) feeding in the forests of western Costa Rica, Whitehead (1986) found that infants ate leaves only of those plant species that they had observed adult group members feed upon and that infants often fed from the same branch on which their mother was feeding. On the other hand, infants frequently began to ingest unfamiliar fruits before adult group members did so, and when eating fruit, seldom fed from the same branch on which an adult was feeding. Thus, the food choices of young howler monkeys selecting among leaves (presumably relatively rich in toxic secondary compounds) appear to be open to maternal effects, while their choice of fruit is independent of adult influence.

King's (1994) observations of differences in the extent of co-feeding between adult and juvenile savanna baboons (*Papio cynocephalus*) as a function of the type of food fed upon provides a second example of selective co-feeding. Juvenile ate together with adults more frequently when eating corms and at a lower rate when feeding on leaves than when feeding on fruits or seeds.

"Muzzle-muzzle"

After a naïve rat (an observer) interacts with a conspecific that has recently eaten a food unfamiliar to the observer, the observer exhibits an enhanced preference for whatever food its demonstrator ate (Galef & Wigmore 1983; Strupp & Levitsky 1984). Such effects of demonstrators on their observers' food preferences are long lasting. For example, when a 1-month interval was imposed between demonstrator rats interacting with observers and testing of observers for their diet preferences, effects of demonstrators on observers' diet preferences were as strong as when observers were tested immediately after interacting with their demonstrators (Galef & Whiskin 2003b). Demonstrator effects on observers' food preferences are also surprisingly powerful, able to reverse an aversion to a food established by associating that food with illness (Galef 1986) or a congenital flavor aversion, for example, to a piquant food (Galef 1989).

Demonstrator influence on observer food choices depends on observers extracting olfactory information from demonstrators during periods when demonstrator and observer are close to one another and face to face (Galef & Wigmore 1983; Galef & Stein 1985), a position that King (1994) labeled "muzzle-muzzle" in discussing interactions between juvenile and adult baboons in Amboseli National Park in Kenya.

Possibly similar cases of olfactory communication of information concerning foods have been reported in a number of mammalian species (e.g., hyenas: Yoerg 1991; voles: Solomon et al. 2002; mice: Valsecchi et al. 1993; Mongolian gerbils: Choleris et al. 1998; roof rats: Chou et al. 2002; fruit bats: Ratcliffe & ter Hofstede 2005; hamsters: Lupfer et al. 2003). Although, in general, it is not known whether weanlings' food choices following muzzle-muzzle are more likely to result from their interactions with their mothers than with other adults, there is one case in which exclusive transmission from mothers to young has been clearly demonstrated.

Lupfer et al. (2003) conducted a study in which both young golden hamsters (*Mesocricetus auratus*) and young dwarf hamsters (*Phodopus campbelli*) interacted with a demonstrator of their own species that had recently eaten a distinctively flavored food. Golden and dwarf hamsters were selected as subjects because of differences in the social organizations of the two species. In the wild, golden hamsters are relatively solitary, and adults are highly aggressive towards one another, whereas dwarf hamsters are a moderately social species, with males and females sometimes sharing burrows and engaging in biparental care.

When allowed to choose between a flavor eaten by a conspecific demonstrator and another unfamiliar food, juveniles of the more social dwarf hamsters preferred the food that their demonstrators had eaten regardless of the demonstrator's age or relatedness to the observer, whereas juveniles of the relatively solitary golden hamster did not show a similar preference if the demonstrator hamster was either an unrelated adult or a littermate. However, when the demonstrator was the mother of an observer golden hamster, the juvenile exhibited a preference for the diet that its mother had eaten.

MATERNAL EFFECTS ON FOOD EXTRACTION

One of the more interesting possible consequences of adults and young feeding together is that, while in the presence of feeding adults, young have opportunities to observe whatever techniques adults use to extract potential foods that are, in some way, difficult to access. There is an emerging body of data suggesting that, especially in primates and cetaceans, young animals may acquire extractive skills, at least in part, as a consequence of observing the behavior of adults that have developed those skills. In many species, the adult whose behavior is most frequently observed is the mother of the observer.

Rats in Israel

Although such social learning of relatively complex feeding skills is most frequently reported in mammals with relatively large brains, even rodents appear to engage in functionally similar forms of behavior acquisition that can have profound effects on their dietary repertoires. For example, in what is probably the best-documented case of social acquisition of a complex feeding skill in any species, Terkel and colleagues have analyzed a traditional extractive technique used by roof rats (*Rattus rattus*) living in the pine forests of Israel. These rats living there subsist on a diet that consists solely of pine seeds and water (Aisner & Terkel 1992; Zohar & Terkel 1992).

Laboratory studies have revealed that for rats to gain more energy in eating pine seeds than they use to extract them from beneath the scales of pine cones the rats must take advantage of the physical structure of pinecones in removing the tough scales that protect the seeds. A rat must first remove scales from the cone's base and then remove, in succession, the spiral of scales running around the cone's shaft to its apex.

Only 6 of 222 naïve young rats left in the presence of pinecones for months while maintained at 85 percent of their normal body weight learned to open pinecones efficiently, and no adult rat trapped outside a pine forest learned to do so even when confined for months in a cage with both pinecones and a knowledgeable adult rat that demonstrated the efficient method of opening the cones. To the contrary, essentially all young rats learned to open pinecones efficiently if they were reared either by their own mother or by a foster mother that, while in the juveniles' presence, efficiently stripped scales from pinecones. Clearly some aspect of the postnatal interaction between adults and young is needed for transmission of the desirable method of stripping pinecones of their scales to occur (Aisner & Terkel 1992; Zohar & Terkel 1992).

Further experiments demonstrated that the experience of young rats in completing stripping pinecones that had been started appropriately by an experienced adult rat (or by an experimenter using a pair of pliers to imitate the pattern of scale removal used by efficient adults) enabled more than 70% of young rats to learn to feed efficiently on pine seeds (Terkel 1996). The relative frequency with which mothers and other adult rats contribute partially stripped pinecones to juveniles in natural settings is not known.

Dolphins in Australia

Perhaps the most intriguing example of behavioral transmission from mothers to offspring of a complex extractive technique concerns use of sponges by bottle-nosed dolphins (*Tursiops* spp.) foraging in deep channels at Shark Bay, Australia (Mann & Sargeant 2003). Fragmentary observations of sponge use by dolphins in the murky, shark-infested waters where sponge carrying has been observed suggest that dolphins use the sponges to protect their rostrums while grubbing in the substrate in search of prey. Sponge use appears to be adaptive in that sponge users have higher calving success than sympatric females that do not use sponges (Mann et al. in press).

Analysis of mitochondrial DNA of spongers and nonspongers showed that sponging is found only in a single matriline. Further, because both spongers and nonspongers forage in the same areas, it is unlikely that ecological conditions alone cause some animals to use sponges while foraging, and DNA analyses led to rejection of the hypothesis that sponging females are a mutant subset of the dolphin population at Shark Bay (Krutzen et al. 2005). Observations of a strong general correspondence between the foraging styles of mothers and their calves (Mann & Sargeant 2003), as well as anecdotal and laboratory evidence of dolphins' ability to learn by observation, are consistent with the results from analysis of sponging in suggesting powerful maternal effects on foraging behavior of dolphins.

Chimpanzees in Africa

Studies of six geographically separate chimpanzee populations have identified at least 15 patterns of tool use in foraging that are found in some populations but not others (Whiten et al. 2001). For example, chimpanzees in the Tai Forest in West Africa use stone hammers and anvils to crack nuts, whereas no chimpanzees in East Africa do so (Boesch & Boesch 1990; Boesch 1991);

conversely, chimpanzees at Gombe National Park in East Africa use twigs as probes to fish for termites (Goodall 1986), whereas chimpanzees at Tai do not (McGrew 1992).

Although many such differences in behavior of members of the various populations of chimpanzees have been described, almost nothing is known of their development. However, results of a recent long-term study of termite fishing by young chimps at Gombe suggest an important role for maternal influence on development of this foraging technique. Lonsdorf's (2005; Lonsdorf et al. 2004) 4-year study of the ontogeny of termite fishing revealed that mother chimpanzees showed different frequency distributions of the depths to which they inserted tools into termite nests while fishing. Daughters' patterns of depth of tool insertion, but not those of sons, were significantly correlated with those of their mothers.

The sex difference in similarity between mothers and offspring in termite fishing did not reflect a difference in either frequency of social interaction between mothers and their sons and daughters or tolerance of mothers for their male and female offspring. However, young females spent more time watching their mothers fish for termites than did young males. Daughters' use of their mothers' pattern of fishing for termites is probably as clean a case of maternal influences on extractive techniques as is likely to result from purely observational studies of free-living chimpanzees.

FINAL DISCUSSION AND CONCLUSIONS

As the preceding review makes clear, mammalian mothers can profoundly influence the food choices and feeding behaviors of their young. The behavioral processes supporting such maternal effects vary from species to species. They range from exclusively maternal influences on dietary preferences of weanlings mediated by flavors of foods that mothers ingest passing into their amniotic fluid or milk to influences that all adults can have on young, but in which mothers are likely to predominate because of their physical proximity to their own young during weaning.

Adaptive Value of Maternal Influences on Feeding Behavior

Regardless of the mechanism or mechanisms involved in maternal influences on the food choices of juveniles, the functions of maternal influence are probably similar, increasing the probability that young wean to an adequate diet of safe foods before exhausting their internal reserves of any nutrient and without ingesting deleterious quantities of toxins. Although the potential benefits of maternal influence on food choice seem obvious, there is little empirical evidence that either maternal effects or influences of any other adult on the food choices of young mammals result in facilitation of weaning to an adequate diet in free-living populations (Galef et al. 2004). Even in the laboratory, demonstrations of the adaptive value of social influences on food choice are rare. Beck & Galef (1989) have shown that weanling Norway rats offered a choice among four distinctively flavored foods, only one of which contained sufficient protein for survival, could not learn by trial and error to eat enough of the adequate diet to remain healthy, but focused their feeding on the protein-rich diet and grew normally if caged with an adult trained to eat that diet. Similarly, Terkel's (1996) laboratory studies of pinecone stripping by black rats (described above) strongly suggest that rats' survival in the pine forests of Israel is dependent on feeding behaviors learned from adults.

Mothers in predatory species that bring wounded or dead prey to their offspring to kill or feed upon seem likely to facilitate development of the predatory behavior in developing young. However, until very recently there was no unambiguous evidence of such effects even in laboratory settings. For example, studies of development of mouse killing in young domestic cats, *Felis catus*, have both found (Caro 1980a, b) and failed to find (Baerends-van Roon & Baerends 1979) effects of interacting with adult cats that kill mice on kittens' age at initiation of predation on mice. Similarly, although Flandera & Novakova (1975) found accelerated development of mouse killing in rat pups whose mothers killed mice in their presence, Galef (1996b) did not.

Thornton & McAuliffe (2006) recently reported the first compelling evidence of social facilitation of the development of effective predatory behavior and, most unexpectedly, they did so in free-living animals. In brief, Thornton and McAuliffe found that young meerkats that they had provided with disabled scorpions similar to those that adult provisioners bring young in response to juveniles' begging calls were more successful at holding on to a healthy scorpion and less likely to be stung or pincered by it than siblings that had not been artificially provisioned with disabled scorpions.

Extrapolation from Laboratory to Field

It may seem strange that after more than 50 years of study of interactions between mammalian mothers and their weaning young there is so little conclusive evidence of adaptive consequences of maternal effects on the development of food choices and feeding behaviors of mammals living in natural environments. There are two major difficulties in providing such evidence. First, in most studies of free-living animals, even when the frequency and nature of contacts between mothers and young can be measured with some precision, determining the consequences of maternal interaction with off-spring on the subsequent feeding behavior of the young is all but impossible. For example, in a particularly careful study, Caro (1994) described a changing pattern of mother cheetahs' (*Acinonyx jubatus*) release of prey in the presence of their cubs allowing them to "practice" killing both hare and young antelope. However, whether such behavior facilitated development of cubs' hunting skills could not be determined.

Second, in the laboratory, although experiments to measure effects of interaction with a mother on the subsequent feeding behavior of her offspring can be conducted easily, conditions of captivity may, as noted previously, dramatically increase the time young spent in contact with their mothers or decrease the frequency of contact of young with other adults (or both) and therefore may exaggerate maternal effects that would occur in free-living animals.

Third, changes in the behavior of young that appear to be adaptive in the protected environment of the laboratory might look quite different in the more complex and challenging natural environment. Clearly, there is much work to be done before the role of maternal effects in the development of adaptive feeding repertoires of free-living mammals is understood.

A Suggestion for the Future

In any mammalian species where juveniles and adults exploit different foods or use different feeding techniques, juvenile acquisition of the food preferences or foraging behaviors of conspecific adults could prove maladaptive. Young should be resistant to acquiring the feeding behaviors of their adult caregivers. Even laboratory evidence that maternal effects seen in species where mothers and young exploit similar resources are not found in species where mothers and young exploit different resources would provide valuable support for hypotheses concerning the adaptive value of maternal influences on the development of feeding behaviors.

ACKNOWLEDGMENTS

Preparation of this chapter was facilitated by funds from the Natural Sciences and Engineering Research Council of Canada.

REFERENCES

Aisner, R. & Terkel, J. 1992. Ontogeny of pine-cone opening behaviour in the black rat (*Rattus rattus*). *Animal Behaviour*, 44, 327–336.

Baerends-van Roon, J. M. & Baerends, G. P. 1979. *The Morphogenesis of the Behaviour of the Domestic Cat with Special Emphasis on the Development of Prey Catching*. Amsterdam: North Holland.

Beck, M. & Galef, B. G., Jr. 1989. Social influences on the selection of a protein-sufficient diet by Norway rats. *Journal of Comparative Psychology*, 103, 132–139.

Bilko, A. Altbacker, V. & Hudson, R. 1994. Transmission of food preferences in the rabbit: the means of information transfer. *Physiology & Behavior*, 56, 907–912.

Birch, L. L. 1980. The relationship between children's food preferences and those of their parents. *Journal of Nutrition Education*, 12, 14–18.

Boinski, S. & Fragaszy, D. M. 1989. The ontogeny of foraging in squirrel monkeys, *Saimiri oerstedi*. *Animal Behaviour*, 37, 415–428.

Boesch, C. 1991. Teaching among chimpanzees. *Animal Behaviour*, 41, 530–532.

Boesch, C. & Boesch, H. 1990. Tool use and making in wild chimpanzees. *Folia Primatologica*, 54, 86–99.

Booth, D. A. 1985. Food conditioned eating preferences and aversions with interoceptive elements: conditioned appetites and satieties. *Annals of the New York Academy of Sciences*, 443, 22–41.

Boyd, R. & Richerson, P. J. 1985. *Culture and the Evolutionary Process*. Princeton: Princeton University Press.

Brown, G. R., Almond, R. E. A. & van Bergen, Y. 2004. Begging, stealing and offering: functional perspectives on food transfer in non-human primates. *Advances in the Study of Behavior*, 34, 265–295.

Campbell, R. G. 1976. A note on the use of food flavor to stimulate the food intake of weaner pigs. *Animal Production*, 23, 417–419.

Caro, T. M. 1980a. Predatory behaviour in domestic cat mothers. *Behaviour*, 74, 128–147.

Caro, T. M. 1980b. Effects of the mother, object play and adult experience on predation in cats. *Behavioral and Neural Biology*, 29, 29–51.

Caro, T. M. 1994. *Cheetahs of the Serengeti Plains*. Chicago: University of Chicago Press.

Chou, L.-S., Marsh, R. E. & Richerson, P. 2002. Constraints on social transmission of food selection by roof rats, *Rattus rattus*. *Acta Zoologica Taiwanica*, 11, 95–109.

Choleris, E., Valsecchi, P., Wang, Y., Ferrari, P. Kavaliers. M. & Mainardi, M. 1998. Social learning of food preference in male and female Mongolian gerbils is facilitated by anxiolytic chlordiazepoxide. *Pharmacology, Biochemistry and Behavior*, 60, 575–584.

Coussi-Korbel, S. & Fragaszy, D. M. 1995. On the relation between social dynamics and social learning. *Animal Behaviour*, 50, 1441–1453.

DiBattista, D. 2002. Preference for novel flavors in adult golden hamsters (*Mesocricetus auratus*). *Journal of Comparative Psychology*, 116, 63–92.

Distel, R. A. & Provenza, F. D. 1991. Experience early in life affects voluntary intake of blackbrush by goats. *Journal of Chemical Ecology*, 16, 431–450.

Dominguez, H. D., Lopez, M. F. & Molina, J. C. 1998. Neonatal responsiveness to alcohol odor and infant alcohol intake as a function of alcohol experience during late gestation. *Alcohol*, 16, 109–117.

Ewer, R. F. 1968. *The Ethology of Mammals*. London: Logos Press.

Ewer, R. F. 1969. The "instinct to teach." *Nature*, 222, 698.

Flandera, V. & Novakova, V. 1975. Effect of the mother on the development of aggressive behavior in rats. *Developmental Psychobiology*, 8, 49–54.

Fragaszy, D. M. Fuerstein, J. M. & Mitra, D. 1997. Transfer of foods from adults to infants in tufted capuchins (*Cebus apella*). *Journal of Comparative Psychology*, 111, 194–200.

Galef, B. G., Jr. 1979. Investigations of the functions of coprophagy in juvenile rats. *Journal of Comparative and Physiological Psychology*, 93, 295–305.

Galef, B. G., Jr. 1981. The ecology of weaning: parasitism and the achievement of independence by altricial mammals. In: *Parental Care in Mammals* (Ed. by D. J. Gubernick & P. H. Klopfer), pp. 211–241. New York: Plenum Press.

Galef, B. G., Jr. 1986. Social interaction modifies learned aversions, sodium appetite, and both palatability and handling-time induced dietary preference in rats (*R. norvegicus*). *Journal of Comparative Psychology*, 100, 432–439.

Galef, B. G., Jr. 1989. Enduring social enhancement of rats' preferences for the palatable and the piquant. *Appetite*, 13, 81–92.

Galef, B. G., Jr. 1991. A contrarian view of the wisdom of the body as it relates to food selection, *Psychological Review*, 98, 218–224.

Galef, B. G., Jr. 1996a. Food selection: problems in understanding how we choose foods to eat. *Neuroscience and Biobehavioral Reviews*, 20, 67–73.

Galef, B. G, Jr. 1996b. Traditions in animals: field observations and laboratory analyses. In: *Readings in Animal Cognition* (Ed. by M. Bekoff & D. Jamieson), pp. 91–106. Cambridge: MIT Press.

Galef, B. G., Jr. & Beck, M. 1990. Diet selection and poison avoidance by mammals individually and in social groups. In: *Handbook of Neurobiology*. Vol. 10. *Neurobiology of Food and Fluid Intake* (Ed. by E. M. Stricker), pp. 329–349. New York: Plenum Press.

Galef, B. G., Jr. & Clark, M. M. 1971a. Parent-offspring interactions determine time and place of first ingestion of solid food by wild rat pups. *Psychonomic Science*, 25, 15–16.

Galef, B. G., Jr. & Clark, M. M. 1971b. Social factors in the poison avoidance and feeding behavior of wild and domesticated rat pups. *Journal of Comparative and Physiological Psychology*, 75, 341–357.

Galef, B. G., Jr. & Henderson, P. W. 1972. Mother's milk: a determinant of the feeding preferences of weaning rat pups. *Journal of Comparative and Physiological Psychology*, 78, 213–219.

Galef, B. G., Jr. & Sherry, D. F. 1973. Mother's milk: a medium for transmission of cues reflecting the flavor of mother's diet. *Journal of Comparative and Physiological Psychology*, 83, 374–378.

Galef, B. G., Jr. & Whiskin, E. E. 2003a. Preference for novel flavors in adult Norway rats (*Rattus norvegicus*). *Journal of Comparative Psychology*, 117, 96–100.

Galef, B. G., Jr. & Whiskin, E. E. 2003b. Socially transmitted food preferences can be used to study long-term memory in rats. *Learning & Behavior*, 31, 160–164.

Galef, B. G., Jr. & Whiskin, E. E. 2005. Differences between golden hamsters and Norway rats in preference for the sole diet that they are eating. *Journal of Comparative Psychology*, 119, 8–13.

Galef, B. G., Jr. & Wigmore, S. W. 1983. Transfer of information concerning distant foods: a laboratory investigation of the information-centre hypothesis. *Animal Behaviour*, 31, 748–758.

Galef, B. G., Jr. & Stein, M. 1985. Demonstrator influence on observer diet preference: analyses of critical social interactions and olfactory signals. *Animal Learning & Behavior*, 3, 31–38.

Galef, B. G., Jr., Marczinski, C. A., Murray, K. A. & Whiskin, E. E. 2001. Studies of food stealing by young Norway rats. *Journal of Comparative Psychology*, 115, 16–21.

Galef, B. G., Jr., Whiskin, E. E. & Dewar, G. 2004. A new way to study teaching in animals: despite demonstrable benefits, rat dams do not teach their young what to eat. *Animal Behaviour*, 70, 91–96.

Garcia, J. & Hankins, W. G. 1975. The evolution of bitter and the acquisition of toxiphobia. In: *Olfaction and Taste*. Vol. 5 (Ed. by D. A. Denton & J. P. Coghan), pp. 39–46. New York: Academic Press.

Garcia, J. & Koelling, R. A. 1966. The relationship of cue to consequence in avoidance learning. *Psychonomic Science*. 5, 123–124.

Goodall, J. 1986. *The Chimpanzees of Gombe: Patterns of Behavior*. Cambridge: Belknap Press.

Green, G. C., Elwin, R. L., Mottershead, B. E., Keogh, R. G. & Lynch, J. J. 1984. Long-term effects of early experience to supplementary feeding in sheep. *Proceedings of the Australian Society of Animal Production*, 15, 373–375.

Hepper, P. G. 1988. Adaptive fetal learning: prenatal exposure to garlic affects postnatal preference. *Animal Behaviour*, 36, 935–936.

Hiraiwa-Hasegawa, M.{12} 1990. Role of food sharing between mother and infant in the ontogeny of feeding behavior. In: *The Chimpanzees of the Mahale Mountains: Sexual and Life History Strategies*. (Ed. by T. Nishida), pp. 277–283. Tokyo: University of Tokyo Press.

Honey, P. L. & Galef, B. G., Jr. 2003. Ethanol consumption by rat dams during gestation, lactation and weaning enhances voluntary ethanol consumption by their adolescent young. *Developmental Psychobiology*, 42, 252–260.

King, B. J. 1994. *The Information Continuum*. Sante Fe: SAR Press.

Krutzen, M., Mann, J. Heithaus, M. R., Connor, R. C., Bejder, L.& Sherwin, W. B. 2005. Cultural transmission of tool use in bottlenose dolphins. *Proceedings of the National Academy of Sciences USA*, 102, 8939–8943.

Laland, K. 2004. Social learning strategies. *Learning & Behavior*, 32, 4–14.

Lonsdorf, E. V. 2005. Sex differences in the development of termite-fishing skills in wild chimpanzees, *Pan troglodytes schweinfurthii*, of Gombe National Park, Tanzania. *Animal Behaviour*, 70, 673–683.

Lonsdorf, E. V., Eberly, L. E. & Pusey, A. E. 2004. Sex differences in learning in chimpanzees. *Nature*, 428, 715–716.

Lupfer, G., Frieman, J. & Coonfield, D. 2003. Social transmission of flavor preferences in two species of hamster (*Mesocricetus auratus* and *Phodopus campbelli*). *Journal of Comparative Psychology*, 117, 449–455.

Lynch, J. J., Keogh, R. G., Elwin, R. L., Green, G. C. & Mottershead, B. E. 1983. Effects of early experience on the post weaning acceptance of whole grain wheat by fine-wool Merino lambs. *Animal Production*, 36, 175–183.

Mainardi, M., Poli, M. & Valsecchi, P. 1989. Ontogeny of dietary selection in weaning mice: effects of early experience and mother's milk. *Biology of Behaviour*, 14, 185–194.

Mann, J. & Sargeant, B. L. 2003. Like mother like calf: the ontogeny of foraging traditions in wild Indian Ocean bottlenosed dolphins (*Tursiops* sp.). In: *The Biology of Traditions: Models and Evidence* (Ed by D. M. Fragaszy & S. Perry), pp. 236–266. Cambridge: Cambridge University Press.

Mann, J., Sargeant, B. L., Watson-Capps, J. J., Connor, R. C. & Heithaus, M. R. in press. Tool-use in wild bottlenose dolphins: function, fitness and development. *Behavioral Ecology and Sociobiology*.

McFadyen-Ketchum, S. A. & Porter, R. H. 1989. Transmission of food preferences in spiny mice (*Acomys cahirinus*) via mouth-nose interaction between mothers and weanlings. *Behavioral Ecology and Sociobiology*, 24, 59–62.

McGrew, W. C. 1992. *Chimpanzee Material Culture*. Cambridge. Cambridge University Press.

Mennella, J. A. & Beauchamp, G. K. 1999. Mother's milk modifies the infant's acceptance of flavored cereal. *Developmental Psychobiology*, 35, 197–203.

Mennella, J. A., Jagnow, C. P. & Beauchamp, G. K. 2001. Prenatal and postnatal flavor learning by human infants. *Pediatrics*, 107, 88–98.

Morrill, J. L. & Dayton, A. D. 1978. Effect of feed flavor in milk and calf starter on feed consumption and growth. *Journal of Dairy Science*, 61, 229–232.

Mueller, K. L., Hoon, M. A., Erlenbach, I., Chandrashekar, J., Zuker, C. S. & Ryba, N. J. 2005. The receptor and coding logic for bitter taste. *Nature*, 434, 225–229.

Nolte, D. L. & Provenza, F. D. 1992. Food preferences in lambs after exposure to flavors in milk. *Applied Animal Behavior Science*, 32, 381–389.

Nolte, D. L., Provenza, F. D. & Balph, D. F. 1990. The establishment and persistence of food preferences in lambs exposed to selected foods. *Journal of Animal Science*, 68, 998–1002.

Pliner, P. & Pelchat, M. L. 1986. Similarities in food preferences between children and their siblings and parents. *Appetite*, 7, 333–342.

Price, E. C. & Feistner, A. T. C. 1993. Food sharing in lion tamarins: tests of three hypotheses. *American Journal of Primatology*, 31, 211–221.

Provenza, F. 1994. Ontogeny and social transmission of food selection in domesticated ruminants. In: *Behavioral Aspects of Feeding* (Ed. by B. G. Galef, Jr., M. Mainardi & P. Valsecchi), pp. 147–164. Chur: Harwood Academic.

Ratcliffe, J. M. & ter Hofstede, H. M. 2005. Roosts as information centres: social learning of food preferences in bats. *Biology Letters*, 1, 72–74.

Ratcliffe, J. M., Fenton, M. B. & Galef, B. G., Jr. 2003. An exception to the rule: vampire bats do not learn flavour-aversions. *Animal Behaviour*, 65, 385–389.

Reed, D. R., Tanake, T. & McDaniel, A. H. 2006. Diverse tastes: genetics of sweet and bitter perception. *Physiology & Behavior*, 88, 215–226.

Rozin, P. & Schulkin, J. 1990. Food selection. In: *Handbook of Neurobiology*. Vol. 10. *Neurobiology of Food and Fluid Intake* (Ed. by E. M. Stricker), pp. 297–328. New York: Plenum Press.

Rozin, P., Fallon, A. E. & Mandell, R. 1984. Family resemblances in attitudes towards foods. *Developmental Psychology*, 20, 309–314.

Schaal, B., Marlier, L. & Soussignan, R. 2000. Human fetuses learn odours from their pregnant mother's diet. *Chemical Senses*, 25, 729–737.

Schaal, B. Orgeur, P. & Arnould, C. 1995. Olfactory preferences in newborn lambs: possible influence of prenatal experience. *Behaviour*, 132, 351–365.

Sclafani, A. 1995. How food preferences are learned: laboratory animal models. *Proceedings of the Nutrition Society*, 54, 419–427.

Scott, T. R. 1990. Gustatory control of food selection. In: *Handbook of Neurobiology*. Vol. 10. *Neurobiology of Food and Fluid Intake*. (Ed. by E. M. Stricker), pp. 243–264. New York: Plenum Press.

Silk, J. 1979. Feeding, foraging and food sharing behavior of immature chimpanzees. *Folia Primatologica*, 31, 123–142.

Smotherman, W. P. 1982. In utero chemosensory experience alters taste preferences and corticosterone responsiveness. *Behavioral and Neural Biology*, 36, 61–68.

Solomon, N. G., Yaeger, C. S. & Beeler, L. A. 2002. Social transmission and memory of food preferences in pine voles (*Microtus pinetorum*). *Journal of Comparative Psychology*, 116, 35–38.

Spear, N. E. & Molina, J. C. 2005. Fetal or infantile exposure to ethanol promotes ethanol ingestion in adolescence and adulthood: a theoretical review. *Alcoholism: Clinical & Experimental Research*, 29, 906–929.

Steiniger, von F. 1950. Beiträge zur Soziologie und sonstigen Biologie der Wanderratte. *Zeitschrift für Tierpsychologie*, 7, 356–370.

Strupp, B. J. & Levitsky, D. E. 1984. Social transmission of food preferences in adult hooded rats (*Rattus norvegicus*). *Journal of Comparative Psychology*, 98, 384–389.

Terkel, J. 1996. Cultural transmission of feeding behavior in the black rat (*Rattus rattus*). In: *Social Learning in Animals: The Roots of Culture* (Ed. by C. M. Heyes & B. G. Galef, Jr.), pp. 17–48. San Diego: Academic Press.

Thornton, A. & McAuliffe, K. 2006. Teaching in wild meerkats. *Nature*, 313, 227–229.

Trivers, R. L. 1974. Parent-offspring conflict. *American Zoologist*, 14, 249–264.

Valsecchi, P., Moles, A. & Mainardi, M. 1993. Does mother's diet affect food selection of weanling wild mice? *Animal Behaviour*, 46, 827–828.

Wells, D. L. & Hepper, P. G. 2006. Prenatal olfactory learning in the domestic dog. *Animal Behaviour*, 72, 681–686.

Whitehead, J. M. 1986. Development of feeding selectivity in mantled howler monkeys (*Alouatta palliata*). In: *Primate Ontogeny, Cognition and Social Behavior* (Ed by J. G. Else & P. C. Lee), pp. 105–117. Cambridge: Cambridge University Press.

Whiten, A., Goodall, J., McGrew, W. C., Nishida, T., Reynolds, V., Sugiyama, Y., Tutin, C. E. G., Wrangham, R. W., Boesch, C. 2001. Charting cultural variation in chimpanzees. *Behaviour*, 138, 1481–1516.

Wyrwicka, W. 1974. Eating banana in cats for brain stimulation reward. *Physiology & Behavior*, 12, 1063–1066.

Wyrwicka, W. 1978. Imitation of mother's inappropriate food preference in weanling kittens. *Pavlovian Journal of Biological Science*, 13, 55–72.

Wyrwicka, W. 1981. *The Development of Food Preferences: Parental Influences and the Primacy Effect*. Springfield: Charles C. Thomas.

Wyrwicka, W. & Long, A. M. 1980. Observations on the initiation of eating new food by weanling kittens. *Pavlovian Journal of Biological Science*, 15, 115–122.

Yoerg, S. I. 1991. Social feeding reverses flavor aversions in spotted hyena (*Crocuta crocuta*). *Journal of Comparative Psychology*, 105, 185–189.

Zohar, O. & Terkel, J. 1992. Acquisition of pinecone stripping behavior in black rats (*Rattus rattus*). *International Journal of Comparative Psychology*, 5, 1–6.

9

The Trans-Generational Influence of Maternal Care on Offspring Gene Expression and Behavior in Rodents

FRANCES A. CHAMPAGNE AND JAMES P. CURLEY

INTRODUCTION

In mammals, mother-infant interactions promote infant growth and survival within the immediate environment but can also mediate long-term changes in gene expression and behavior that persist into adulthood. Maternal effects have been studied extensively in rodents and illustrate the profound impact of disruptions or variations in maternal care on offspring development. Early life experiences have been found to alter numerous neuroendocrine systems associated with stress responsivity, cognition, and social and reproductive behavior. These maternal effects involve epigenetic modifications of DNA, which are associated with the quality of mother-infant interactions occurring during the postnatal period. Moreover, maternal effects can shift patterns of offspring maternal care resulting in a transmission of the consequences of mother-infant interactions across generations. This nongenomic transmission of variations in gene expression and behavior is an example of a Lamarckian-type inheritance pattern in which experiences occurring in one generation can alter the development of subsequent generations. This chapter will review evidence from laboratory studies with rodents that demonstrate the mediating role of maternal care in shaping gene expression and behavior of offspring and explore the potential mechanisms through which this influence is mediated as well as the consequences of mother-infant interactions for future generations.

MATERNAL CARE IN RODENTS

The interaction between mother and infant occurs over an extended period in mammals and is characterized initially by the transfer of resources across the placental barrier. During the prenatal period the energetic demands of fetal development are reliant on maternal responsiveness to endocrine signals which increase maternal circulation and food intake (Weissgerber & Wolfe 2006). Throughout gestation, increasing levels of estrogen and prolactin prepare maternal physiology for parturition and lactation and prime the maternal brain resulting in the emergence of behavioral patterns that will promote care of offspring during the postpartum period (Rosenblatt 1975; Bridges 1994; Bridges et al. 1996). In the laboratory, lactating rats and mice display typical patterns of postpartum behavior including nest building, retrieval, nursing, licking and grooming of pups, and maternal defense of the nest (Rosenblatt 1967; 1975; Fleming & Rosenblatt 1974; Gammie 2005; Shoji & Kato 2006). Although the overall frequency of these behaviors varies between strains and between individuals, it is clear that these aspects of care are essential to the survival of these altricial offspring. The retrieval and grouping of pups into the nest is necessary for thermoregulation, allowing the female to crouch over the litter and provide pups with ventral heat (Croskerry et al. 1978; Leon et al. 1978), and this positioning allows pups to gain access to the nipples to permit suckling (Stern & Johnson 1990). Licking and grooming, particularly of the anogenital region of pups, are necessary for stimulating urination and defecation (Rosenblatt & Lehrman 1963), increase the motor activity of the pups, which enhances their ability to attach to the nipples, and also serve to regulate brain and body temperature (Sullivan et al. 1988a, b). For the mother, licking and grooming pups provide a mechanism to reclaim salt and water that have been lost through lactation (Gubernick & Alberts 1983; 1985). Social behavior of lactating females is also altered, resulting in aggressive behavior towards intruders and defense of the litter (Lonstein & Gammie 2002; Gammie 2005). Over the course of the preweaning period there are dramatic reductions in the frequency of time mothers spend in contact and licking/grooming (LG) pups corresponding to increased growth and maturation of pups (Gubernick & Alberts 1983; Champagne et al. 2003a). Overall, the behavior of the mother provides care needed by the pups to survive and allows the mother to meet the physiological demands of prolonged care of young.

MATERNAL DEPRIVATION AND OFFSPRING DEVELOPMENT

Beyond the basic needs of survival, it is evident that some aspects of maternal care are necessary for offspring to thrive within and beyond the postnatal period. This has been demonstrated in studies examining the long-term impact of complete maternal deprivation. Much like infant rhesus macaques in Harlow's artificial rearing paradigm (Seay & Harlow 1965; Harlow & Suomi 1971; 1974), rodents can be completely deprived of maternal contact during the postnatal period. Although pups reared under these conditions are provided with adequate levels of nutrients and warmth, and are stimulated with a paintbrush to promote urination and defecation, as juveniles and adults these offspring display behavioral and neuroendocrine abnormalities (Gonzalez et al. 2001; Lovic et al. 2001; Gonzalez & Fleming 2002; Lovic & Fleming 2004). They are more fearful, engaging in fewer open-arm entries in an elevated plus maze, display elevated locomotor activity and cognitive impairments related to attentional shifting, and are impaired on measures of social behavior, including maternal care (Gonzalez et al. 2001; Lovic et al. 2001; Gonzalez & Fleming 2002). Females raised under these conditions display deficits in maternal LG and contact toward their own pups (Fleming et al. 2002) and may be less responsive to hormonal priming of maternal behavior (Novakov & Fleming 2005). Although peer rearing (in which pups are raised with one peer) can ameliorate the deficits normally observed, the pronounced effect of being maternally deprived is still evident (Melo et al. 2006).

Although complete maternal deprivation is a highly artificial procedure, prolonged periods of separation between mother and infant may not be uncommon in many mammalian species. A methodology that has been used extensively in this regard involves physical separation of mother and pups for varying periods of time. The maternal separation paradigm, involving hours of daily mother-infant separation, was initially used to manipulate the responsivity of the hypothalamic-pituitary-adrenal (HPA) axis (Rosenfeld et al. 1992; Plotsky & Meaney 1993; Plotsky et al. 2005). These rearing conditions lead to behavioral and neurobiological changes in the offspring including decreased exploration, behavioral inhibition, increased CRH mRNA in the brain paraventricular nucleus (PVN), increased corticosterone response to stress, and decreased levels of hippocampal glucocorticoid receptor (GR) mRNA (Plotsky & Meaney 1993; Meaney et al. 1996; Lehmann et al. 1999). The short-term stress response to separation from the mother, indicated by increased corticosterone release and ultrasonic vocalizations produced by pups, is accompanied by long-term changes in stress responsivity (Hofer et al. 1993; Lehmann et al. 2002b). Neurocognitive function is also modified by early experience, as indicated by increased latencies on the Morris water maze, decreased hippocampal synaptophysin levels, and increased apoptosis (Lehmann et al. 2002a). There is also growing evidence that maternal separation reduces social behavior in both males and females. For example, female rats separated from their mothers for 5 hours per day during the preweaning period showed deficits in maternal LG toward their own offspring later in life (Fleming et al. 2002).

In these experimental paradigms it is difficult to pinpoint the particular aspect of maternal separation that is critical for inducing changes in offspring development. Certainly the lengthy duration of separation is an important factor. Brief periods of separation, in which pups are removed from the home cage for 10–20 minutes, are referred to as neonatal handling (Levine 1957). Neonatal handing typically has the opposite effect of longer periods of maternal separation, resulting in an attenuated response to stress, increased hippocampal GR mRNA, improved cognitive ability, and increased frequency of social behavior (Meaney et al. 1985; 1989; 1991; Lehmann et al. 2002a). These short-duration separations are thought to increase the responsivity of females and stimulate maternal care when reintroduced to pups (Lee & Williams 1974; Liu et al. 1997) whereas longer periods of separation are associated with increased pup corticosterone (Lehmann et al. 2002b) and reduced maternal care following separation and reunion (Boccia et al. 2006). However, if pups are treated with anxiolytic drugs during the period of separation, the long-term detrimental consequences of long separations for offspring development are ameliorated. For example, pups treated with oxytocin or allopregnalone exhibit reduced levels of separation-induced ultrasonic vocalizations and an attenuated stress response and as adults do not exhibit the behavioral and physiological stress reactivity associated with early maternal separation (Zimmerberg et al. 1994; Pedersen & Boccia 2002). These studies suggest that the quality of the mother-infant interaction is altered by long periods of maternal separation, and this change in the quality of care may modulate offspring phenotype.

Maternal separation and handling manipulations provide useful methodologies for studying the impact of early experiences on development. However, more ecologically relevant techniques have been developed to study these effects in which a variable foraging demand is used to alter the frequency and quality of mother-infant contact. Through varying the accessibility of food, the foraging effort of mothers can be adjusted to be high (high foraging demand, or HFD, when food availability is consistently low), low (low foraging demand, or LFD, when food availability is consistently high) or unpredictable (variable foraging demand, or VFD, when food availability alternates randomly between high and low) during the early postnatal period (Rosenblum & Paully 1984). Following these manipulations, mothers are expected to adjust their caregiving budget accordingly, providing offspring with differential care (but see Maestripieri & Wallen 2003 for a critique of this paradigm). This methodology was originally developed for use in bonnet ma-caques (*Macaca radiata*) by Rosenblum and colleagues (Rosenblum & Paully 1984), but has since been adapted for use in rodents (Bredy et al. 2007; Macri & Wurbel 2007). Early studies in bonnet macaques revealed that it is those infant monkeys reared by VFD mothers that are particularly affected; these offspring are less affiliative, autonomous, and exploratory than those reared under the other foraging conditions (Rosenblum & Paully 1984). These in-fants also cling to mothers excessively, are more socially timid and subordi-nate, and are excessively fearful when separated from the mother (Andrews & Rosenblum 1994). These behavioral changes are related to altered levels of a number of hormones and neurotransmitters, including corticotropin releasing factor, cortisol, dopamine, serotonin, and growth hormones, in the CSF of infants reared by VFD mothers (Coplan et al. 2000; 2001; 2006; but see Maestripieri & Wallen 2003). In rats, mothers that are required to leave the home cage to forage for food in a separate cage (HFD) are away from their offspring for longer periods than mothers that do not have to leave the home cage to forage (LFD), but display a more active and efficient nursing style (Macri & Wurbel 2007). Offspring reared by these HFD mothers have lower HPA activity and are less fearful than offspring born to LFD mothers. Congruent with primate studies, rat offspring reared by VFD mothers were found to be more fearful and have higher HPA activity than offspring reared by either the HFD or LFD mothers (Bredy et al. 2007; Macri & Wurbel 2007). Thus, it is not the level of demand but rather the variability of the demand that can profoundly alter mother-infant interactions.

DEVELOPMENTAL IMPACT OF NATURAL VARIATIONS IN MATERNAL CARE

Although studies investigating the effects of maternal deprivation on off-spring development are suggestive of maternal effects, these studies do not necessarily point to a particular aspect of the mother-infant interaction that is critical for determining the long-term consequences of these manipu-lations (Macri & Wurbel 2007). More direct evidence for the role of post-

partum behavior in shaping patterns of offspring behavioral and physiological development comes from detailed characterization of naturally occurring
variations in mother-infant interactions. During the first week postpartum,
lactating female rats display high levels of nursing contact with pups accompanied by bouts of licking and grooming. The frequency of these behaviors
varies both within and between strains and has been implicated in shaping offspring phenotype. Myers et al. (1989b) reported that spontaneously
hypertensive (SHR) and Wistar Kyoto (WKY) rats exhibited differences in
maternal behavior during the postpartum period. Furthermore, they demonstrated that strain differences in adult blood pressure between offspring
of SHR and WKY rats were related to differences in maternal LG, retrieval of
pups, and nursing posture exhibited by these two strains. Strain differences
in maternal care of mice have also been implicated as a possible source of
strain differences in behavior and physiology (Francis et al. 2003; Priebe
et al. 2005). BALB/c mouse females display lower levels of postpartum LG
compared to C57BL/6 mice, and this may explain why BALB/c mice have an
elevated physiological and behavioral response to stress and perform poorly
on learning and memory tasks compared to mice of the other strain (Francis
et al. 2003).

The role of individual differences in maternal LG in modulating offspring
gene expression, physiology, and behavior has been explored extensively in
Long-Evans rats (Meaney 2001). In a large cohort of Long-Evans rat females,
the frequency of LG behavior observed during the first week postpartum
is normally distributed (Figure 9-1). Females can be characterized as being
high or low in LG depending on the frequency of LG behavior exhibited,
and this characterization is highly stable within each female (Champagne et
al. 2003a). Thus, comparisons can be made between offspring of high- and
low-LG mothers to determine the effects of this particular aspect of postpartum maternal care on offspring development. Early studies demonstrated
an association between levels of LG and stress responsivity, with the adult
male offspring of high-LG females being more exploratory in a novel environment, having reduced plasma adrenocorticotropin and corticosterone
in response to stress, elevated hippocampal glucocorticoid receptor mRNA,
decreased hypothalamic CRH mRNA, and increased density of benzodiazepine receptors in the amygdala compared to the offspring of low-LG mothers (Liu et al. 1997; 2000a; Francis et al. 1999; Caldji et al. 2000). Offspring
of high-LG mothers also exhibit enhanced performance on tests of spatial
learning and memory, elevated hippocampal brain derived neurotrophic factor (BDNF) mRNA, and increased hippocampal choline acetyltransferase

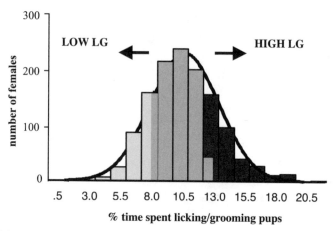

Figure 9-1. Normal distribution of licking/grooming (LG) behavior in a large cohort (*n* = 400) of lactating females. Individual differences in LG are highly stable and females can be characterized as being low (1 standard deviation below the cohort mean) or high (1 standard deviation above the cohort mean).

and synaptophysin (Liu et al. 2000b). GABA subunit expression is altered by maternal LG with implications for benzodiazepine binding (Caldji et al. 2003). Neuronal survival in the hippocampus is increased and apoptosis is decreased among the offspring of low-LG mothers, in association with elevated levels of fibroblast growth factor (Weaver et al. 2002; Bredy et al. 2003). Dopaminergic release associated with stress responsivity in males and reward in females is also altered as a function of LG (Champagne et al. 2004; Zhang et al. 2005).

Although these studies are strongly suggestive of the role of maternal care in shaping offspring development, it is critical in these paradigms to establish the environmental mediation of these effects. In the case of individual differences in LG, this was accomplished with cross-fostering between high- and low-LG mothers (Figure 9-2) in which pups were removed from the genetic mother at birth and fostered to a mother that had been previously characterized as having a high- or low-LG phenotype (Champagne et al. 2003a). Cross-fostering studies have demonstrated that the differences in the offspring behavior and physiology previously described are related to the level of postpartum care received rather than to genetic or prenatal factors (Francis et al. 1999; Champagne et al. 2003a). Thus, the offspring of high-LG mothers cross-fostered to low LG mothers are indistinguishable in phenotype from the genetic offspring of non-cross-fostered low-LG mothers. Conversely, the offspring of low-LG mothers when reared by high-LG moth-

biological mother   adoptive mother

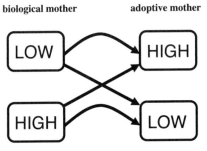

Figure 9-2. Design of a cross-fostering study. Offspring born to low-LG females are taken at birth and fostered to a high-LG female, whereas offspring born to a high-LG female are taken at birth and fostered to a low-LG female. Within-group fostering of high-LG to high-LG and low-LG to low-LG can be used to control for the effects of the cross-fostering manipulation.

ers resemble the genetic offspring of non-cross-fostered high-LG mothers on measures of both gene expression and behavior. This is consistent with earlier studies in which cross-fostering offspring between WKY and SHR strains resulted in a shift in offspring phenotype such that pups cross-fostered from a WKY mother to a SHR mother exhibited a phenotype similar to that of a genetic offspring of a non-cross-fostered SHR mother (Myers et al. 1989a). In mice, cross-fostering has also demonstrated the importance of the prenatal environment in shaping offspring phenotype. C57BL/6 embryos that are transferred prenatally to BALB/c mothers and reared by a BALB/c female develop characteristics similar to BALB/c mice including decreased exploration of a novel environment indicating increased anxiety (Francis et al. 2003). This shift to a BALB/c phenotype does not occur if embryos are exposed to a C57BL/6 prenatal environment or cross-fostered at birth to a C57BL/6 female. Thus, in some species it may be important to consider mother-infant interactions during both the gestational and postnatal periods when exploring the origins of variation in offspring phenotypic development.

MECHANISMS OF MATERNAL INFLUENCE

Maternal care provided during the first week postpartum clearly has long-term consequences for offspring; however, the question that emerges pertains to the mechanism through which these maternal effects are sustained. This question has led to the exploration of epigenetic regulation of gene expression in the offspring of high- and low-LG mothers. In the cell nucleus, DNA is wrapped around a complex of histone proteins, and it is clusters of these DNA/histone complexes that form chromatin (Turner 2001). However, to be expressed, DNA must come into contact with RNA polymerase and transcription factors. Thus, gene expression can occur only when DNA is in an active state where it is unwrapped from the histone proteins and the nucleic acids sequences are exposed (Russo et al. 1996; Turner 2001). Our knowledge of these processes is advancing rapidly and hence the term "epigenetic" has come to refer to the changes in chromatin and DNA structure that alter gene expression and hence phenotype, without involving changes to the sequence of DNA.

The molecular mechanisms involved in the epigenetic regulation of the genome are numerous and complex, yet one particular mechanism produces stable changes in gene expression and thus may be essential to understanding the maternal effects described in rodents. Within the DNA sequence, there are specific sites where a methyl group can attach to cytosine through an enzymatic reaction (Razin 1998; Turner 2001). The sites where this can occur reside primarily within the regulatory regions of a gene, in the promoter area upstream from the transcription start site. At a functional level, methylation prevents access of transcription factors and RNA polymerase to DNA, resulting in silencing of the gene. In addition to the gene silencing that occurs in the presence of DNA methylation, these methyl groups attract other protein complexes that promote histone deacetylation, further inhibiting the likelihood of gene expression (Strathdee & Brown 2002). The bond between the cytosine and methyl group is very strong, resulting in a stable yet potentially reversible change in gene expression. DNA methylation patterns are maintained after cell division and thus passed from parent to daughter cells; it is through this form of epigenetic modification that cellular differentiation occurs. Though several examples of environmentally induced changes in DNA methylation have been demonstrated (Jaenisch & Bird 2003; Anway et al. 2005; Waterland 2006), the question of whether the changes in gene expression associated with postnatal mother-infant interactions result from these epigenetic mechanisms has only recently been addressed.

The first investigation of epigenetic regulation of phenotypes associated with levels of maternal care explored the levels of hippocampal GR mRNA observed in the offspring of high- and low-LG mothers (Weaver et al. 2004a). The differential levels of expression of this receptor are thought to be critical to mediating the differences in stress responsivity between the offspring of high- and low-LG females due to the negative feedback relationship between hippocampal GR and the HPA axis (Sapolsky et al. 1985). Analysis of the level of DNA methylation within the GR 1_7 promoter region suggests that elevated levels of maternal LG are associated with decreased GR 1_7 methylation corresponding to the elevated levels of receptor expression observed in the hippocampus. Site-specific analysis of the methylation pattern in this region indicated that the NGF1-A (nerve growth factor) binding site is differentially methylated in the offspring of high- and low-LG mothers; subsequent analysis indicated that the binding of NGF1-A to this region is reduced in hippocampal tissue taken from the offspring of low-LG mothers. Thus, the differential methylation of the GR promoter prevents the binding of factors necessary for increased expression of the receptor. A temporal analysis of the methylation of the GR 1_7 promoter indicated that differences between the offspring of high- and low-LG mothers emerge during the postpartum period and are sustained at weaning and into adulthood (Weaver et al. 2004a). Thus, the differences in gene expression and behavior that are observed in the adult offspring associated with the quality of early maternal care may be mediated by the long-term silencing of gene expression achieved through differential methylation. Moreover, cross-fostering studies confirm that these effects are indeed mediated by the quality of the postnatal environment.

The mediating role of epigenetic modifications in the relationship between maternal care and offspring phenotype is supported by findings that the increased anxiety, elevated stress-induced corticosterone, and decreased hippocampal GR mRNA expression observed in the offspring of low-LG mothers can be altered by pharmacologically targeting the epigenome (Weaver et al. 2004a, b; 2006). Central administration of trichostatin-A, a histone deacetylase inhibitor that promotes demethylation, to the adult offspring of low-LG mothers reverses the effects of maternal care and produces a phenotype that is indistinguishable from that of the offspring of high-LG mothers (Weaver et al. 2004a). Conversely, central administration of methionine, a methyl donor that promotes methylation, to the adult offspring of high-LG mothers results in an increased anxiety, increased corticosterone response to stress, decreased GR mRNA, and decreased binding of NGF1-A to the hippocampal GR 1_7 promotor region (Weaver et al. 2005; 2006). Thus, epigenetic regulation of gene expression is critical for shaping offspring brain and behavior, and maternal care experienced in infancy is associated with the epigenetic status of genes in adulthood.

MATERNAL EFFECTS ACROSS GENERATIONS

In many mammalian species, the quality of the early environment can affect the development of social and reproductive behavior. In rodents, pups exposed to adverse maternal environments and impaired maternal behavior exhibit impaired maternal care toward their offspring. For example, reducing the normal exposure of female mouse pups to maternal interactions through early weaning is associated with lower levels of LG and nursing toward their own pups (Kikusui et al. 2005). Female rat pups that are either separated from their mothers for short repeated periods (Lovic et al. 2001) or experience complete maternal deprivation (Gonzalez et al. 2001) exhibit impaired maternal care, such as retrieving fewer pups when the nest is disrupted and exhibiting reduced LG and crouching behaviors. Natural variations in maternal care are also transmitted across generations. The offspring of high-LG rat mothers exhibit high levels of maternal LG toward their own offspring, whereas the offspring of low-LG mothers are themselves low in LG (Francis et al. 1999; Champagne & Meaney 2001; Gonzalez et al. 2001; Fleming et al. 2002; Champagne et al. 2003a). Moreover, cross-fostering female offspring between high- and low-LG mothers has confirmed the role of postnatal care in mediating this intergenerational transmission (Francis et al. 1999; Champagne et al. 2003a).

The mechanisms through which the maternal care of one generation can influence the maternal behavior of subsequent generations may involve sensitivity to hormones that normally serve to up-regulate gene expression and promote the physiological and behavioral aspects of maternal care. In rats, complete maternal deprivation in infancy is associated with reduced maternal behavior in response to estrogen and progesterone treatment (Novakov & Fleming 2005). Offspring of low-LG mothers exhibit decreased estrogen-mediated up-regulation of oxytocin receptor binding and c-fos immunoreactivity in hypothalamic regions implicated in maternal care, such as the medial preoptic area (MPOA; Champagne et al. 2001). This sensitivity may be mediated by lower levels of estrogen receptor α (ERα) mRNA in the MPOA that are found in offspring of low- than in those of high-LG mothers (Champagne et al. 2003b). Differential levels of ERα mRNA are observed in infancy and maintained into adulthood, suggesting a long-term suppression of gene expression in response to low levels of LG. Analysis of MPOA levels of DNA methylation within the ERα promoter indicated that low levels of maternal LG are associated with high levels of ERα methylation, whereas high levels of LG are associated with low levels of ERα methylation among female offspring (Champagne et al. 2006). Thus, LG is associated with epigenetic regulation of genes in female offspring that mediate maternal behavior, and this may be one of the mechanisms underlying the transmission of maternal care across generations.

Evidence for the nongenomic transmission of patterns of maternal behavior from mother to offspring is not limited to rodents (see Maestripieri, this volume). Abusive parenting styles of rhesus macaques (*Macaca mulatta*) modulate subsequent maternal behavior of offspring such that over 50% of offspring who had received abusive parenting during the first 6 months of life exhibit abusive parenting themselves as adults (Maestripieri 1998; 2005; Maestripieri & Carroll 1998). Infants cross-fostered from an abusive female to a nonabusive female are not abusive to their own offspring, suggesting the role of early experience and specific neuroendocrine mechanisms in mediating these effects (Maestripieri 2005; Maestripieri et al. 2006). Such an intergenerational transmission of abuse was suggested by long-term studies of studies of rhesus and pigtail macaque (*Macaca nemestrina*) captive populations, in which infant abuse is highly concentrated within certain matrilines and among closely related females (Maestripieri et al. 1997; Maestripieri & Carroll 1998). Among captive vervet monkeys (*Chlorocebus aethiops*), the best predictor of the frequency of mother-infant contact is the level of contact the adult female received from her mother during the first six months of life (Fairbanks 1989). In rhesus macaques, there are differences among matrilines in the frequency of certain maternal behaviors (Simpson & Howe 1986), and there is evidence that maternal rejection rates are transmitted across generations through nongenetic mechanisms (Berman 1990; Maestripieri et al. 2007; Maestripieri, this volume).

In humans, the trans-generational continuity of child abuse is striking. It is currently estimated that up to 70% of abusive parents were themselves abused, whereas 20–30% of abused infants are likely to become abusers (Egeland et al. 1987; Chapman & Scott 2001). Women reared in institutional settings without experiencing appropriate parental care display less sensitivity and are more confrontational towards their children (Dowdney et al. 1985). An intergenerational transmission of maternal care and overprotection as rated by the Parental Bonding Index (PBI) has also been shown between women and their daughters (Miller et al. 1997); this transmission of parental style has been determined to be independent of socioeconomic status, and maternal or daughter's temperament or depression. A mother's own attachment to her mother, as assessed by the retrospective Adult Attachment Interview, is a strong predictor of her infant's attachment, especially for secure and disorganized patterns of attachment (Main & Solomon 1990; Benoit & Parker 1994; van IJzendoorn 1995; Pederson et al. 1998). Sroufe and colleagues have also reported preliminary results from a prospective study suggesting evidence for the transmission of attachment classifications from mother to daughter and to granddaughter (Sroufe 2005; Sroufe et al. 2005)

TACTILE STIMULATION AND DEVELOPMENT

The behavioral transmission of variation in mother-infant interactions described in the previous section is apparent in species that differ considerably in the form of postnatal care provided. However, a common function of these behaviors may be to provide the infant with tactile stimulation. Evidence from maternal separation and artificial rearing studies in rodents suggests that tactile stimulation can have long-term consequences for brain development and behavior. The immediate effects of maternal separation include decreases in serum growth hormone and ornithine decarboxylase (ODC) in the brain and periphery (Kuhn et al. 1978). ODC activity is critical to normal growth and development (Slotkin & Bartolome 1986). However, if pups are stroked with a paintbrush during the period of maternal separation, levels of growth hormone and ODC return to normal (Evoniuk et al. 1979; Pauk et al. 1986). In artificial rearing paradigms, providing pups with high levels of tactile stimulation (stroking with a paintbrush) during the postnatal period improves maternal responsivity and lessens fear-related behaviors (Gonzalez et al. 2001). McCarthy et al. (1997) demonstrated that tactile stimulation of pups on postnatal day 3 resulted in a rapid induction of Fos immunoreactivity in the ventral MPOA. Tactile stimulation of neonates also increases serum lactate, a major source of energy for the metabolic needs of the developing brain (Alasmi et al. 1997), and increases Fos immunoreactivity in oxytocin neurons in the PVN (Caba et al. 2003). In humans, touch during the postnatal period results in increased weight gain and improved performance on development tasks by premature and low birth-weight babies (Solkoff & Matuszak 1975; Field et al. 1986; Kuhn et al. 1991). Extended contact during the postpartum period also increases maternal responsiveness to infants (Kennell et al. 1974). Secure infant attachment is thought to be dependent on physical contact between mother and infant (Ainsworth et al. 1978), and Main & Solomon (1990) reported that infants of mothers with insecure attachments showed an aversion to physical contact. Thus, there is converging evidence to suggest that tactile stimulation may be one important behavioral mechanism through which maternal effects on social and emotional responsiveness occur in mammals.

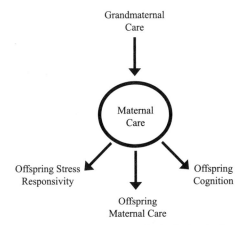

Figure 9-3. Transmission of maternal care across generations has implications for the transmission of maternal effects on stress responsivity and cognition of offspring.

CONCLUSIONS

The transmission of patterns of maternal care from one generation to the next has implications for the transmission of stress responsivity, cognition, and the neurobiological consequences for offspring of the maternal effects that have been described in the previous sections (Figure 9-3). This behavioral transmission of gene expression and behavior challenges our understanding of the concept of inheritance and suggests that maternal effects can be as pervasive as direct genetic effects in transmitting information across generations. Maternal care exhibits a high degree of plasticity in response to environmental conditions and thus the quality of the environment is conferred to offspring by the level of care received, or perhaps more specifically, by the level of tactile stimulation provided. This source of communication between mother and infant provides a simple yet elegant mechanism for shaping offspring development that is conserved across species. The adaptive significance of these maternal effects is a matter of speculation within the context of laboratory studies, but it is clear that the quality of maternal care provided within the postnatal environment can have stable effects on the neurobiological substrates of stress responsivity. Stressful environments, characterized by unpredictable food availability, predation, or reduced social contact may shift patterns of maternal care to induce increased stress responsivity in offspring. Thus mothers may "prepare" offspring for the environmental conditions in which they will reside. The involvement of epigenetic rather than genomic variation in these maternal effects allows for transmission of information about the environment from mothers to offspring and may result in enhanced adaptation to the environment in the offspring. However, as explained by Cheverud & Wolf (this volume) and Wade et al. (this volume), induced phenotypic changes that do not permanently modify the genome may attenuate the strength of selection. Therefore, the epigenetic effects reviewed in this chapter may influence evolutionary processes through multiple mechanisms.

REFERENCES

Ainsworth, M., Blehar, M., Waters, E. & Wall, S. 1978. *Patterns of Attachment: A Psychological Study of the Strange Situation.* Hillsdale, NJ: Erlbaum.

Alasmi, M. M., Pickens, W. L. & Hoath, S. B. 1997. Effect of tactile stimulation on serum lactate in the newborn rat. *Pediatric Research*, 41, 857–861.

Andrews, M. W. & Rosenblum, L. A. 1994. The development of affiliative and agonistic social patterns in differentially reared monkeys. *Child Development*, 65, 1398–1404.

Anway, M. D., Cupp, A. S., Uzumcu, M. & Skinner, M. K. 2005. Epigenetic transgenerational actions of endocrine disruptors and male fertility. *Science*, 308, 1466–1469.

Benoit, D. & Parker, K. C. 1994. Stability and transmission of attachment across three generations. *Child Development*, 65, 1444–1456.

Berman, C. M. 1990. Intergenerational transmission of maternal rejection rates among free-ranging rhesus monkeys on Cayo Santiago. *Animal Behaviour*, 39, 329–337.

Boccia, M. L., Razzoli, M., Prasad Vadlamudi, S., Trumbull, W., Caleffie, C. & Pedersen, C. A. 2006. Repeated long separations from pups produce depression-like behavior in rat mothers. *Psychoneuroendocrinology*, 2, 65–71.

Bredy, T. W., Grant, R. J., Champagne, D. L. & Meaney, M. J. 2003. Maternal care influences neuronal survival in the hippocampus of the rat. *European Journal of Neuroscience*, 18, 2903–2909.

Bredy, T. W., Brown, R. E. & Meaney, M. J. 2007. Effect of resource availability on biparental care, and offspring neural and behavioral development in the California mouse (*Peromyscus californicus*). *European Journal of Neuroscience*, 25, 567–575.

Bridges, R. S. 1994. The role of lactogenic hormones in maternal behavior in female rats. *Acta Paediatrica Supplement*, 397, 33–39.

Bridges, R. S., Robertson, M. C., Shiu, R. P., Friesen, H. G., Stuer, A. M. & Mann, P. E. 1996. Endocrine communication between conceptus and mother: placental lactogen stimulation of maternal behavior. *Neuroendocrinology*, 64, 57–64.

Caba, M., Rovirosa, M. J. & Silver, R. 2003. Suckling and genital stroking induces Fos expression in hypothalamic oxytocinergic neurons of rabbit pups. *Developmental Brain Research*, 143, 119–128.

Caldji, C., Diorio, J. & Meaney, M. J. 2000. Variations in maternal care in infancy regulate the development of stress reactivity. *Biological Psychiatry*, 48, 1164–1174.

Caldji, C., Diorio, J. & Meaney, M. J. 2003. Variations in maternal care alter GABA(A) receptor subunit expression in brain regions associated with fear. *Neuropsychopharmacology*, 28, 1950–1959.

Champagne, F. A. & Meaney, M. J. 2001. Like mother, like daughter: evidence for non-genomic transmission of parental behavior and stress responsivity. *Progress in Brain Research*, 133, 287–302.

Champagne, F. A., Diorio, J., Sharma, S. & Meaney, M. J. 2001. Naturally occurring variations in maternal behavior in the rat are associated with differences in estrogen-inducible central oxytocin receptors. *Proceedings of the National Academy of Sciences USA*, 98, 12736–12741.

Champagne, F. A., Francis, D. D., Mar, A. & Meaney, M. J. 2003a. Variations in maternal care in the rat as a mediating influence for the effects of environment on development. *Physiology & Behavior*, 79, 359–371.

Champagne, F. A., Weaver, I. C., Diorio, J., Sharma, S. & Meaney, M. J. 2003b. Natural variations in maternal care are associated with estrogen receptor alpha expression and estrogen sensitivity in the medial preoptic area. *Endocrinology*, 144, 4720–4724.

Champagne, F. A., Chretien, P., Stevenson, C. W., Zhang, T. Y., Gratton, A. & Meaney, M. J. 2004. Variations in nucleus accumbens dopamine associated with individual differences in maternal behavior in the rat. *Journal of Neuroscience*, 24, 4113–4123.

Champagne, F. A., Weaver, I. C., Diorio, J., Dymov, S., Szyf, M. & Meaney, M. J. 2006. Maternal care associated with methylation of the estrogen receptor-alpha1b promoter and estrogen receptor-alpha expression in the medial preoptic area of female offspring. *Endocrinology*, 147, 2909–2915.

Chapman, D. & Scott, K. 2001. The impact of maternal intergenerational risk factors on adverse developmental outcomes. *Developmental Review*, 21, 305–325.

Cheverud, J. M. & Wolf, J. B. 2009. The genetics and evolutionary consequences of maternal effects. In: *Maternal Effects in Mammals* (Ed. by D. Maestripieri & J. M. Mateo), pp. 11–37. Chicago: University of Chicago Press.

Coplan, J. D., Smith, E. L., Trost, R. C., Scharf, B. A., Altemus, M., Bjornson, L., Owens, M. J., Gorman, J. M., Nemeroff, C. B. & Rosenblum, L. A. 2000. Growth hormone response to clonidine in adversely reared young adult primates: relationship to serial cerebrospinal fluid corticotropin-releasing factor concentrations. *Psychiatry Research*, 95, 93–102.

Coplan, J. D., Smith, E. L., Altemus, M., Scharf, B. A., Owens, M. J., Nemeroff, C. B., Gorman, J. M. & Rosenblum, L. A. 2001. Variable foraging demand rearing: sustained elevations in cisternal cerebrospinal fluid corticotropin-releasing factor concentrations in adult primates. *Biological Psychiatry*, 50, 200–204.

Coplan, J. D., Smith, E. L., Altemus, M., Mathew, S. J., Perera, T., Kral, J. G., Gorman, J. M., Owens, M. J., Nemeroff, C. B. & Rosenblum, L. A. 2006. Maternal-infant response to variable foraging demand in nonhuman primates: effects of timing of stressor on cerebrospinal fluid corticotropin-releasing factor and circulating glucocorticoid concentrations. *Annals of the New York Academy of Sciences*, 1071, 525–533.

Croskerry, P. G., Smith, G. K. & Leon, M. 1978. Thermoregulation and the maternal behaviour of the rat. *Nature*, 273, 299–300.

Dowdney, L., Skuse, D., Rutter, M., Quinton, D. & Mrazek, D. 1985. The nature and qualities of parenting provided by women raised in institutions. *Journal of Child Psychology and Psychiatry*, 26, 599–625.

Egeland, B., Jacobvitz, D. & Papatola, K. 1987. Intergenerational continuity of abuse. In: *Child Abuse and Neglect: Biosocial Dimensions* (Ed. by R. J. Gelles & J. B. Lancaster), pp. 255–276. New York: Aldine.

Evoniuk, G. E., Kuhn, C. M. & Schanberg, S. M. 1979. The effect of tactile stimulation on serum growth hormone and tissue ornithine decarboxylase activity during maternal deprivation in rat pups. *Communications in Psychopharmacology*, 3, 363–370.

Fairbanks, L. A. 1989. Early experience and cross-generational continuity of mother-infant contact in vervet monkeys. *Developmental Psychobiology*, 22, 669–681.

Field, T. M., Schanberg, S. M., Scafidi, F., Bauer, C. R., Vega-Lahr, N., Garcia, R., Nystrom, J. & Kuhn, C. M. 1986. Tactile/kinesthetic stimulation effects on preterm neonates. *Pediatrics*, 77, 654–658.

Fleming, A. S. & Rosenblatt, J. S. 1974. Maternal behavior in the virgin and lactating rat. *Journal of Comparative and Physiological Psychology*, 86, 957–972.

Fleming, A. S., Kraemer, G. W., Gonzalez, A., Lovic, V., Rees, S. & Melo, A. 2002. Mothering begets mothering: the transmission of behavior and its neurobiology across generations. *Pharmacology, Biochemistry and Behavior*, 73, 61–75.

Francis, D., Diorio, J., Liu, D. & Meaney, M. J. 1999. Nongenomic transmission across generations of maternal behavior and stress responses in the rat. *Science*, 286, 1155–1158.

Francis, D., Szegda, K., Campbell, G., Martin, W. D. & Insel, T. R. 2003. Epigenetic sources of behavioral differences in mice. *Nature Neuroscience*, 6, 445–446.

Gammie, S. C. 2005. Current models and future directions for understanding the neural circuitries of maternal behaviors in rodents. *Behavioral and Cognitive Neuroscience Reviews*, 4, 119–135.

Gonzalez, A. & Fleming, A. S. 2002. Artificial rearing causes changes in maternal behavior and c-fos expression in juvenile female rats. *Behavioral Neuroscience*, 116, 999–1013.

Gonzalez, A., Lovic, V., Ward, G. R., Wainwright, P. E. & Fleming, A. S. 2001. Intergenerational effects of complete maternal deprivation and replacement stimulation on maternal behavior and emotionality in female rats. *Developmental Psychobiology*, 38, 11–32.

Gubernick, D. J. & Alberts, J. R. 1983. Maternal licking of young: resource exchange and proximate controls. *Physiology & Behavior*, 31, 593–601.

Gubernick, D. J. & Alberts, J. R. 1985. Maternal licking by virgin and lactating rats: water transfer from pups. *Physiology & Behavior*, 34, 501–506.

Harlow, H. F. & Suomi, S. J. 1971. Social recovery by isolation-reared monkeys. *Proceedings of the National Academy of Sciences USA*, 68, 1534–1538.

Harlow, H. F. & Suomi, S. J. 1974. Induced depression in monkeys. *Behavioral Biology*, 12, 273–296.

Hofer, M. A., Brunelli, S. A. & Shair, H. N. 1993. Ultrasonic vocalization responses of rat pups to acute separation and contact comfort do not depend on maternal thermal cues. *Developmental Psychobiology*, 26, 81–95.

Jaenisch, R. & Bird, A. 2003. Epigenetic regulation of gene expression: how the genome integrates intrinsic and environmental signals. *Nature Genetics*, 33 Suppl, 245–254.

Kennell, J. H., Jerauld, R., Wolfe, H., Chesler, D., Kreger, N. C., McAlpine, W., Steffa, M. & Klaus, M. H. 1974. Maternal behavior one year after early and extended post-partum contact. *Developmental Medicine and Child Neurology*, 16, 172–179.

Kikusui, T., Isaka, Y. & Mori, Y. 2005. Early weaning deprives mouse pups of maternal care and decreases their maternal behavior in adulthood. *Behavioural Brain Research*, 162, 200–206.

Kuhn, C. M., Butler, S. R. & Schanberg, S. M. 1978. Selective depression of serum growth hormone during maternal deprivation in rat pups. *Science*, 201, 1034–1036.

Kuhn, C. M., Schanberg, S. M., Field, T., Symanski, R., Zimmerman, E., Scafidi, F. & Roberts, J. 1991. Tactile-kinesthetic stimulation effects on sympathetic and adrenocortical function in preterm infants. *Journal of Pediatrics*, 119, 434–440.

Lee, M. & Williams, D. 1974. Changes in licking behaviour of rat mother following handling of young. *Animal Behaviour*, 22, 679–681.

Lehmann, J., Pryce, C. R., Bettschen, D. & Feldon, J. 1999. The maternal separation paradigm and adult emotionality and cognition in male and female Wistar rats. *Pharmacology, Biochemistry and Behavior*, 64, 705–715.

Lehmann, J., Pryce, C. R., Jongen-Relo, A. L., Stohr, T., Pothuizen, H. H. & Feldon, J. 2002a. Comparison of maternal separation and early handling in terms of their neurobehavioral effects in aged rats. *Neurobiology of Aging*, 23, 457–466.

Lehmann, J., Russig, H., Feldon, J. & Pryce, C. R. 2002b. Effect of a single maternal separation at different pup ages on the corticosterone stress response in adult and aged rats. *Pharmacology, Biochemistry and Behavior*, 73, 141–145.

Leon, M., Croskerry, P. G. & Smith, G. K. 1978. Thermal control of mother-young contact in rats. *Physiology & Behavior*, 21, 790–811.

Levine, S. 1957. Infantile experience and resistance to physiological stress. *Science*, 126, 405.

Liu, D., Diorio, J., Tannenbaum, B., Caldji, C., Francis, D., Freedman, A., Sharma, S., Pearson, D., Plotsky, P. M. & Meaney, M. J. 1997. Maternal care, hippocampal glucocorticoid receptors, and hypothalamic-pituitary-adrenal responses to stress. *Science*, 277, 1659–1662.

Liu, D., Caldji, C., Sharma, S., Plotsky, P. M. & Meaney, M. J. 2000a. Influence of neonatal rearing conditions on stress-induced adrenocorticotropin responses and norepinepherine release in the hypothalamic paraventricular nucleus. *Journal of Neuroendocrinology*, 12, 5–12.

Liu, D., Diorio, J., Day, J. C., Francis, D. D. & Meaney, M. J. 2000b. Maternal care, hippocampal synaptogenesis and cognitive development in rats. *Nature Neuroscience*, 3, 799–806.

Lonstein, J. S. & Gammie, S. C. 2002. Sensory, hormonal, and neural control of maternal aggression in laboratory rodents. *Neuroscience and Biobehavioral Reviews*, 26, 869–888.

Lovic, V. & Fleming, A. S. 2004. Artificially-reared female rats show reduced prepulse inhibition and deficits in the attentional set shifting task: reversal of effects with maternal-like licking stimulation. *Behavioural Brain Research*, 148, 209–219.

Lovic, V., Gonzalez, A. & Fleming, A. S. 2001. Maternally separated rats show deficits in maternal care in adulthood. *Developmental Psychobiology*, 39, 19–33.

Macri, S. & Wurbel, H. 2007. Effects of variation in postnatal maternal environment on maternal behaviour and fear and stress responses in rats. *Animal Behaviour*, 73, 171–184.

Maestripieri, D. 1998. Parenting styles of abusive mothers in group-living rhesus macaques. *Animal Behaviour*, 55, 1–11.

Maestripieri, D. 2005. Early experience affects the intergenerational transmission of infant abuse in rhesus monkeys. *Proceedings of the National Academy of Sciences USA*, 102, 9726–9729.

Maestripieri, D. 2009. Maternal influences on offspring growth, reproduction, and behavior in primates. In: *Maternal Effects in Mammals* (Ed. by D. Maestripieri & J. M. Mateo), pp. 256–291. Chicago: University of Chicago Press.

Maestripieri, D. & Carroll, K. A. 1998. Child abuse and neglect: usefulness of the animal data. *Psychological Bulletin*, 123, 211–223.

Maestripieri, D. & Wallen, K. 2003. Nonhuman primate models of developmental psychopathology: problems and prospects. In: *Neurodevelopmental Mechanisms in Psychopathology* (Ed. by D. Cicchetti & E. Walker), pp. 187–214. Cambridge: Cambridge University Press.

Maestripieri, D., Wallen, K. & Carroll, K. A. 1997. Infant abuse runs in families of group-living pigtail macaques. *Child Abuse and Neglect*, 21, 465–471.

Maestripieri, D., Higley, J. D., Lindell, S. G., Newman, T. K., McCormack, K. M. & Sanchez, M. M. 2006. Early maternal rejection affects the development of monoaminergic systems and adult abusive parenting in rhesus macaques. *Behavioral Neuroscience*, 120, 1017–1024.

Maestripieri, D., Lindell, S. G. & Higley, J. D. 2007. Intergenerational transmission of maternal behavior in rhesus monkeys and its underlying mechanisms. *Developmental Psychobiology*, 49, 165–171.

Main, M. & Solomon, J. 1990. Procedures for identifying infants as disorganized/disoriented during the Ainsworth Strange Situation. In: *Attachment in the Preschool Years* (Ed. by M. Greenberg, D. Cicchetti & E. Cummings), pp. 121–160. Chicago: University of Chicago Press.

McCarthy, M. M., Besmer, H. R., Jacobs, S. C., Keidan, G. M. & Gibbs, R. B. 1997. Influence of maternal grooming, sex and age on Fos immunoreactivity in the preoptic area of neonatal rats: implications for sexual differentiation. *Developmental Neuroscience*, 19, 488–496.

Meaney, M. J. 2001. Maternal care, gene expression, and the transmission of individual differences in stress reactivity across generations. *Annual Review of Neuroscience*, 24, 1161–1192.

Meaney, M. J., Aitken, D. H., Bodnoff, S. R., Iny, L. J. & Sapolsky, R. M. 1985. The effects of postnatal handling on the development of the glucocorticoid receptor systems and stress recovery in the rat. *Progress in Neuropsychopharmacology and Biological Psychiatry*, 9, 731–734.

Meaney, M. J., Aitken, D. H., Viau, V., Sharma, S. & Sarrieau, A. 1989. Neonatal handling alters adrenocortical negative feedback sensitivity and hippocampal type II glucocorticoid receptor binding in the rat. *Neuroendocrinology*, 50, 597–604.

Meaney, M. J., Aitken, D. H., Bhatnagar, S. & Sapolsky, R. M. 1991. Postnatal handling attenuates certain neuroendocrine, anatomical, and cognitive dysfunctions associated with aging in female rats. *Neurobiology of Aging*, 12, 31–38.

Meaney, M. J., Diorio, J., Francis, D., Widdowson, J., LaPlante, P., Caldji, C., Sharma, S., Seckl, J. R. & Plotsky, P. M. 1996. Early environmental regulation of forebrain glucocorticoid receptor gene expression: implications for adrenocortical responses to stress. *Developmental Neuroscience*, 18, 49–72.

Melo, A. I., Lovic, V., Gonzalez, A., Madden, M., Sinopoli, K. & Fleming, A. S. 2006. Maternal and littermate deprivation disrupts maternal behavior and social-learning of food preference in adulthood: tactile stimulation, nest odor, and social rearing prevent these effects. *Developmental Psychobiology*, 48, 209–219.

Miller, L., Kramer, R., Warner, V., Wickramaratne, P. & Weissman, M. 1997. Intergenerational transmission of parental bonding among women. *Journal of the American Academy of Child and Adolescent Psychiatry*, 36, 1134–1139.

Myers, M. M., Brunelli, S. A., Shair, H. N., Squire, J. M. & Hofer, M. A. 1989a. Relationships between maternal behavior of SHR and WKY dams and adult blood pressures of cross-fostered F1 pups. *Developmental Psychobiology*, 22, 55–67.

Myers, M. M., Brunelli, S. A., Squire, J. M., Shindeldecker, R. D. & Hofer, M. A. 1989b. Maternal behavior of SHR rats and its relationship to offspring blood pressures. *Developmental Psychobiology*, 22, 29–53.

Novakov, M. & Fleming, A. S. 2005. The effects of early rearing environment on the hormonal induction of maternal behavior in virgin rats. *Hormones and Behavior*, 48, 528–536.

Pauk, J., Kuhn, C. M., Field, T. M. & Schanberg, S. M. 1986. Positive effects of tactile versus kinesthetic or vestibular stimulation on neuroendocrine and ODC activity in maternally-deprived rat pups. *Life Science*, 39, 2081–2087.

Pedersen, C. A. & Boccia, M. L. 2002. Oxytocin links mothering received, mothering bestowed and adult stress responses. *Stress*, 5, 259–267.

Pederson, D. R., Gleason, K. E., Moran, G. & Bento, S. 1998. Maternal attachment representations, maternal sensitivity, and the infant-mother attachment relationship. *Developmental Psychobiology*, 34, 925–933.

Plotsky, P. M. & Meaney, M. J. 1993. Early, postnatal experience alters hypothalamic corticotropin-releasing factor (CRF) mRNA, median eminence CRF content and stress-induced release in adult rats. *Molecular Brain Research*, 18, 195–200.

Plotsky, P. M., Thrivikraman, K. V., Nemeroff, C. B., Caldji, C., Sharma, S. & Meaney, M. J. 2005. Long-term consequences of neonatal rearing on central corticotropin-releasing factor systems in adult male rat offspring. *Neuropsychopharmacology*, 30, 2192–2204.

Priebe, K., Romeo, R. D., Francis, D. D., Sisti, H. M., Mueller, A., McEwen, B. S. & Brake, W. G. 2005. Maternal influences on adult stress and anxiety-like behavior in C57BL/6J and BALB/cJ mice: a cross-fostering study. *Developmental Psychobiology*, 47, 398–407.

Razin, A. 1998. CpG methylation, chromatin structure and gene silencing: a three-way connection. *EMBO Journal*, 17, 4905–4908.

Rosenblatt, J. S. 1967. Nonhormonal basis of maternal behavior in the rat. *Science*, 156, 1512–1514.

Rosenblatt, J. S. 1975. Prepartum and postpartum regulation of maternal behaviour in the rat. *Ciba Foundation Symposia*, 33, 17–37.

Rosenblatt, J. S. & Lehrman, D. S. 1963. Maternal behavior in the laboratory rat. In: *Maternal Behavior in Mammals* (Ed. by H. L. Rheingold), pp. 8–57. New York: Wiley.

Rosenblum, L. A. & Paully, G. S. 1984. The effects of varying environmental demands on maternal and infant behavior. *Child Development*, 55, 305–314.

Rosenfeld, P., Wetmore, J. B. & Levine, S. 1992. Effects of repeated maternal separations on the adrenocortical response to stress of preweanling rats. *Physiology & Behavior*, 52, 787–791.

Russo, E., Martienssen, R. & Riggs, A. D. 1996. *Epigenetic Mechanisms of Gene Regulation*. Plainview, NY: Cold Spring Harbor Lab.

Sapolsky, R. M., Meaney, M. J. & McEwen, B. S. 1985. The development of the glucocorticoid receptor system in the rat limbic brain. III. Negative-feedback regulation. *Brain Research*, 350, 169–173.

Seay, B. & Harlow, H. F. 1965. Maternal separation in the rhesus monkey. *Journal of Nervous and Mental Disease*, 140, 434–441.

Shoji, H. & Kato, K. 2006. Maternal behavior of primiparous females in inbred strains of mice: a detailed descriptive analysis. *Physiology & Behavior*, 89, 320–328.

Simpson, M. & Howe, S. 1986. Group and matriline differences in the behaviour of rhesus monkey infants. *Animal Behaviour*, 34, 444–459.

Slotkin, T. A. & Bartolome, J. 1986. Role of ornithine decarboxylase and the polyamines in nervous system development: a review. *Brain Research Bulletin*, 17, 307–320.

Solkoff, N. & Matuszak, D. 1975. Tactile stimulation and behavioral development among low-birthweight infants. *Child Psychiatry and Human Development*, 6, 33–37.

Sroufe, L. A. 2005. Attachment and development: a prospective, longitudinal study from birth to adulthood. *Attachment and Human Development*, 7, 349–367.

Sroufe, L. A., Egeland, B., Carlson, E. & Collins, W. 2005. *The Development of the Person: The Minnesota Study of Risk and Adaptation from Birth to Adulthood*. New York: Guildford Press.

Stern, J. M. & Johnson, S. K. 1990. Ventral somatosensory determinants of nursing behavior in Norway rats. I. Effects of variations in the quality and quantity of pup stimuli. *Physiology & Behavior*, 47, 993–1011.

Strathdee, G. & Brown, R. 2002. Aberrant DNA methylation in cancer: potential clinical interventions. *Expert Review of Molecular Medicine*, 2002, 1–17.

Sullivan, R. M., Shokrai, N. & Leon, M. 1988a. Physical stimulation reduces the body temperature of infant rats. *Developmental Psychobiology*, 21, 225–235.

Sullivan, R. M., Wilson, D. A. & Leon, M. 1988b. Physical stimulation reduces the brain temperature of infant rats. *Developmental Psychobiology*, 21, 237–250.

Turner, B. 2001. *Chromatin and Gene Regulation*. Oxford: Blackwell Science Ltd.

van IJzendoorn, M. H. 1995. Adult attachment representations, parental responsiveness, and infant attachment: a meta-analysis on the predictive validity of the Adult Attachment Interview. *Psychological Bulletin*, 117, 387–403.

Wade, M. J., Priest, N. K. & Cruickshank, T. E. 2009. A theoretical overview of genetic maternal effects: evolutionary predictions and empirical tests with mammalian data. In: *Maternal Effects in Mammals* (Ed. by D. Maestripieri & J. M. Mateo), pp. 38–63. Chicago: University of Chicago Press.

Waterland, R. A. 2006. Assessing the effects of high methionine intake on DNA methylation. *Journal of Nutrition*, 136, 1706S-1710S.

Weaver, I. C., Grant, R. J. & Meaney, M. J. 2002. Maternal behavior regulates long-term hippocampal expression of BAX and apoptosis in the offspring. *Journal of Neurochemistry*, 82, 998–1002.

Weaver, I. C., Cervoni, N., Champagne, F. A., D'Alessio, A. C., Sharma, S., Seckl, J. R., Dymov, S., Szyf, M. & Meaney, M. J. 2004a. Epigenetic programming by maternal behavior. *Nature Neuroscience*, 7, 847–854.

Weaver, I. C., Diorio, J., Seckl, J. R., Szyf, M. & Meaney, M. J. 2004b. Early environmental regulation of hippocampal glucocorticoid receptor gene expression: characterization of intracellular mediators and potential genomic target sites. *Annals of the New York Academy of Sciences*, 1024, 182–212.

Weaver, I. C., Champagne, F. A., Brown, S. E., Dymov, S., Sharma, S., Meaney, M. J. & Szyf, M. 2005. Reversal of maternal programming of stress responses in adult offspring through methyl supplementation: altering epigenetic marking later in life. *Journal of Neuroscience*, 25, 11045–11054.

Weaver, I. C., Meaney, M. J. & Szyf, M. 2006. Maternal care effects on the hippocampal transcriptome and anxiety-mediated behaviors in the offspring that are reversible in adulthood. *Proceedings of the National Academy of Sciences USA*, 103, 3480–3485.

Weissgerber, T. L. & Wolfe, L. A. 2006. Physiological adaptation in early human pregnancy: adaptation to balance maternal-fetal demands. *Applied Physiology Nutrition and Metabolism*, 31, 1–11.

Zhang, T. Y., Chretien, P., Meaney, M. J. & Gratton, A. 2005. Influence of naturally occurring variations in maternal care on prepulse inhibition of acoustic startle and the medial prefrontal cortical dopamine response to stress in adult rats. *Journal of Neuroscience*, 25, 1493–1502.

Zimmerberg, B., Brunelli, S. A. & Hofer, M. A. 1994. Reduction of rat pup ultrasonic vocalizations by the neuroactive steroid allopregnanolone. *Pharmacology, Biochemistry and Behavior*, 47, 735–738.

10

Effects of Intrauterine Position in
Litter-Bearing Mammals

JOHN G. VANDENBERGH

INTRODUCTION

As pointed out throughout this volume, maternal care is the hallmark of "mammalhood." The developing embryo not only receives half of its genes from its mother, it is an integral part of the mother for the most important part of its development as an individual, i.e., the prenatal development period. Mothers communicate in many ways with their offspring in utero during development. As pointed out by Champagne & Curley and in other chapters in this volume, strong influences from the mother continue during the early postnatal days. While such postnatal maternal effects are the focus of this volume, this is not the whole story. Here we will examine the interactions that occur among fetuses in utero that have long-lasting effects on later anatomical, physiological, and behavioral characteristics.

Many anatomical, physiological, and behavioral characteristics are sexually dimorphic in mammals. In addition to the variability seen between sexes, a great deal of variability exists among individuals within each sex. The origins of this variability lie in complex interactions between an individual's genome and its environment. In this chapter I will present evidence that the location of a fetus in relation to other fetuses in the uterus of litter-bearing species can contribute to phenotypic variation among individuals during gestation, at birth, during early development, and in adulthood. Further, I will suggest that the findings from work on prenatal influences play an important role in animal tests of endocrine disruptors and perhaps in determining the consequences of human exposure to such substances.

Many mammalian species produce more than one offspring at a time and are referred to as litter-bearing or polytocous species. Littermates may begin interacting with their siblings early in fetal development with long-lasting consequences for their anatomy, physiology, and behavior. This intrauterine communication among fetuses operates within the context of the maternal environment and contributes to the epigenetic variability seen within a litter.

The possibility that the development of a fetus in utero could be affected by neighboring fetuses emerged from a finding by Clemens & Coniglio (1971) and Clemens et al. (1978) that the anogenital distance of rat pups varied not only between sexes, as was well known, but also among the females. Delivering pups by cesarean section and recording their position in the uterus as related to siblings allowed the investigators to relate the intensity of the rat's adult sexual behavior to its prior intrauterine position (IUP). In pursuing the correlates of IUP, Clemens and coworkers discovered that female pups flanked by males in utero had long anogenital distances but still shorter than males. Female pups not flanked by males had shorter anogenital distances. The finding was further explored by vom Saal & Bronson (1978) in mice. They found that female mice gestated between two males were masculinized in morphological, physiological, and behavioral characteristics.

Interest in the IUP phenomenon stimulated a number of laboratories to explore its role in explaining some of the variability among individuals. Identifying and explaining the sources of variability is an essential quest in biology. Clearly, genetic sources of variability are high on the list of explanatory mechanisms. However, the genetic background interacts with environmental factors to result in variability in expression. In recent years, it has become clear that fetal development is an important period for such interaction to take place. Other chapters in this volume explore maternal influences during and after fetal development. Here the focus is on endocrine communication among siblings in the uterus. The findings not only provide some explanation for the natural variability among pups in a litter but also suggest that there is an interaction between the hormones resulting from male proximity and endocrine disruptor compounds (EDCs) to which the mother is exposed.

Much of this chapter is a detailed review of the IUP effect with a focus on opportunities for future research, on an analysis of the importance of IUP, and on the research it has spawned on EDCs and human biology (see also Ryan & Vandenbergh 2002).

INTRAUTERINE POSITION

The rodent uterus is composed of left and right uterine horns, called a bicornate uterus; pups can be implanted in both horns of the uterus and, with perhaps one exception, are randomly distributed with regard to sex. Clark & Galef (1990) found that in gerbils there was a preponderance of male fetuses in the right horn of the uterus and female fetuses in the left, just as Hippocrates suggested. No mechanism for such a distribution has been uncovered. Random distribution between the uterine horns seems to be likely in most species and, assuming random distribution of sexes between right and left horns of the uterus, three possibilities exist. A fetus can lie between the two members of the same sex, be flanked by a different sex on each side, or be positioned at either end of the uterus and thus have only one near neighbor. Various abbreviations have been applied to this distribution but the one used here is 2M, 1M, or 0M for females and males, as shown in Figure 10-1. Fetuses at the ends of the uterine horns can have a wombmate of the same or different sex but are still considered 0M or 1M.

In rodents and other placental mammals, males produce testosterone at relatively high concentrations early in development. In rats, for example, testosterone production begins about midway during gestation and remains high until shortly after birth (Vreeburg et al. 1983). Females, on the other hand, produce small amounts of androgens or estrogens early in development and only later do estrogen titers rise in females at puberty. This developmental exposure to testosterone sets sexual differentiation into motion. Masculinization includes development of the Wolffian duct system, external genitalia, and those CNS structures that influence behavior and physiological characteristics associated with maleness. The testosterone produced by a male, which directs its own development, can also affect nearby fetuses. Since females are not significantly exposed to self-produced androgen, any androgens coming from nearby males can have a masculinizing effect on them. Male neighbors, on the other hand, produce their own testosterone, so any amount of this hormone coming from a neighbor can only add to what they are producing, possibly reaching a ceiling level in the male.

To produce an effect, testosterone must move from a male to adjacent fetuses. Because testosterone is a steroid hormone, it is lipophilic and can

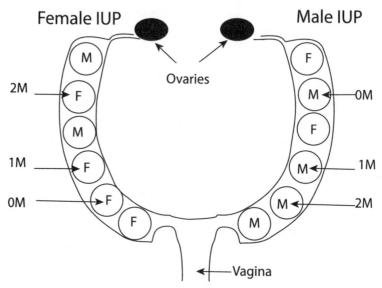

Figure 10-1. A schematic representation of the gravid uterus in the house mouse showing the intrauterine position effect. Embryos implant in either horn of the uterus in a random fashion with the result that female fetuses can be in utero between two males (2M), next to one male (1M), or not be adjacent to any males (0M). The figure was modified from Ryan and Vandenbergh (2002).

diffuse through amniotic membranes and amniotic fluid. Using this information, vom Saal et al. (1990) demonstrated that diffusion rather than circulation was the most likely mechanism of testosterone transfer between pups since 2M female mice had higher blood concentrations of testosterone than did 0M females. In mice it is unlikely that maternal circulation could be responsible for the effect of testosterone on adjacent fetuses (vom Saal & Dhar 1992). In laboratory rats, however, testosterone could be transported via maternal circulation. In rats, circulation to the uterus flows from the caudal to the cephalic end of each uterine horn, i.e., from the vagina to the ovaries. Thus, fetuses located at the caudal end of the uterus would be exposed to higher concentrations of maternal testosterone than those at the caudal end (Gorodeski et al. 1995). This finding is supported by studies showing that cocaine injected into the mother reached the caudal fetuses first and the cephalic fetuses last and at lower concentration (Lipton et al. 1998). It is likely that some of the variability seen in studies of maternally administered drugs could be due to this transport mechanism. Many investigators classify rat fetuses by the number of males in the litter because this seems to fit the circulatory mechanism of maternal transmission. In rats, two factors can influence the effects of testosterone produced by male fetuses. In addition to the mouse-like IUP effect, i.e., the location of a female fetus between two males, the number of male fetuses in a uterine horn may also be important. However, regardless of how interfetal transmission of testosterone occurs, the result is a variable concentration of testosterone present during critical periods for sexual differentiation. This may underlie the variable degrees of masculinization noted in litters.

ANATOMICAL CORRELATES OF IUP

The original anatomical clue to the presence of an IUP effect was an observation that the distance between the anus and the genital papilla varied within a sex as well as between sexes in the newborn rat (Clemens & Coniglio 1971; Clemens et al. 1978). This distance is easy to measure with a ruler or calipers and has become an important index of fetal masculinization (Vandenbergh & Huggett 1995). In mice, the anogenital distance (AGD) can be useful to detect prenatal androgen exposure in females when measured at the time of weaning as well as at birth. In gerbils it seems that the IUP effect is more pronounced in males than females (Clark et al. 1990) and in swine no effect of the IUP was found on AGD (Rohde Parfet et al. 1990); however, in swine a correlation does exist between AGD and the number of males in the litter (Drickamer et al. 1997). To date, the possibility of an IUP effect has been examined only in a few litter-bearing mammalian species (Ryan & Vandenbergh 2002) and thus we cannot assume it is present in other mammals.

In earlier studies, the IUP was noted upon cesarean section followed by cross-fostering of the pups. This procedure yielded an accurate identification of IUP but was tedious and has been largely replaced by measures of anogenital distance. One of the issues complicating use of the AGD is that the length of the tissue is affected by the body size of the subject. Several standardizing measures can be taken such as body length, tail length, and body mass. The length measures would be more isometric but body mass allows for allometry (Steel et al. 1995). Similarly, several statistical transformations can be used to make appropriate comparisons. We have performed both isometric and allometric adjustments in our laboratory and found that length measures of body, tail, and hind foot did not yield sufficient variability to accurately reflect the size of a mouse. Thus length measures did not improve the reliability of the AGD to correctly identify IUP (Hotchkiss

& Vandenbergh 2005). We also conducted an extensive statistical analysis on the IUP of 152 female mice from known intrauterine positions that had been delivered by cesarean section to determine the most useful transformation (Hotchkiss & Vandenbergh 2005). We found that AGD simply divided by body weight termed an anogenital distance index (AGDI) and the residual log transformation are the most accurate means to predict the IUP of the pup.

In addition to AGD, the IUP affects several other morphological traits. Male mice, for example, have larger seminal vesicles and smaller prostate glands if they are from the 2M position than from the 0M position (Nonneman et al. 1992). In rats (van der Hoeven et al. 1992) and gerbils (Clark et al. 1990; 1993) 2M males have heavier testes than 0M males, but in swine no IUP effect on testes weight was found (Rohde Parfet et al. 1990). The sexually dimorphic ventral scent gland is heavier in 2M male gerbils than in their 0M siblings (Clark et al. 1990). The size of a number of reproductive organs varies with IUP, as will be described below.

Adult sexual dimorphism within the central nervous system (CNS) is well documented to be correlated with perinatal steroid hormone concentrations. Thus, it is no surprise that the IUP has consequences for sexually dimorphic CNS structures. The sexually dimorphic nucleus of the preoptic area in the hypothalamus (SDN-POA) is larger in male rats than in females (Forger 1999). Faber & Hughes (1992) found that the size of the SDN-POA varies directly with the AGD of female rats. Furthermore, 2M female gerbils have an increased number of motoneurons in the spinal nucleus of the bulbocavernosus muscle (Forger et al. 1996). Female rats with a longer AGD have a larger sexually dimorphic nucleus in the SDN-POA. Pei et al. (2006), using the calbindin-D28k technique as a specific marker for identification of the SDN-POA (Sickel & McCarthy 2000), were able to use three-dimensional reconstruction to determine more accurate estimates of the SDN-POA volume. The volume of the SDN-POA in male rats was two to four times as large as in females, and in 2M males the SDN-POA was approximately twice the size of that in 0M males. However, no significant difference was found between 2M and 0M females. In contrast, the IUP does not appear to influence the sexually dimorphic hippocampus in gerbils (Sherry et al. 1996) nor the corpus callosum in mice (Bulman-Fleming & Wahlsten 1991). Additional descriptions of anatomical differences can be found in Ryan & Vandenbergh (2002). The consequences of such structural variability in the CNS will be explored when the behavioral effects of IUP are discussed later in this chapter.

PHYSIOLOGICAL EFFECTS OF IUP

The single most important effect in rodents of having contiguous fetuses of the same or different sex results from the exposure to testosterone produced by male fetuses. Vom Saal & Bronson (1980) and vom Saal et al. (1981) found that 2M female mouse fetuses have higher concentrations of testosterone than do 0M females in both blood and amniotic fluid (vom Saal & Bronson 1980; vom Saal et al. 1990). In an attempt to determine the distribution of steroid hormones in rat fetuses, Pei et al. (2006) measured testosterone and estradiol in diencephalic tissue and found differences on gestation day 21 only in a few comparisons. Both testosterone and estradiol were higher in 2M than in 0M males. Concentrations of both hormones were low and quite variable, however, making it difficult to document a significant difference between 2M and 0M females.

There are a number of biochemical events in the developing fetus that appear to be affected by IUP. In one study, the AGD of weanling mice was found to correlate with several measures of blood chemistry and with activity-related behavior (Kerin et al. 2003). Furthermore, in 0M female rats, testosterone 5alpha reductase activity is higher than in 2M females (van der Hoeven et al. 1992), and 0M male mice have low 5α reductase in the seminal vesicles and prostate (Nonneman et al. 1992). The enzyme 5α reductase converts testosterone to dihydrotestosterone, which has effects on many androgen-sensitive tissues. Aromatase, an enzyme that converts testosterone to 17β estradiol, does not appear to vary with IUP when measured in the rat forebrain (Tobet et al. 1985). Higher cytochrome oxidase reactivity, a measure of metabolism, is elevated in the anterior hypothalamus of 2M gerbils (Jones et al. 1997). A number of other enzymatic effects, summarized in Ryan & Vandenbergh (2002), suggest that the IUP may be an interesting variable to include in tests of putative therapeutic drugs or toxicological substances.

EFFECTS OF IUP ON REPRODUCTION

Since the effects of IUP occur as a result of the transfer of testosterone from male fetuses, it is no surprise that IUP can have important consequences on reproductive parameters such as puberty and adult fertility. 0M female mice and gerbils display earlier vaginal opening and age at first estrus than do 2M females (McDermott et al. 1978; Clark & Galef 1988; vom Saal 1989b; Clark et al. 1991; Vandenbergh & Huggett 1995). 0M female mice also produce more potent puberty-delaying pheromone(s) than do 2M female mice, but they do not differ in sensitivity to the male puberty accelerating pheromone (vom Saal & Bronson 1980; vom Saal et al. 1981; vom Saal 1989b; Vandenbergh & Huggett 1995).

IUP effects continue to influence reproductive characteristics into adulthood. Using AGD as an indicator of IUP, Drickamer (1996) showed that wild female mice with a short AGD are more likely to become pregnant than females with a long AGD. In the laboratory, 2M female mice produce fewer viable litters and become infertile earlier than 0M females (vom Saal & Moyer 1985). Studies with swine, which may have some practical importance for food production, show that insemination and subsequent pregnancy are more successful if the sow is from a female-biased litter and thus likely to be less exposed to testosterone than other females (Drickamer et al. 1997).

Clark and her colleagues have demonstrated that gerbils show a strong IUP effect among males, quite unlike mice. An adult male derived from the 2M position (termed a "stud" male) is more attractive to females, inseminates more females, and sires more pups per litter than a 0M male (Clark et al. 1992; 1998). The 0M males with the lowest testosterone concentrations (termed "dud" males) show little sexual behavior and higher amounts of parental behavior (Clark et al. 1992; Clark & Galef 2000).

One of the effects of IUP on reproduction that has defied explanation to date is that the sex ratio produced by 2M females is modified. When 2M female gerbils are raised to adulthood and allowed to mate, they produce about 60% male pups, whereas 0M females produce about 40% males and 1M females produce the expected 50% males (Clark & Galef 1990; Clark et al. 1993). Overall, gerbils generate the expected 50:50 sex ratio but the variability among litters can be partially explained by the IUP of the mother. The results with gerbils prompted us to study the effects of IUP on sex ratios in mice. Using mice derived by cesarean section, we found that 2M females produced 58% males in their first litter, 1M females produced 51% and 0M females produced 42% males (Vandenbergh & Huggett 1994), very similar to the results of Clark & Galef's (1995; 2000) studies with gerbils. In the mouse study, all males used were from the 1M position and thus the IUP of the male was probably irrelevant. The mice were mated again, and the second litter produced a similar sex ratio effect. Pup loss during gestation or postnatal cannibalism is an unlikely explanation because the number of placental scars correlated well with litter size and no postpartum loss was detected. It is possible, but not yet tested, that the timing of insemination during the female's estrous cycle may be related to the sex ratio shifts. Female rats inseminated close to the time of ovulation produce more females than females inseminated before or after ovulation, which produce more males (Hendricks & McClintock 1990). A similar phenomenon may be present in humans (James 1987).

IUP EFFECTS IN THE NATURAL ENVIRONMENT

To test whether IUP is an important variable in wild mice, we examined the survival, reproduction, and home range size of wild mice from known intrauterine positions (Zielinski et al. 1992). Testing a truly wild population of mice would present great challenges, so a more confined, but naturalistic, population was utilized. Wild house mice were captured and allowed to mate in the laboratory. As the females approached term, cesarean sections were performed, permitting identification of the location of the pups. Pups were then cross-fostered to other females and, when mature, were taken to the field to populate "highway islands." Highway islands are segments of inter-state highway exchanges, or cloverleaves. The ones we chose had typical old field vegetation and were approximately 1 hectare in area. See Massey (1982) for more details on the ecology of highway islands.

Each highway island was cleared of native rodents and populated with about 24 0M and 2M females and 40 1M males. Populations were then moni-tored by periodic live-trapping over the next seven weeks. Measures of sur-vivorship and reproduction indicated no significant differences between 0M and 2M females but the re-trapping data showed that the 2M females had 40% larger home ranges than the 0M females, quite likely giving them a competitive advantage. This effect, as will be noted below, may be due to the increased levels of aggression shown by 2M females. These results suggest that the IUP effect has ecological consequences on the use of space in the field. With the more recent information that the IUP can be estimated by measure of the anogenital distance, it may now be possible to include an anogenital distance index (AGDI) as a variable in rodent field studies.

The absence of an effect on survival and reproduction in the highway island populations may be due to the short duration of the experiment. Drickamer (1996) conducted a longer experiment, seven months, using mice dwelling in large outdoor enclosures. In this study, IUP was not known by cesarean section but AGD was used as an indicator of prior IUP. The results showed that male mice with longer AGDs were more aggressive and had larger home ranges that did males with shorter AGDs. Survival was not cor-related with either AGD or sex. Female mice with shorter AGDs, however, were more likely to reproduce and had more pregnancies than 2M females.

This study further supports the notion that IUP may play a role in rodent population dynamics.

Females derived from wild populations have also been tested for the IUP effect on reproductive parameters in the laboratory. In a study of the interaction between IUP and social stress induced by introducing unfamiliar individuals to a cage containing the test female, Zielinski et al. (1991) showed that IUP influenced sexual phenotype and that social stress induces masculinization of female pups in wild mice as it does in laboratory mice (vom Saal & Bronson 1978). Further, Zielinski & Vandenbergh (1991) found no effect of IUP on age at first reproduction but did find that 0M females living in grouped conditions showed earlier vaginal opening. The results of such studies with wild house mice support the suggestion of vom Saal (1984) and Vandenbergh (2003) that the IUP effect and changes due to prenatal stress can have a role in reproductive parameters underlying rodent population dynamics (see also Cowell et al. 1998). The important concept here is that the presence of an androgen, e.g., testosterone, during gestation either due to proximity of a male fetus or due to stress can influence female reproductive performance.

Several litter-bearing species that are not commonly used in the laboratory have also have been shown to display IUP effects. While these effects have usually been measured in the laboratory, they may be relevant to field populations since the subjects have not undergone the intense selection for reproduction and docility seen in laboratory rodents. In California mice (*Peromyscus californicus*), the AGD of females born in male-biased litters are longer than in non–male biased litters (Cantoni et al. 1999). Ferrets (*Mustela furo*) are also likely to display IUP effects since females located downstream from males in the uterus contain higher concentrations of testosterone than other females (Krohmer & Baum 1989). In one study that may have implications for population ecology, grey-sided voles (*Clethrionomys rufocanus*) showed a greater tendency to disperse if derived from male-biased litters than from female-biased litters (Ims 1989; 1990). As noted above, support for the effect of fetal androgen exposure on dispersal behavior was also found by Drickamer (1996) using wild mice in large outdoor enclosures. Many other domestic and wild litter-bearing species remain to be examined for IUP effects.

BEHAVIORAL EFFECTS OF IUP

Since IUP effects result from the transfer of gonadal hormones during fetal development and such hormones are known to play a major role in organizing later sexual behavior and sexually dimorphic behaviors (Goy & McEwen 1980; Naftolin et al. 1988), it is not surprising that the IUP influences such behaviors. In fact, the IUP effect was discovered because of an attempt to explain differences in female mounting behavior (Clemens & Coniglio 1971). Female rats downstream from two or more males show a decreased lordosis quotient, as measured by the arch of the female's back during copulation when mounted (Houtsmuller & Slob 1990). Using the AGD as an index of testosterone exposure, Zehr et al. (2001) found that shorter AGDs were related to weaker lordosis responses induced by manual palpation. Treatment with neonatal testosterone, gestating at a distal intrauterine position, or being in a 2M position results in female rats exhibiting more female-female mounting behavior (Meisel & Ward 1981; Houtsmuller & Slob 1990). Most of the reported IUP effects in rats are in females, but one study demonstrated that males may also be affected. When given a choice, 2M males sniff females more than do 0M males (Hernandez-Tristan et al. 1999).

IUP effects on reproductive behavior have been more frequently noted in mice than in rats. Female mice from the 0M position show a higher lordosis quotient and are more likely to receive the male's first ejaculation upon mating than 2M females (Rines & vom Saal 1984). Further, when tested in a choice chamber, 0M females are more likely than 2M females to be chosen, mounted, and attacked when rejecting a male's advances (vom Saal & Bronson 1978; 1980). Male mice are also more attracted to a 0M female's pheromones than to those of 2M females (Drickamer et al. 2001). Demonstrating that 2M male's behavioral phenotype can also be affected by IUP, Drickamer et al. (2001) found that females prefer the odor of 2M males over that of 0M males. Thus, in mice, studies strongly suggest the IUP plays an important role in female behavior. In both rats and mice there are only a few instances of an important effect on males.

In contrast, studies on the Mongolian gerbil present even stronger evidence of the IUP effect on reproductive behavior in both sexes. 2M male gerbils attract females more readily and show shorter latency to mount than do 0M males (Clark et al. 1992). Further, 2M males are less likely to spend time with pups than 0M males (Clark et al. 1998). It is possible that some of the IUP effects seen could be modulated by parental care received soon after birth. Both male and female gerbil pups from the 2M position receive higher rates of anogenital licking from their parents than do 0M males and females (Clark et al. 1989). This may be important because the quantity of anogenital licking correlates with the number of motor neurons in the sexually dimorphic spinal nucleus of the bulbocavernosus muscle of 2M female gerbils (Forger et al. 1996). As with most studies of maternal effects, cross-fostering of the pups might remove the confound of differential licking.

Studies on both sexes suggest that an individual mouse's IUP influences its response to androgens later in life. Several studies have focused on aggressive behavior, a clear sexually dimorphic trait. Both female and male mice from a 2M position become aggressive sooner in response to testosterone injections than do animals from a 0M position (Gandelman et al. 1977; vom Saal et al. 1983). Parental behavior, another sexually dimorphic trait, is shown more frequently by 2M males than 0M males after castration (vom Saal 1983; 1989a). Also in mice, 2M females are more likely to show intense aggression when pregnant than 0M females. Similarly, in swine, 2M females are more likely to initiate fights than 0M females (Rohde Parfet et al. 1990). A behavior closely related to aggression in rodents is urine marking, often associated with territoriality. Studies have shown a relationship between IUP and urine marking. Males mark more frequently in the presence of 2M females than 0M females (Politch & Herrenkohl 1984) and 2M females mark at a higher rate than 0M females (vom Saal & Bronson 1980). Observations on gerbil marking behavior are even more compelling. Clark et al. (1990; 1992) have found that male gerbils, whether gonadally intact, castrated, or testosterone replaced, scent mark more frequently if derived from the 2M position than those from the 0M position. The finding that adult castration does not alter the IUP effect suggests that the testosterone derived from adjacent males has an organizing effect on females and, in some cases, on the male's expression of sexually dimorphic behaviors. The developing brain is permanently altered by the presence of small quantities of testosterone during critical periods. In most cases this is detected by responsivity to testosterone later in life, but with the studies by Clark and colleagues on gerbils, there is evidence that the effects of prenatal testosterone can be expressed in the absence of additional testosterone.

Dietary preferences in rodents may also be influenced by IUP. Mice and rats show sexual dimorphism in preference for sweet items, with females having a stronger preference for saccharin solutions than males (Valenstein et al. 1967). In mice, 0M females prefer higher concentrations of saccharin than do 2M females (Bushong & Mann 1994).

IUP EFFECTS IN HUMANS

Humans show an array of sexually dimorphic characteristics (Wizemann & Pardue 2001) and it is possible that male-female dizygotic twins may show some effects of the intrauterine environment. For example, humans show dental asymmetry with males having larger teeth in the right jaw than females, but in dizygotic twins this difference disappears (Boklage 1985). Another difference found is in the production of spontaneous otoacoustic emissions, continuous tonal sounds produced naturally in the cochlea. Although the functional significance of otoacoustic signals is not clear, the frequency of such cochlear emissions is higher in females than in males. However, this difference is reduced in dizygotic twins (McFadden 1993). The sexual dimorphism was not affected significantly in a follow-up study of click-induced otoacoustic emissions, sounds made in the cochlea in response to a click stimulus (McFadden et al. 1996). Although these sexual dimorphisms (see Miller 1994 for a review) do not seem to be related to reproductive success or have any clear survival value, some other apparent consequences of being a dizygotic twin are suggestive.

In a large study of Australian twins, Loehlin & Martin (1998) found only a few minor differences in reproductive parameters among opposite sex twins. In response to a questionnaire, women with a female co-twin were compared to women with a male co-twin. No significant differences emerged but puberty onset and age at first pregnancy were later as would be expected after male exposure in utero. However, in a second large study of 5,679 twins in Finland focusing on female reproductive traits, women having a male twin showed no differences from the other twins in reproductive traits such as age at menarche, fertility, fecundity, or self-reported feminine traits (Rose et al. 2002). Some studies provide modest support that a twin brother can influence his sister's development in utero. In a study of adult behavior among dizygotic human twins, Resnick et al. (1993) found that females, when adult, showed higher levels of sensation-seeking behaviors if they had a twin brother than if they had a twin sister. Measures of sensation seeking were based on a questionnaire to assess participation in dangerous sports, activities associated with high speeds, seeking potentially dangerous experiences through travel, use of consciousness-altering drugs, and indications of restlessness and intolerance of the routine. Miller (1994) similarly found more masculine attitudes among females with a male co-twin than with a same-sex co-twin. Postnatal experiences may confound these findings and cannot be excluded from these studies. Recent work on human twins born prior to the availability of assisted reproduction techniques may shed new light on the issue of male effects on female co-twin. Lummaa et al. (2007) studying a population of mixed-sex twins born from 1734 to 1888, long before assisted reproduction was available to humans, showed that in the presence of a male co-twin, the female co-twin can have significantly reduced lifetime reproductive success. This finding suggests that differences in exposure to hormones or hormone mimics during fetal development are likely to have important fitness consequences for humans.

As demonstrated in a large number of studies reviewed above, testosterone from the adjacent male or from other sources during intrauterine development has long-lasting effects on anatomy, physiology, and behavior. In rodent studies, the AGD was demonstrated to be a valid index of prior exposure to androgens. If the AGD at birth could index hormone exposure in human fetuses, it may be possible to use it as an index of early exposure to androgenic hormones, hormone mimics, or hormone blockers. Using the AGD of newborn infants as an index of such androgen exposure, a recent study (Swan et al. 2005) strongly suggests that the AGD in humans could be so used. The AGD in humans (the distance from the anus to the base of the scrotum in males or to the base of the clitoris in females) is twice as long in males as in females (Salazar-Martinez et al. 2004) and thus could be used as an indicator of previous androgen exposure.

In toxicological studies of EDCs, the AGD has been used as a measure of prenatal antiandrogen exposure in rats (Rhees et al. 1997; Hotchkiss et al. 2004). For example, Hotchkiss et al. (2004) showed that neonatal AGD was shortened by exposure to a mixture of antiandrogens, linuron, a fungicide, and butyl benzyl phthalate, a component of some plastics and other items used by humans. The demasculinizing effects continued into adulthood in the rats. So it is reasonable to expect that the AGD of humans can reveal any effects of a known EDC. The human AGD was described and called the "anogenital ratio" in a study of fetal virilization of female newborn infants (Callegari et al. 1987). In a recent study of 781 newly delivered male infants, Romano-Riquer et al. (2007) have provided extensive information on the measurement of human AGD and its statistical reliability. In their study, birth weight and gestational length were strongly related to AGD, but penile length was not.

Recent studies have begun to explore the role of an EDC, phthalates, on human AGD. Phthalates are commonly used in industry and commerce in a variety of personal care products, soaps, plastics, and some pesticides. Humans are known to have extensive exposure to phthalates and significant circulating and urinary concentrations of phthalate esters in human females have been reported (Colón et al. 2000; Swan et al. 2005). In rat studies, phthalate esters are a demonstrated antiandrogen (Parks et al. 2000) and are known to have a demasculinizing effect on AGD and reproductive organs (Nagao et al. 2000; Barlow & Foster 2003; Hotchkiss et al. 2004; Tyl et al. 2004). In the first examination of the possible role of an EDC on human AGD, Swan et al. (2005) compared variation in prenatal phthalate exposure (as measured by urinary metabolites in pregnant women) to the AGDI of their male offspring at or shortly after birth. A significant negative correlation was found between prenatal urinary phthalate concentration and the boys' AGDI. The findings were consistent with the phthalate induction of demasculinization demonstrated in rodent studies.

One concern with the Swan et al. (2005) study is whether the AGDI of infants was appropriately adjusted for body size. As described above in the mouse, simple body weight adjustment is the preferred adjustment (Hotchkiss & Vandenbergh 2005; see also Houtsmuller et al. 1997 for regression approaches to AGD in rats). Swan et al. (2005) found that simple body weight is the preferred method to standardize the measure of anogenital distance in human males. Use of body weight to calculate the AGDI in mice has been examined in a population of subjects with known IUP and compared with a variety of other possible standardizing measures (Hotchkiss & Vandenbergh 2005). Adjustment by body weight was found to equal or exceed all others in accurately detecting mice from different IUPs. Earlier, Gallavan et al. (1999) using a data set of 1,501 rat pup AGDs concluded that analysis of covariance is a preferred procedure to simply using body weight. Perhaps species differences may explain these contrasting conclusions.

Although more work needs to be done to explore the use of the AGD in human studies of endocrine disruption, it seems clear that preliminary studies, such as that of Swan et al. (2005), provide very provocative data. Their findings may stimulate searches for other possible correlates of the AGDI in humans with EDCs. Rodent studies demonstrate a robust relationship between AGDI and a variety of androgen exposures during fetal development and are used to identify putative EDCs (Gray et al. 1999; Mylchreest et al.; 1998 McIntyre et al. 2002; Wilson et al. 2002; Hotchkiss et al. 2004). The data now available suggest that the AGDI measured during infancy can signal prior exposure to such substances.

SUMMARY AND CONCLUSIONS

Research on the intrauterine position effect stimulated by the early studies of Clemens & Coniglio (1971) in rats and vom Saal & Bronson (1978) in mice has inspired a variety of studies. A summary of these findings on mice and rats is shown in Table 10-1. It has become clear that developing in proximity to males or, as in the rat, downstream from males, has long-term consequences for females. Most findings relate to transfer of testosterone from male to female fetuses but in at least one species, Mongolian gerbils, males can have a robust effect on adjacent male siblings as well. Some investigators have explored the long-term consequences of being near a male in utero on anatomical, physiological, behavioral, and ecological variables. In most species the results clearly demonstrated a long-term effect of being near a male fetus. Others have investigated the neuroendocrine mechanisms underlying hormone transfer between fetuses.

Few studies have explored the IUP effect in other litter-bearing species; this remains an area ripe for investigation. It may be particularly relevant to study potential AGD effects in domestic species such as the dog and cat as well as additional studies in swine. This review is a selective analysis of the literature and an attempt to uncover areas needing more research. Further, it is possible that research on the IUP can influence our understanding of the variability in human phenotypic traits. Specifically, whether the animal findings related to IUP can apply to characteristics of females in which there is male exposure during fetal development and whether the findings of IUP research can have relevance to our understanding of the consequences of endocrine and endocrine disruptor exposure during fetal development. With these thoughts in mind, the following general conclusions can be derived from this review of the literature on intrauterine effects in mammals.

First, fetuses in litter-bearing mammals have the possibility of being adjacent to the same-sex or different-sex siblings. This position can have developmental consequences because males produce relatively high concentrations of testosterone in the prenatal period that can transfer either through fetal membranes or the maternal circulation to adjacent siblings.

Second, the effects of intrauterine position contribute to variability seen in a number of traits in litter-bearing species at birth and continuing into adulthood. Consequently, as part of the "environment-gene" interaction, the perinatal period should be considered as a critical environmental period in the development of sexually dimorphic traits. Intrauterine position can be considered as an epigenetic effect.

Table 10-1. Major anatomical, physiological, and behavioral correlates of intrauterine position in female mice and rats (modified and updated from Table 1 in Ryan & Vandenbergh 2002). 0M and 2M designations can be based either on cesarean sections or on an index of the anogenital distance of the female. Mice and rats were chosen for this analysis because they are common animal models used in studies of endocrine disruptors.

Anatomical Effects:	0M	2M	References
Anogenital distance	shorter	longer	Clemens et al. 1978
Testis weight	lighter	heavier	van der Hoeven 1992
Accessory reproductive glands	smaller	larger	Nonneman et al. 1992
Brain SDN-POA	smaller	larger	Faber & Hughes 1992

Physiological Effects:	0M	2M	References
Fetal testosterone concentrations (m)	lower	higher	vom Saal et al. 1990
Postnatal sensitivity to testosterone (m)	less	more	vom Saal et al. 1983
Postnatal hypothalamic GABA (r)	less	more	Hernandez-Tristan et al. 1999
Sensitivity to bisphenol A	more	less	Howdeshell & vom Saal 2000
Impregnation	earlier	later	vom Saal 1989b

Behavioral Effects:	0M	2M	References
Female-female mounting behavior	less	more	Houtsmuller & Slob 1990
Lordosis	weaker	stronger	Houtsmuller & Slob 1990
Attractiveness to males	higher	lower	Rines & vom Saal 1984
Aggressive behavior	lower	higher	vom Saal et al. 1983
Rate of scent marking	lower	higher	vom Saal & Bronson 1980
Saccharin preference	lower	higher	Bushong & Mann 1994
Home range size in wild	smaller	larger	Zielinski et al. 1992
Home range size in enclosures	smaller	larger	Drickamer 1996
Fertility in enclosures	higher	lower	Drickamer 1996

Third, the factors regulating population dynamics, especially of rodents, should include consideration of the intrauterine position. The IUP can explain some of the variability contributing to distribution of individuals, reproductive performance, sex ratio, and, perhaps, survival.

Fourth, recent research on the IUP effect is relevant to toxicological studies in two ways. First, the studies suggest that developing females are exposed to varying amounts of fetal androgens, and this can interact with EDCs, especially antiandrogens. To reduce variability in rodent assays of putative EDCs, mice or rats can be selected on the basis of AGD to be either more (2M-like) or less (0M-like) sensitive to antiandrogens. Alternatively, 1M-like individuals could be chosen as subjects to eliminate potential outliers. Second, the primary measure of androgenization used in rodent studies is the anogenital distance; this measure may prove to be a simple but informative parameter for discovering endocrine disruptor effects in humans.

Finally, sex differences in disease susceptibility and treatment have been explored intensively in recent years. For an important report on sex differences in human biology and health from the Institute of Medicine of the National Academies of Science see Wizemann & Pardue (2001). It is clear that a complex of genetic and environmental causes underlie these differences; among the environmental factors, we must consider natural exposure to hormones from the fetus's own developing gonads, the maternal circulation, and exposure to hormone mimics during pregnancy and the nursing period. The IUP effect provides an animal model that could benefit research in this area.

ACKNOWLEDGMENTS

I thank Drs. Bryce C. Ryan and Andrew K. Hotchkiss for informative laboratory discussions and for providing editorial comments on this paper. Barbara Vandenbergh and several anonymous reviewers provided editorial comments to greatly improve the manuscript. Support was provided by the W. M. Keck Center for Behavioral Biology at North Carolina State University.

REFERENCES

Barlow, N. J. & Foster, P. M. D. 2003. Pathogenesis of male reproductive tract lesions from gestation through adulthood following in utero exposure to di(n-butyl) phthalate. *Toxicologic Pathology*, 31, 397–410.

Boklage, C. E. 1985. Interactions between opposite-sex dizygotic fetuses and the assumptions of Weinberg difference method epidemiology. *American Journal of Human Genetics*, 37, 591–605.

Bulman-Fleming, B. & Wahlsten, D. 1991. The effects of intrauterine position on the degree of corpus callosum deficiency in two substrains of BALB/c mice. *Developmental Psychobiology*, 24, 395–412.

Bushong, M. E. & Mann, M. A. 1994. Gender and intrauterine position influence saccharin preference in mice. *Hormones and Behavior*, 28, 207–218.

Callegari, C., Everett, S., Ross, M. & Brasel, J. A. 1987. Anogenital ratio: measure of fetal virilization in premature and full-term newborn infants. *Journal of Pediatrics*, 111, 240–243.

Cantoni, D., Glaizot, O. & Brown, R. E. 1999. Effects of sex composition of the litter on anogenital distance in California mice (*Peromyscus californicus*). *Canadian Journal of Zoology*, 77, 124–131.

Champagne, F. & Curley, J. P. 2009. The trans-generational influence of maternal care on offspring gene expression and behavior in rodents. In: *Maternal Effects in Mammals* (Ed. by D. Maestripieri & J. M. Mateo), pp. 182–202. Chicago: University of Chicago Press.

Clark, M. M. & Galef, B. G., Jr. 1988. Effects of uterine position on rate of sexual development in female Mongolian gerbils. *Physiology & Behavior*, 42, 15–18.

Clark, M. M & Galef, B. G., Jr. 1990. Sexual segregation in the left and right horns of the gerbil uterus: "The male embryo is usually on the right, the female on the left" (Hippocrates). *Developmental Psychobiology*, 23, 29–37.

Clark, M. M. & Galef, B. G., Jr. 1995. A gerbil dam's fetal intrauterine position affects the sex ratios of litters she gestates. *Physiology & Behavior*, 57, 297–299.

Clark, M. M. & Galef, B. G., Jr. 2000. Why some male Mongolian gerbils may help at the nest: testosterone, asexuality and alloparenting. *Animal Behaviour*, 59, 801–806.

Clark, M. M., Bone, S. & Galef, B. G., Jr. 1989. Uterine positions and schedules of urination: correlates of differential maternal anogenital stimulation. *Developmental Psychobiology*, 22, 389–400.

Clark, M. M., Malenfant, S. A., Winter, D. A. & Galef, B. G., Jr. 1990. Fetal uterine position affects copulation and scent marking by adult male gerbils. *Physiology & Behavior*, 47, 301–305.

Clark, M. M., Crews, D. & Galef, B. G., Jr. 1991. Concentrations of sex steroid hormones in pregnant and fetal Mongolian gerbils. *Physiology & Behavior*, 49, 239–243.

Clark, M. M., Tucker, L. & Galef, B. G., Jr. 1992. Stud males and dud males: intra-uterine position effects on the reproductive success of male gerbils. *Animal Behaviour*, 43, 215–221.

Clark, M. M., Karpuik, P. & Galef, B. G., Jr. 1993. Hormonally mediated inheritance of acquired characteristics in Mongolian gerbils. *Nature*, 364, 712.

Clark, M. M., Vonk, J. M. & Galef, B. G., Jr. 1998. Intrauterine position, parenting, and nest-site attachment in male Mongolian gerbils. *Developmental Psychobiology*, 32, 177–181.

Clemens, L. G. & Coniglio, L. 1971. Influence of prenatal litter composition on mounting behavior of female rats. *American Zoologist*, 11, 617–618.

Clemens, L. G., Gladue, B. A. & Coniglio, L. P. 1978. Prenatal endogenous androgenic influences on masculine sexual behavior and genital morphology in male and female rats. *Hormones and Behavior*, 10, 40–53.

Colón, I., Caro, D., Bourdony, C. J. & Rosario, O. 2000. Identification of phthalate esters in the serum of young Puerto Rican girls with premature breast development. *Environmental Health Perspectives*, 108, 895–900.

Cowell, L. G., Crowder, L. B. & Kepler, T. B. 1998. Density-dependent prenatal androgen exposure as an endogenous mechanism for the generation of cycles in small mammal populations. *Journal of Theoretical Biology*, 190, 93–106.

Drickamer, L. C. 1996. Intra-uterine position and anogenital distance in house mice: consequences under field conditions. *Animal Behaviour*, 51, 925–934.

Drickamer, L. C., Arthur, R. D. & Rosenthal, T. L. 1997. Conception failure in swine: importance of the sex ratio of a female's birth litter and tests of other factors. *Journal of Animal Science*, 75, 2192–2196.

Drickamer, L. C., Robinson, A. S. & Mossman, C. A. 2001. Differential responses to same and opposite sex odors by adult house mice are associated with anogenital distance. *Ethology*, 107, 509–519.

Faber, K. A. & Hughes, C. L., Jr. 1992. Anogenital distance at birth as a predictor of volume of the sexually dimorphic nucleus of the preoptic area of the hypothalamus and pituitary responsiveness in castrated adult rats. *Biology of Reproduction*, 46, 101–104.

Forger, N. G. 1999. Psychological sexual differentiation. In: *Encyclopedia of Reproduction.* Vol. 4 (Ed. by E. Knobil & J. D. Neill), pp. 421–430. New York: Academic Press.

Forger, N. G., Galef, B. G., Jr. & Clark, M. M. 1996. Intrauterine position affects motoneuron number and muscle size in sexually dimorphic neuromuscular system. *Brain Research*, 735, 119–124.

Gallavan, R. H., Jr., Holson, J. F., Stump, D. G., Knapp, J. F. & Reynolds, V. L. 1999. Interpreting the toxicologic significance of alterations in anogenital distance: potential for confounding effects of progeny body weights. *Reproductive Toxicology*, 13, 383–390.

Gandelman, R., vom Saal, F. S. & Reinisch, J. M. 1977. Contiguity to male fetuses affects morphology and behavior of female mice. *Nature*, 266, 722–724.

Gorodeski, G. I., Sheean, L. A. & Utian, W. H. 1995. Sex hormone modulation of flow velocity in the parametrial artery of the pregnant rat. *American Journal of Physiology*, 268, R614–624.

Goy, R. W. & McEwen B. S. 1980. *Sexual Differentiation of the Brain*. Cambridge, MA: MIT Press.

Gray, L. E., Jr., Wolf, C., Lambright, C., Mann, P., Price, M., Cooper, R. L. & Ostby, J. 1999. Administration of potentially antiandrogenic pesticides (procymidone, linuron, iprodione, chlozolinate, p,p'-DDE, and ketoconazole) and toxic substances (dibutyl- and diethylhexyl phthalate, PCB 169, and ethane dimethane sulphonate) during sexual differentiation produces diverse profiles of reproductive malformations in the male rat. *Toxicology and Industrial Health*, 15, 94–118.

Hendricks, C. & McClintock, M. K. 1990. Timing of insemination is correlated with the secondary sex ratio of Norway rats. *Physiology & Behavior*, 48, 625–632.

Hernandez-Tristan, R., Arevalo, C. & Canals, S. 1999. Effects of prenatal uterine position on male and female rats' sexual behavior. *Physiology & Behavior*, 67, 401–408.

Hotchkiss, A. K. & Vandenbergh, J. G. 2005. The anogenital distance index of mice (*Mus musculus domesticus*): an analysis. *Contemporary Topics in Laboratory Animal Science*, 44, 46–48.

Hotchkiss, A. K., Parks-Saldutti, L. G., Ostby, J. S., Lambright, C., Furr, J., Vandenbergh, J. G. & Gray, L. E., Jr. 2004. A mixture of the "antiandrogens" linuron and butyl benzyl phthalate alters sexual differentiation of the male rat in a cumulative fashion. *Biology of Reproduction*, 71, 1852–1861.

Houtsmuller, E. J. & Slob, A. K. 1990. Masculinization and defeminization of female rats by males located caudally in the uterus. *Physiology & Behavior*, 48, 555–560.

Houtsmuller, E. J., Thornton, J. A. & Rowland, D. L. 1997. Using a regression approach to study the influence of male fetuses on the genital morphology of neonatal female rats. *Multivariate Behavioral Research*, 32, 77–94.

Howdeshell, K. I. & vom Saal, F. S. 2000. Developmental exposure to bisphenol A: interaction with endogenous estradiol during pregnancy in mice. *American Zoologist*, 40, 429–437.

Ims, R. A. 1989. Kinship and origin effects on dispersal and space sharing in *Clethrionomys rufocanus*. *Ecology*, 70, 607–616.

Ims, R. A. 1990. Determinants of natal dispersal and space use in grey-sided voles, *Clethrionomys rufocanus*: a combined field and laboratory experiment. *Oikos*, 57, 106–113.

James, W. H. 1987. The human sex ratio. I. A review of the literature. *Human Biology*, 59, 721–752.

Jones, D, Gonzalez-Lima, F., Crews, D., Galef, Jr. B. G. & Clark, M. M. 1997. Effects of intrauterine position on the metabolic capacity of the hypothalamus of female gerbils. *Physiology & Behavior*, 61, 513–519.

Kerin, T. K., Vogler, G. P., Blizard, D. A., Stout, J. T., McClearn, G. E. & Vandenbergh, D. J. 2003. Anogenital distance measured at weaning is correlated with measures of blood chemistry and behaviors in 450-day-old female mice. *Physiology & Behavior*, 78, 697–702.

Krohmer, R. W. & Baum, M. J. 1989. Effect of sex, intrauterine position and androgen manipulation on the development of brain aromatase activity in fetal ferrets. *Journal of Neuroendocrinology*, 1, 265–271.

Lipton, J. W., Robie, H. C., Ling, Z., Weese-Mayer, D. E. & Carvey, P. M. 1998. The magnitude of brain dopamine depletion from prenatal cocaine exposure is a function of uterine position. *Neurotoxicology and Teratology*, 20, 373–381.

Loehlin, J. C. & Martin, N. G. 1998. A comparison of adult female twins from opposite-sex and same-sex pairs on variables related to reproduction. *Behavior Genetics*, 28, 21–27.

Lummaa, V., Pettay, J. R. & Russell, A. F. 2007. Male twins reduce fitness of female co-twins in humans. *Proceedings of the National Academy of Sciences USA*, 104, 10915–10920.

Massey, A. 1982. Ecology of house mouse populations confined by highways. *Journal of the Elisha Mitchell Scientific Society*, 98, 135–143.

McDermott, N. J., Gandelman, R. & Reinisch, J. M. 1978. Contiguity to male fetuses influences ano-genital distance and time of vaginal opening in mice. *Physiology & Behavior*, 20, 661–663.

McFadden, D. 1993. A masculinizing effect on the auditory systems of human females having male co-twins. *Proceedings of the National Academy of Sciences USA*, 90, 11900–11904.

McFadden, D., Loehlin, J. C. & Pasanen, E. G. 1996. Additional finding on heritability and prenatal masculinization of cochlear mechanisms: click-evoked otoacoustic emissions. *Hearing Research*, 97, 102–119.

McIntyre, B. S., Barlow, N. J. & Foster, P. M. D. 2002. Male rats exposed to linuron *in utero* exhibit permanent changes in anogenital distance, nipple retention, and epididymal malformations that result in subsequent testicular atrophy. *Toxicological Sciences*, 65, 62–70.

Meisel, R. L. & Ward, I. L. 1981. Fetal female rats are masculinized by male littermates located caudally in the uterus. *Science*, 213, 239–242

Miller, E. M. 1994. Prenatal sex hormone transfer: a reason to study opposite-sex twins. *Personality and Individual Differences*, 17, 511–529.

Mylchreest, E., Cattley, R. C. & Foster, P. M. D. 1998. Male reproductive tract malformations in rats following gestational and lactational exposure to di(*n*-butyl) phthalate: an antiandrogenic mechanism? *Toxicological Sciences*, 43, 47–60.

Naftolin, F., MacLusky, N. J., Leranth, C. Z., Sakamoto, H. S. & Garcia-Segura, I. M. 1988. The cellular effects of estrogens on neuroendocrine tissues. *Journal of Steroid Biochemistry*, 30, 195–207.

Nagao, T., Ohta, R., Marumo, H., Shindo, T., Yoshimura, S. & Ono. H. 2000. Effect of butyl benzyl phthalate in Sprague-Dawley rats after gavage administration: a two-generation reproductive study. *Reproductive Toxicology*, 14, 513–532.

Nonneman, D. J., Ganjam, V. K., Welshons, W. V. & vom Saal, F. S. 1992. Intrauterine position effects on steroid metabolism and steroid receptors of reproductive organs in male mice. *Biology of Reproduction*, 47, 723–729.

Parks, L. G., Ostby, J. S., Lambright, C. R., Abbott, B. D., Klinefelter, G. R., Barlow, N. J. & Gray, L. E., Jr. 2000. The plasticizer diethylhexyl phthalate induces malformations by decreasing testosterone synthesis during sexual differentiation in the male rat. *Toxicological Sciences*, 58, 339–349.

Pei, M., Matsuda, K., Sakamoto, H. & Kawata, M. 2006. Intrauterine proximity to male fetuses affects the morphology of the sexually dimorphic nucleus of the preoptic area of the adult rat brain. *European Journal of Neuroscience*, 23, 1234–1240.

Politch, J. A. & Herrenkohl, L. R. 1984. Effects of prenatal stress on reproduction in male and female mice. *Physiology & Behavior*, 32, 95–99.

Resnick, S. M., Gottesman, I. I. & McGue, M. 1993. Sensation seeking in opposite-sex twins: an effect of prenatal hormones? *Behavior Genetics*, 23, 323–329.

Rhees, R. W., Kirk, B. A., Sephton, S. & Lephart, E. D. 1997. Effects of prenatal testosterone on sexual behavior, reproductive morphology and LH secretion in the female rat. *Developmental Neuroscience*, 19, 430–437.

Rines, J. P. & vom Saal, F. S. 1984. Fetal effects on sexual behavior and aggression in young and old female mice treated with estrogen and testosterone. *Hormones and Behavior*, 18, 117–129.

Rohde Parfet, K. A., Ganjam, V. K., Lamberson, W. R., Rieke, A. R., vom Saal, F. S. & Day, B. N. 1990. Intrauterine position effects in female swine: subsequent reproductive performance, and social and sexual behavior. *Applied Animal Behaviour Science*, 26, 349–362.

Romano-Riquer, S. P., Hernandez-Avila, M., Gladen, B. C., Cupul-Uicab, L. A. & Longnecker, M. P. 2007. Reliability and determinants of anogenital distance and penis dimensions in male newborns from Chiapas, Mexico. *Paediatric and Perinatal Epidemiology*, 21, 219–228.

Rose, R. J., Kaprio, J., Winter, T., Dick, D. M., Viken, R. J., Pulkkinen, L. & Koskenvuo, M. 2002. Femininity and fertility in sisters with twin brothers: prenatal androgenization? Cross-sex socialization? *Psychological Science*, 13, 263–267.

Ryan, B. C. & Vandenbergh, J. G. 2002. Intrauterine position effects. *Neuroscience and Biobehavioral Reviews*, 26, 665–678.

Salazar-Martinez, E., Romano-Riquir, P., Yanez-Marquez, E., Longnecker, M. P. & Hernandez-Avila, M. 2004. Anogenital distance in human male and female newborns: a descriptive, cross-sectional study. *Environmental Health*, 3, 8–20.

Sherry, D. F., Galef, B. G., Jr. & Clark, M. M. 1996. Sex and intrauterine position influence the size of the gerbil hippocampus. *Physiology & Behavior*, 60, 1491–1494.

Sickel, M. J. & McCarthy, M. M. 2000. Calbindin-D28k immunoreactivity is a marker for a subdivision of the sexually dimorphic nucleus of the preoptic area of the rat: developmental profile and gonadal steroid modulation. *Journal of Neuroendocrinology*, 12, 397–402.

Steel, R. G. D., Torrie, J. H. & Dickey, D. A. 1995. *Principles and Procedures of Statistics: A Biometrical Approach*. New York: McGraw-Hill.

Swan, S. H., Main, K. M., Liu, F., Stewart, S. L., Kruse, R. L., Calafat, A. M., Mao, C. S., Redmon, J. B., Ternand, C. L., Sullivan, S. & Teague, J. L. 2005. Decrease in anogenital distance among male infants with prenatal phthalate exposure. *Environmental Health Perspectives*, 113, 1056–1061.

Tobet, S. A., Baum, M. J., Tang, H. B., Shim, J. H. & Canick, J. A. 1985. Aromatase activity in the perinatal rat forebrain: effects of age, sex and intrauterine position. *Developmental Brain Research*, 23, 171–178.

Tyl, R. W., Myers, C. B., Marr, M. C., Fail, P. A., Seely, J. C., Brine, D. R., Barter, R. A. & Butala, J. H.. 2004. Reproductive toxicity evaluation of dietary butyl benzyl phthalate (BBP) in rats. *Reproductive Toxicology*, 18, 241–264.

Valenstein, E. S., Kakolewski, J. W. & Cox, V. C. 1967. Sex differences in taste preference for glucose and saccharin solutions. *Science*, 156, 942–943.

Vandenbergh, J. G. 2003. Prenatal hormone exposure and sexual variation. *American Scientist*, 91, 218–225

Vandenbergh, J. G. & Huggett, C. L. 1994. Mother's prior intrauterine position affects the sex ratio of her offspring in house mice. *Proceedings of the National Academy of Sciences USA*, 91, 11055–11059.

Vandenbergh, J. G. & Huggett, C. L. 1995. The anogenital distance index: a predictor of the intrauterine position effects on reproduction in female house mice. *Laboratory Animal Science*, 45, 567–573.

van der Hoeven, T., Lefevre, R. & Mankes, R. 1992. Effects of intrauterine position on the hepatic microsomal polysubstrate monooxygenase and cytosolic glutathione S-transferase activity, plasma sex steroids and relative organ weights in adult male and female Long-Evans rats. *Journal of Pharmacology and Experimental Therapeutics*, 263, 32–39.

vom Saal, F. S. 1983. Variation in infanticide and parental behavior in male mice due to prior intrauterine proximity to female fetuses: elimination by prenatal stress. *Physiology & Behavior*, 30, 675–681.

vom Saal, F. S. 1984. The intrauterine position phenomenon: effects on physiology, aggressive behavior and population dynamics in house mice. In: *Biological Perspectives on Aggression* (Ed. by K. J. Flannelly, R. J. Blanchard & D. C. Blanchard), pp. 135–179. New York: Alan R. Liss.

vom Saal, F. S. 1989a. Perinatal testosterone exposure has opposite effects on intermale aggression and infanticide in mice. In: *House Mouse Aggression* (Ed. by P. F. Brain, M. Mainardi & S. Parmigiani), pp. 179–204. New York: Harwood Academic.

vom Saal, F. S. 1989b. The production of and sensitivity to cues that delay puberty and prolong subsequent oestrous cycles in female mice are influenced by prior intrauterine position. *Journal of Reproduction and Fertility*, 86, 457–471.

vom Saal, F. S. & Bronson, F. H. 1978. In utero proximity of female mouse fetuses to males: effects on reproductive performance during later life. *Biology of Reproduction*, 19, 842–853.

vom Saal, F. S. & Bronson, F. H. 1980. Sexual characteristics of adult female mice are correlated with their blood testosterone levels during prenatal development. *Science*, 208, 597–599.

vom Saal, F. S. & Dhar, M. G. 1992. Blood flow in the uterine loop artery and loop vein is bidirectional in the mouse: implications for transport of steroids between fetuses. *Physiology & Behavior*, 52, 163–171.

vom Saal, F. S. & Moyer, C. L. 1985. Prenatal effects on reproductive capacity during aging in female mice. *Biology of Reproduction*, 32, 1116–1126.

vom Saal, F. S., Pryor, S. & Bronson, F. H. 1981. Effects of prior intrauterine position and housing on oestrous cycle length in adolescent mice. *Journal of Reproduction and Fertility*, 62, 33–37.

vom Saal, F. S., Grant, W. M., Mc Mullen, C. W. & Laves, K. S. 1983. High fetal estrogen concentrations: correlation with increased sexual activity and decreased aggression in male mice. *Science*, 220, 1306–1309.

vom Saal, F. S., Quadagno, D. M., Even, M. D., Keisler, L. W., Keisler, D. H. & Khan, S. 1990. Paradoxical effects of maternal stress on fetal steroids and postnatal reproductive traits in female mice from different intrauterine positions. *Biology of Reproduction*, 43, 751–761.

Vreeburg, J. T., Groeneveld, J. O., Post, P. E. & Ooms, M. P. 1983. Concentrations of testosterone and androsterone in peripheral and umbilical venous plasma of fetal rats. *Journal of Reproduction and Fertility*, 68, 171–175.

Wilson, V. S., Lambright, C., Ostby, J. & Gray, L. E., Jr. 2002. In vitro and in vivo effects of 17β-trenbolone: a feedlot effluent contaminant. *Toxicological Sciences*, 70, 202–211.

Wizemann, T. M. & Pardue, M.-L. (Eds.). 2001. *Exploring the Biological Contributions to Human Health: Does Sex Matter?* Institute of Medicine, National Academy Press, Washington, D.C.

Zehr, J. L., Gans, S. E. & McClintock, M. K. 2001. Variation in reproductive traits is associated with short anogenital distance in female rats. *Developmental Psychobiology*, 38, 229–238.

Zielinski, W. J. & Vandenbergh, J. G. 1991. Effect of intrauterine position and social density on age of first reproduction in wild-type female house mice (*Mus musculus*). *Journal of Comparative Psychology*, 105, 134–139.

Zielinski, W. J., Vandenbergh, J. G. & Montano, M. M. 1991. Effects of social stress and intrauterine position on sexual phenotype in wild-type house mice (*Mus musculus*). *Physiology & Behavior*, 49, 117–123.

Zielinski, W. J., vom Saal, F. S. & Vandenbergh, J. G. 1992. The effect of intrauterine position on the survival, reproduction and home range size of female house mice (*Mus musculus*). *Behavioral Ecology and Sociobiology*, 30, 185–191.

11

Maternal Effects in Fissiped Carnivores

KAY E. HOLEKAMP AND STEPHANIE M. DLONIAK

INTRODUCTION

Most mammalian carnivores give birth to altricial young whose survival depends upon parental care for an extended period; in some species, the period of offspring dependence on the mother may last several years (e.g., in various bears, Ramsay & Stirling 1988). Production of helpless offspring with poor motor coordination is particularly characteristic of the fissiped carnivores (Bininda-Emonds & Gittleman 2000). "Fissiped" literally means carnivores having toes separated to the base; these include all the extant mammalian carnivores except seals, sea lions, and the walrus, all of which have toes joined by webbing to transform their feet into paddles (the "pinnipeds"; see Bowen, this volume). Although most fissipeds are terrestrial, this group includes otters, which are members of the same family that contains badgers and weasels. In addition, fissipeds also include cats, dogs, bears, hyenas, mongooses, civets, and raccoons and their relatives.

Young fissiped carnivores generally cannot defend themselves against predators, forage on their own, or acquire prey. They are usually maintained for the first several weeks or months of life in a den, creche, or burrow (Caro 1994). When effects of body mass are controlled, fissipeds have a longer period between weaning and puberty (Read & Harvey 1989), and a longer period between weaning and independence from parental care (Caro 1994) than any other mammals except primates. Although adult males and variable numbers of helpers or "allo-mothers" participate in offspring care in some mongooses and most canids, parental care in virtually all other carnivores is undertaken exclusively by females (Ewer 1973; Rasa 1986). In most fissipeds, the mother is responsible for selection and preparation of the birth site, as well as grooming, guarding, warming, retrieving, and suckling the young (e.g., Hofer & East 1996). In addition, the mother is usually also responsible for providing solid foods for weaned offspring until they can forage independently. Most young carnivores remain with their mother for several weeks or months after weaning, and laboratory data suggest that extended interactions with the mother are critical for normal health and behavioral development. For example, puppies and kittens removed from their mothers just a few weeks before the normal age of mother-infant separation exhibit loss of appetite and weight, as well as increased distress, aggression, and susceptibility to disease and mortality compared with youngsters that remain with their mothers (Seitz 1959; Slabbert & Rasa 1993).

The critical role played by the mother in caring for and interacting with young throughout the extended period before young carnivores can survive independently suggests that maternal variables might have particularly important effects in shaping morphology, survivorship, and behavior of offspring in this group of mammals. Maternal effects occur when maternal phenotype influences offspring phenotype independent of offspring genotype (Bernardo 1996; Mousseau & Fox 1998). Such effects might occur either before or after birth, and they might potentially be mediated by maternal physiology, behavior, social status, or some combination of these variables. Furthermore, maternal effects should theoretically be particularly important in mammalian carnivores because each species occurs under a wide array of environmental conditions throughout its distributional range. Under such varied environmental conditions, maternal effects should permit whatever specific local environment is experienced by the mother to affect the phenotypes of her young and thereby potentially enhance offspring fitness in that environment. Here we review various aspects of the maternal phenotype known or suspected to influence offspring phenotype in fissiped carnivores, independent of offspring genotype. Although data documenting mechanisms by which maternal effects influence offspring phenotype are sparse, we also review those data whenever they are available. Finally, we devote considerable attention to maternal effects associated with social status in spotted hyenas (*Crocuta crocuta*), on which we have gathered a large amount of data through long-term study of known individuals in the wild. The spotted hyena is a particularly interesting species in which to study maternal effects because it occurs under a vast array of different ecological conditions across sub-Saharan Africa, and because of its unusually large and complex societies.

EFFECTS OF MATERNAL AGE, SIZE, CONDITION, AND EXPERIENCE

As in other mammals, a number of physiological and experiential characteristics of female carnivores can have significant effects on their offspring. Infant phenotype and survivorship are known to be influenced in some fissiped carnivores by the mother's age, size, body condition, and experience or parity. Among dwarf mongooses (*Helogale parvula*), older females are more likely than younger females to produce young (Rood 1990), and the success of females' breeding attempts increases with experience (Creel & Waser 1991). Among banded mongooses (*Mungos mungo*), older and larger females bear larger litters containing larger fetuses than do younger or smaller females (Gilchrist et al. 2004). Similarly, maternal mass at the time of conception affects litter size at birth and pup weight at weaning in meerkats (*Suricata suricatta*; Russell et al. 2003); larger females deliver larger litters and produce heavier offspring, both of which are known to correlate positively with measures of female breeding success in meerkats (Russell et al. 2004). Both age and body size in meerkats and banded mongooses often vary with social rank (see below), so effects of these variables may be mediated largely by differential access to food and other resources.

Quantity and quality of resources available to mothers vary among both solitary and gregarious carnivores. Resource availability may affect the body condition of breeding females, and maternal condition in turn is known to affect offspring phenotype and survival in some species under certain ecological conditions. For example, when prey are scarce, the autumn body condition of female European badgers (*Meles meles*) has a strong influence on whether they will produce viable offspring (Woodroffe & Macdonald 1995). Similarly, winter body condition of female wolverines (*Gulo gulo*) affects the likelihood that their young will survive to weaning age (Persson 2005). Poor nutritional condition and stressful environments often inhibit maternal production of antibodies such that females in different conditions may vary in their ability to transfer immunocompetence to their offspring (Roulin & Heeb 1999). In addition, females in poor condition may produce smaller quantities of milk, or milk of poorer quality, as occurs among pinnipeds (Georges et al. 2001).

Whereas multiparous females, those who have borne prior litters, have usually accrued prior experience at parenting, most nulliparous females have not. Thus maternal parity and experience generally covary. Maternal parity affects offspring sex ratios and offspring survivorship in some fissiped carnivores. Maternal parity influences offspring sex in African wild dogs (*Lycaon pictus*); primiparous females produce more sons than daughters, whereas multiparous females produce an excess of daughters (Creel et al. 1998). When parity is controlled, maternal effects on offspring sex ratio have been documented in some fissiped species, and these effects appear to be mediated by the social environment experienced by the mother, presumably because the social environment in turn affects maternal condition. For example, captive female silver foxes (*Vulpes vulpes*) living under highly competitive conditions produce more daughters than sons whereas those living under low-competition conditions produce more sons than daughters (Bakken 1995). In contrast, female spotted hyenas produce more female offspring when population density is low and feeding competition is weak, but produce more sons than daughters when population density is high and feeding competition is relatively intense (Holekamp & Smale 1995). This relationship between intensity of feeding competition and biased offspring sex ratios may be due to the fact that female hyenas are philopatric whereas males disperse; when competition is relaxed, females can afford to produce more daughters as these will inevitably compete with their mothers for food throughout their lives. Competition with sons will be minimal as all males disperse shortly after puberty (Smale et al. 1997). The physiological mechanisms mediating sex ratio adjustment in carnivore litters are not known.

Among Dutch Kooiker dogs (*Canis familiaris*) and spotted hyenas, litters produced by multiparous females tend to be larger than those produced by primiparous females (Mandigers et al. 1994; Holekamp et al. 1996). However, litter size in Dutch Kooiker dogs and dholes (*Cuon alpinus*) also tends to decline with increasing maternal age (Mandigers et al. 1994; Venkataraman 1998), but this does not occur among spotted hyenas (Holekamp et al. 1996) or European badgers (Woodroffe & Macdonald 1995). Neither kitten mortality nor survival time is affected by maternal parity in feral cats (*Felis catus*: Nutter et al. 2004). Among multiparous cheetahs (*Acinonyx jubatus*), Iberian lynx (*Lynx pardinus*), and African lions (*Panthera leo*), maternal age has no effect on offspring survivorship (Packer at al. 1988; Laurenson 1993; Palomares et al. 2005). However, among cheetahs and red foxes (*Vulpes vulpes*), maternal choice of lairs can affect cub survivorship through differential exposure of infants to weather or predators, and maternal experience may influence females' decisions regarding where to keep their young cubs (Henry 1985; Laurenson 1993). Overall, it appears that the effects of maternal age and parity vary considerably among the fissiped species in which such effects have been examined.

In some carnivores, female age, size, condition, and parity may all vary together, so it is difficult to isolate the effects of each variable independently of the others. Overall, however, it seems reasonable to conclude that young carnivores are most likely to fare well if their mothers are relatively large, are in good physical condition, and have already accrued some prior experience at parenting. We would also expect offspring of primiparous females in communally breeding species to fare better if their mothers had previously gained experience serving as helpers or allo-mothers than if their mothers had not served previously in this capacity, although to our knowledge this has never been investigated in fissiped carnivores.

EFFECTS OF MATERNAL DEMEANOR AND STRESS

Pronounced individual differences are commonly observed among female carnivores with respect to the nature of their interactions with their young (e.g., wolves, *Canis lupus*, Fox 1972; domestic dogs, *Canis familiaris*, Rheingold 1963; Wilsson 1984). In some species, such variation in maternal demeanor is associated with variation in offspring behavioral phenotype. For example, German shepherd bitches exhibiting a relatively aggressive "disciplinarian" mothering style during the weaning period appeared to produce offspring that were later more submissive than pups of "nondisciplinarian" mothers (Wilsson 1984).

Epigenetic determination of stress responses through variation in maternal care has not yet been documented in carnivores as it has in captive primates and laboratory rodents (e.g., Levine 1967; Fish et al. 2004; Cameron et al. 2005; Champagne & Curley, Maestripieri, this volume). However, it has been well-established that short periods of daily handling or comparable stimulation can have pronounced long-term effects on the behavioral and physical development of both kittens (Meier 1961) and puppies (Fox 1978). For example, puppies exposed to such daily stimulation from birth to 5 weeks of age are more confident, exploratory, and socially dominant when tested later in strange situations than are unstimulated controls (Fox 1978). Exposing puppies or fox cubs to handling or other mild stressors during the neonatal period tends to produce calm individuals that are not easily stressed or frightened (Fox & Stelzner 1966; Pedersen & Jeppesen 1990; Serpell & Jagoe 1995). Similarly, early handling programs have also been reported to improve subsequent emotional stability and learning capacity in some dog breeds (Fox 1978). These data suggest that felids and canids, and perhaps other carnivores as well, may be sensitive as neonates to some effects of maternal style as well as human handling. It seems reasonable to expect that these long-term effects might be mediated, as they are in rodents (e.g., Fish et al. 2004), by naturally occurring variations in the behaviors female carnivores direct toward their offspring, including grooming and nursing of young. It further seems reasonable to expect that, as occurs in rodents, such variations in maternal behavior modify function of the hypothalamic-pituitary-adrenal (HPA) axis and central nervous system in offspring (e.g., see Champagne & Curley, this volume). However, neither of these hypotheses has yet been tested in carnivores.

Negative effects of maternal stress during the postnatal period are well known in captive carnivores. For example, captive female cheetahs and snow leopards will neglect their cubs when disturbed or stressed (Marma & Yunchis 1986; Laurenson et al. 1995). Similarly, low prey availability in the natural habitat, and perhaps also certain types of anthropogenic disturbances, can cause wild female cheetahs to neglect or abandon their cubs (Laurenson 1993). Anthropogenic disturbance has recently been found to alter patterns of den attendance and other aspects of maternal care in free-living spotted hyenas (Kolowski et al. 2007).

EFFECTS OF NUMBER OF HELPERS OR ALLO-MOTHERS

In many cooperatively breeding carnivores, infants are cared for, not only by their own mothers, but also by other individuals present in the social group (Creel & Creel 1991). Postpubertal older siblings or other adults participate in warming, grooming, protecting, and transporting infants in dwarf mongooses, banded mongooses, and meerkats. In many canid species, helpers perform these caretaking tasks and also routinely provision offspring with food (Moehlman & Hofer 1997). Survivorship of current offspring increases with the number of helpers available in black-backed jackals (*Canis mesomelas*, Moehlman 1979), African wild dogs (Malcolm & Marten 1982; Creel et al. 2004), and meerkats (Russell et al. 2003). We consider these effects of helper number as a subset of maternal effects because, in many respects, number of helpers effectively represents the number of "mothers" participating in offspring care, and many effects of helper number appear to be mediated by the same mechanisms as do more traditional maternal effects. That is, when multiple allo-mothers or helpers provision infants they can potentially transfer more energy or nutrients to the young per unit time than can individual mothers acting alone, and they can also potentially protect infants better than can single mothers. For example, the amount of food fed to meerkat pups increases with the number of helpers (Clutton-Brock et al. 2001), and both growth and survival of pups are directly enhanced by the presence of more helpers (Russell et al. 2002).

In some group-living fissipeds, including lions, brown hyenas (*Parahy-aena brunnea*), various canids, and banded mongooses, adult females breed synchronously, and lactating females allow non-offspring nursing, such that each youngster in each litter born into the social group is nursed by multiple females (Schaller 1972; Rood 1974; Owens & Owens 1979; Gubernick 1981; Mills 1990; Pusey & Packer 1992; Moehlman & Hofer 1997; Macdonald et al. 2004). Furthermore, in various cooperative breeders, although the alpha female is usually the only one to produce surviving young, multiple allo-mothers may also lactate concurrently with the mother, either spontaneously or after loss of their own litters, such that the single surviving litter born into the social group is nursed by multiple females. This occurs in meerkats (Scantlebury et al. 2002) and in several canids including bat-eared foxes (*Otocyon megalotis*, Pauw 2000; Maas & Macdonald 2004), Ethiopian wolves (*Canis simensis*, Sillero-Zubiri et al. 2004), coyotes (*Canis latrans*, Gubernick 1981), and other species (Moehlman & Hofer 1997). In each case, infant ingestion of colostrum and milk produced by multiple mothers might be expected to enhance development of humoral (antibody-mediated) immunity to pathogens in that the array of maternal antibodies (or antibody paratopes to a single pathogen, Roitt et al. 1996) transmitted to young should be larger from multiple lactating females than from single mothers (Roulin & Heeb 1999; Grindstaff et al. 2003). Maternal immunoglobins transferred to youngsters in colostrum and milk are known to enhance postnatal immune system function in domestic cats (*Felis catus*) and dogs (e.g., Omata et al. 1994; Pu et al. 1995; Giger & Casal 1997; Kolb 2003). However, at present very little is known about individual differences within any carnivore species in regard to transmission of maternal antibodies, and no empirical data have yet been gathered documenting how nursing from multiple females affects immune system function in offspring. Thus much remains to be learned in fissipeds about maternal effects that might be mediated by the immune system.

MATERNAL EFFECTS ON HUNTING AND FEEDING

In a diverse array of fissiped carnivores, mothers have important effects on the ability of their offspring to cope with the specific ecological conditions encountered in the natal area. For example, with respect to feeding ecology, female carnivores have been found to affect abilities of their offspring to find prey in the local landscape, select appropriate prey from among those available, and capture prey successfully (e.g., Caro 1994). The species ranges of most fissiped carnivores are quite large, and the prey types available within each range may vary greatly among specific locales. Most fissipeds are highly flexible in their choice of prey, and this flexibility, which appears in many species to be shaped by maternal effects, allows them to take full advantage of whatever prey species are locally available. The important role of maternal effects in the development of offspring foraging competence has been particularly well-documented in felids, otters, and bears.

Early experiments by Kuo (1930; 1938) suggested that female domestic cats ("queens") have a strong influence on prey recognition and prey preferences in their offspring. Domestic cats begin to imitate their mothers' food preferences at 35–56 days of age, and retain this preference long after weaning (Wyrwicka 1978; see Galef, this volume). Kittens presented with a novel food item in the presence of the mother start ingesting it sooner than when the mother is absent (Wyrwicka & Long 1980). A series of experiments by Caro (1980a, b, c) indicated that maternal effects are important determinants of prey recognition and prey capture abilities in domestic cats. First, Caro (1980a) showed that experience with a particular type of prey when kittens are 1–3 months old alters subjects' later responses to that and other prey as adults, and enhances subjects' ability to kill prey. Specifically, adult cats tend to kill and eat more of the particular prey type to which they were exposed as infants than do control cats not exposed to prey during infancy. His work also suggested that the presence of the queen during her infants' exposure to prey may improve her offsprings' ability to catch prey when they are later tested as adults (Caro 1980b). However, Baerends-van Roon and Baerends (1979) found no effect of interacting with adult cats that killed prey on the development of predatory behavior in kittens. Finally, Caro's (1980a, b, c) work documented how predatory behavior in queens changes as their kittens develop, and strongly suggested that specific aspects of the queen's behavior may influence her offsprings' subsequent interactions with prey.

Caro (1980b, c) found that queens vocalize when they bring prey to the lair, and this appears to encourage the kittens to interact with prey. Queens also carry prey directly to kittens, then leave the prey when kittens begin to interact with it. Kittens' interest in prey and time spent interacting with prey both increase when kittens have opportunities to monitor their mothers' interactions with prey. Kittens pay more attention to prey if their mothers also attend to it, and kittens show increased rates of predatory behavior in the presence of their mothers. Caro concluded that the mother plays a critical role in drawing the attention of her offspring to prey and encouraging them to interact with it. Thus the mechanisms by which maternal effects operate here include two simple forms of learning, stimulus enhancement and response facilitation. In stimulus enhancement (Spence 1937), the probability of offspring approaching or contacting a stimulus in the environment is increased by the presence of the mother interacting with that stimulus. In response facilitation (Byrne 1994), the presence of the mother performing an act (particularly one resulting in a reward) increases the probability that the offspring will also perform the same act. Moreover, kittens can learn directly via observation of their mothers (Chesler 1969), so there may also be social learning involved in these maternal effects that makes heavier cognitive demands on youngsters than do these two forms of social facilitation.

Provisioning of young with solid food is often a critical aspect of maternal care in fissiped carnivores; youngsters rely almost entirely on their mothers for solid food from the time they emerge from the natal lair until they separate from the mother. Many species of fissipeds have been observed bringing prey to their young, including domestic cats (Ewer 1969), lions (Schenkel 1966), tigers (*Panthera tigris*, Schaller 1976), cheetahs (Kruuk & Turner 1967; Caro 1994), servals (*Leptailurus serval*, Estes 1991), black-footed cats (*Felis nigripes*, Leyhausen 1979), ermine (*Mustela erminea*, Rasa 1986), dwarf mongooses (Rasa 1973), meerkats (Thornton & McAuliffe 2006), and various otters (Liers 1951; Watt 1993). Domestic cat mothers initially bring dead prey to their offspring in the den or lair and eat it in front of them, but later they bring in prey without eating it themselves (Caro 1980b; 1994). Next they bring live prey to the young, or bring their young to incapacitated prey, and allow offspring to play with it, but mothers often recapture the prey animal if it escapes. Finally, the mother merely initiates movement with her young toward prey, and permits them to capture and kill it (Caro 1980b; 1994). By this sequence of actions during cub development, mothers draw offspring attention to particular prey items, and encourage offspring to interact with prey. Variations on this theme have been reported in other species (e.g., meerkats, Thornton & McAuliffe 2006). For example, solitary nocturnal mongooses do not carry prey to their young while offspring reside in a den, but when young start accompanying the mother on foraging trips, she feeds them on prey that she holds for them in her mouth (Rasa 1986). Similarly, female sea otters (*Enhydra lutris*) find molluscan prey on the sea floor, then return to the surface, where they open and consume the prey while offspring watch. The mother then regularly hands to her cub the mollusk shells from which food has been extracted, and the cub mouths and manipulates these shells (Rasa 1986). Thus, even when resources provisioned to youngsters have little or no nutritive value, the mother is nonetheless providing her offspring with important opportunities to learn about local food resources.

Young river otters (*Lutra lutra*) learn from their mothers which foraging areas are most likely to yield prey and how to handle difficult prey (Watt 1993). Otter cubs initially remain at the surface while the mother forages below, watching her with their faces in the water. Later cubs dive and resurface alongside the mother. When cubs wait at the surface, females provision them with prey. Whereas adults very rarely drop prey, juveniles often do so and thereby lose provisioned food items, so provisioning tends to be costly to mothers. As in felids (Kitchener 1999) and dwarf mongooses (Rasa 1973), female otters often bring prey (in this case, live fish and crayfish) to their young; they carry prey animals to the surface and deposit these for cubs in a shallow pool where cubs are allowed to recapture them (Kruuk 1995). In addition to increasing body size and motor coordination during cub maturation, a great deal of learning appears to be involved in development of efficient foraging in this species. Juvenile otters undergo a 9–10 month period of decreasing dependence on the mother for food as their own foraging efficiency increases, but full nutritional independence does not occur in river otters until approximately one year of age (Watt 1993).

During the long period of mother-cub association in many bear species, youngsters apparently learn a great deal from their mothers about foraging, predators, and other bears (Gilbert 1999). Many bears exploit food resources that can be found only at particular sites or times of year, and by associating closely with their mothers, cubs acquire information about availability of such resources in space and time, as well as about efficient travel routes between foraging sites. In addition, young bears learn methods for prey capture from their mothers. For example, Alaskan brown bears (*Ursus arctos*) utilize up to 28 different techniques for capturing salmon, and learning is extremely important for attaining high rates of salmon capture (Gilbert 1999).

There has been considerable debate in the literature regarding whether or not female carnivores engage in true teaching of their young (see also Galef,

Mateo, this volume). Caro & Hauser (1992) argue that teaching occurs when all three of the following conditions are met: when the teacher incurs some cost (or derives no immediate benefit) as a result of modifying its behavior in the presence of naïve individuals, when the modified behavior of the teacher causes naive individuals to acquire some behavior more rapidly or efficiently than they would otherwise, and when the teacher's behavior changes with the competence of naïve individuals. There can be little doubt that provisioning youngsters with live prey is costly to female carnivores, as they could otherwise simply ingest provisioned prey themselves without risking prey escape. However, evidence is very rare that development of independent prey-capture is accelerated when youngsters interact with provisioned prey.

Thornton & McAuliffe (2006) found, based on observational and experimental data, that wild meerkats teach pups prey-handling skills by providing them with opportunities to interact with live prey. In response to changing pup begging calls, helpers modify their prey-provisioning methods as pups mature. Helpers gain no direct benefits from their provisioning behavior, and in fact they incur costs by giving pups prey animals (scorpions) that are difficult to handle, and that sometimes escape. Furthermore, Thornton & McAuliffe (2006) provided evidence that handling of provisioned prey by pups plays an important role in improving their future prey-handling ability.

Regardless of whether or not the actions of these meerkat helpers and mothers in other carnivore species are considered to be teaching per se, many maternal effects on offspring foraging success are clearly mediated by the opportunities for learning provided by mothers to their offspring. Maternal competence at hunting is likely to affect the number and types of learning opportunities of these sorts that female carnivores can provide for their offspring. In addition, maternal hunger level is known in some species to affect maternal provisioning behavior. For example, cheetah cubs of well-fed mothers get more early practice killing prey than do cubs of hungry mothers because the latter proceed to kill prey themselves immediately so they can consume some of it rather than allowing cubs to gain this valuable experience (Caro 1994).

Pups in several canid species learn specific prey capture techniques from their mothers (e.g., red foxes, Macdonald 1980; dingoes, *Canis lupus dingo*, Corbett 1995; Nel 1999). Schenkel (1966) observed lionesses taking their cubs on "mock hunts" during which stalking techniques were practiced. When the lionesses intended to make serious hunting efforts, cubs were left behind, but on mock hunts, cubs were encouraged to join. Eaton (1970) observed this same phenomenon in cheetahs. In addition to acquiring from their mothers information useful in improving foraging success, young carnivores may also learn from their mothers about which local species are potentially dangerous as predators, and how to avoid them. It is also likely that youngsters learn from their mothers about important features of the natal territory, including border locations and effective sites for scent-marking (Kitchener 1999). Indeed, while their cubs still reside at the communal den, female spotted hyenas often take their cubs on expeditions up to several hundred meters from the den entrance before returning them, and these activities appear to familiarize cubs with the local terrain (Holekamp et al., unpublished data).

Social learning clearly functions as an important mechanism mediating maternal effects on foraging behavior in fissiped carnivores (Kitchener 1999). It has often been suggested (e.g., Bekoff et al. 1984; Gittleman 1986) that the evolution of extended parental care in fissiped carnivores has been promoted by the slow development in youngsters of foraging and prey-capture abilities. Gittleman (1986) noted that, within the canid family, young of carnivorous species reach independence from the parents more slowly than young in omnivorous species. Based on this observation, he hypothesized that carnivorous species may need to acquire more complex skills during early ontogeny than more generalized feeders in order to forage successfully on their own. In addition, constrained development of the feeding apparatus in youngsters of some carnivore species might similarly promote the evolution of extended parental care. Various fissiped and pinniped carnivores that ingest hard foods (e.g., sea otters and walruses, which both feed on hard-shelled invertebrates, or hyenas, which ingest a great deal of bone) wean their offspring much later than do most other carnivores of comparable body size, and they engage in far more extensive maternal care than do other members of their respective clades (Rasa 1986). This suggests that adoption of a durophagous diet (ingestion of hard foods like bone and shells) presents special problems for maturing young in these species. Because consumption of hard foods demands an unusually robust feeding apparatus, this may take longer to develop in durophagous species than in carnivores that feed more exclusively on softer foods. Preliminary data from spotted hyenas (Tanner 2007) suggest that constrained development of the skull and jaws in this species, in which individuals routinely engage in intensive feeding competition, may have set the evolutionary stage for profoundly important maternal effects that permit young hyenas to benefit from the social status of their mothers.

EFFECTS OF MATERNAL DOMINANCE STATUS

In addition to helping youngsters adjust to the ecological conditions encountered in any particular environment, maternal effects also play key roles in some fissiped species by adjusting offspring to the specific social conditions they are likely to experience. In carnivore societies in which priority of access to resources or likelihood of breeding are affected by an individual's social status, maternal rank can have significant effects on offspring phenotype. The societies of most canids and gregarious mongooses are hierarchically organized on the basis of social rank, but in many of these species only the alpha female usually produces any viable young (Creel & Creel 1991; Creel et al. 1992; 1997). However, multiple females sometimes produce young concurrently in some cooperative breeders, including meerkats and banded mongooses (Russell et al. 2002; Gilchrist et al. 2004). In meerkats, pup survival to weaning increases with maternal dominance status (Russell et al. 2002), but maternal status apparently has no such effects in banded mongooses (Gilchrist et al. 2004).

Status-related maternal effects and their mediating mechanisms have been intensively studied in spotted hyenas, and have been found to be critical influences on offspring phenotype. Spotted hyenas are gregarious carnivores that live in stable fission-fusion groups of up to 90 individuals, called clans, that are structured by hierarchical rank relationships (Kruuk 1972; Tilson & Hamilton 1984; Frank 1986; Holekamp & Smale 1990). Virtually every aspect of the life of a spotted hyena is strongly affected by its position in the clan's dominance hierarchy. All adult females are socially dominant to all adult males not born in the clan (Kruuk 1972; Smale et al. 1993); adult females are also heavier and more aggressive than adult immigrant males (Kruuk 1972; Frank 1986; Hamilton et al. 1986; Mills 1990; Szykman et al. 2003). Female hyenas are philopatric, but almost all males disperse from the natal clan between the ages of 2 and 5 years (Frank 1986; Henschel & Skinner 1987; Smale et al. 1997; Van Horn et al. 2003).

Each hyena clan defends a territory containing one or two active communal denning sites, where females rear their offspring (Kruuk 1972; East et al. 1989; Boydston et al. 2006). Male spotted hyenas do not participate in care of young, so all effects of social rank on offspring phenotype in this species are mediated through the behavior and physiology of the mother. A hyena's social rank is not determined by fighting ability or size (Engh et al. 2000), but rather rank is learned in a fashion virtually identical to the associative learning process in primates that has been dubbed "maternal rank inheritance" by primatologists (Frank 1986; Holekamp & Smale 1991; 1993; Smale et al. 1993; see also Maestripieri, this volume). That is, when cubs behave aggressively toward lower-ranking conspecifics, they are joined by the mother and other group members to form coalitions, and mothers also frequently intervene on behalf of their cubs when cubs engage in disputes with clan-mates (Holekamp & Smale 1990; 1993; Smale et al. 1993; Engh et al. 2000). Through this process of associative learning during interactions with conspecifics, young hyenas assume positions in the clan's dominance hierarchy immediately below those of their mothers (Engh et al. 2000). Although this process is called maternal rank "inheritance," no literal inheritance is involved, as it has been possible to rule out any direct genetic influences on offspring rank (Mills 1990; Holekamp et al. 1993; Engh et al. 2000; Hofer & East 2003). Instead, both cubs and other members of the society learn that cubs will be supported in contests by the mother and her allies. High-ranking mothers support their cubs more frequently and more effectively than do low-ranking mothers (Engh et al. 2000). Thus, a key maternal effect in spotted hyenas is the nongenetic transmission of maternal rank to offspring. Individuals of both sexes maintain their maternal ranks as long as they remain in the natal clan, meaning that these effects endure at least until dispersal in males, and throughout the life span in females.

Many aspects of a hyena's behavioral phenotype are strongly affected by maternal rank, and evidence for this appears very early in life. For example, *Crocuta* often fight vigorously with their siblings during the first days or weeks after birth (Frank et al. 1991; Smale et al. 1999; Wachter et al. 2002; Wahaj & Holekamp 2006), and the rates and intensities at which siblings fight decrease as maternal rank increases (Golla et al. 1999; Smale et al. 1999). Subordinate adults are obliged to hunt at higher hourly rates than are dominants, presumably because their priority of access to kills made by other clan members is so low (Holekamp et al. 1997a). High-ranking cubs enjoy a superior ability to win in contests with conspecifics over resources, especially when their mothers or other allies are nearby to help them out in these contests. The time budgets of high- and low-ranking hyenas differ (Kolowski et al. 2007), and low-ranking hyenas exhibit higher rates of vigilance at kills than do dominant individuals (Pangle 2008), most likely because they must beware of many more conspecific competitors than must high-ranking hyenas. Subordinate individuals are more likely than dominants to reconcile with their opponents after fights (Wahaj et al. 2001), perhaps because they have a greater need for their relationships with higher-ranking animals to be in a state of good repair. Dominant females are much more attractive to males as prospective mates than are subordinate females (Szykman et al. 2001), presumably because their offspring are so much more likely to survive to adulthood and to live longer as adults (Watts 2007). Finally, sons of higher-ranking females may find it easier to integrate themselves into a new clan after dispersal (Boydston et al. 2005), may enjoy greater success at securing mating opportunities (Frank 1986), and may even perform elements of the act of mating itself (e.g., mounting or intromission: Drea et al. 2002; Dloniak et al. 2006) more effectively than sons of low-ranking females.

In addition to the social learning described earlier, rank-related maternal effects in spotted hyenas are also mediated in three other ways: by the social environment itself, by nutritional and energetic variables, and by endocrine physiology. First, offspring of high-ranking females experience a very different social environment from that experienced by cubs of low-ranking females, and this environment effectively creates or prohibits opportunities for expression of behavioral phenotypes in youngsters. Dominant cubs associate much more closely with their mothers and other kin than do subordinate cubs, and females from high-ranking matrilines associate more closely than do females of low rank (Holekamp et al. 1997b). Dominant females tend to be more gregarious in general, and they are more attractive social companions, than are subordinate females (Smith et al. 2007), so they occur with their cubs in subgroups of larger mean size than do low-ranking cubs. Because survivorship among dominant animals is so much better than in subordinate hyenas (Watts 2007), dominant animals tend to have many more surviving kin in the population at any given time than do subordinates, and thus they enjoy a much larger network of potential allies, should the need for those arise (Van Horn et al. 2004).

An adult's social status determines its priority of access to food, so rank has profound effects on hyenas' intake of calories and nutrients. A youngster's maternal rank determines its priority of access to food at ungulate kills (Kruuk 1972; Tilson & Hamilton 1984; Frank 1986; Mills 1990). With help from their mothers, cubs of high-ranking females have access to a larger quantity and higher quality of food than do cubs of low-ranking females. When cubs are dependent on the den for shelter, the rates at which females return to dens from kills carrying leftover food items for their cubs increases with maternal rank, and high-ranking females are also more successful than their low-ranking counterparts at ensuring their cubs have access to provisioned foods at the den (Holekamp & Smale 1990). After cubs leave the den, dominant mothers are more successful than subordinate females at helping their cubs gain access to ungulate carcasses (Holekamp & Smale 1990). This

Figure 11-1. Barbara, the larger cub in this photograph, was the daughter of the alpha (highest-ranking) female in our free-ranging study clan in Kenya, whereas Batman, the smaller cub, was the daughter of the 19th-ranked female. These individuals were born 4 days apart in July 1989. This photograph was taken when they were 6 months old, and clearly shows a dramatic difference in growth rate between the two cubs.

rank-related variation in cubs' ability to access food has striking effects on cub growth rates (Hofer & East 1996; 2003), with high-ranking cubs growing much faster than their low-ranking peers (Figure 11-1). Dominant females are also able to wean their cubs at much younger ages than can subordinate females (Frank et al. 1995; Holekamp et al. 1996). However, because all females outrank all adult males, all females can wean their cubs long before development is complete of the skull morphology required for adult levels of feeding performance (Tanner 2007).

Because subordinate females have lower priority of access to food than dominant females, they are often obliged to make the best of a bad situation by avoiding competition with dominants. For example, subordinate females are far less likely than dominant females to forage in the central prey-rich areas of the clan's territory (Boydston et al. 2003). In *Crocuta* populations, where females often hunt migratory antelope outside the boundaries of the clan's territory (e.g., in the Serengeti), low-ranking females need to commute to distant prey much more frequently than do high-ranking females

(Hofer & East 1993a, b). Although female dispersal is rare in *Crocuta*, subordinate females are far more likely to disperse than are dominant females (Holekamp et al. 1993; Höner et al. 2005). Natal dens used by subordinate females tend to be located substantially further from the clan's communal den than those of dominant females (White 2005; Boydston et al. 2006). Probably because they must sacrifice time with their cubs at dens in order to travel further afield to forage, subordinate females attend their cubs at dens much less frequently than do dominant females (Hofer & East 1993c). In addition to the obvious direct effects of low rates of den attendance and poor access to resources on cub growth and survival, these factors may indirectly influence other aspects of cub development. For example, although this has not been studied in carnivores, lack of resources or high rates of conflict with conspecifics might lead to harsh or inconsistent parenting behavior, and this in turn may affect a cub's behavioral and reproductive development. Variation in the amount or quality of parental care might be expected to influence behavioral development and future mothering style in hyenas, as has been argued to occur in primates (Maestripieri 2005), including humans (Belsky et al. 1991).

A number of life-history traits are strongly influenced by maternal rank in spotted hyenas, and these effects are most likely mediated directly by rank-related variation in nutritional state or energy availability as reflected in priority of access to food (Holekamp & Smale 2000; Hofer & East 2003). High-ranking females obtain more resources (Holekamp & Smale 2000), and thus are able to provide better nourishment to their cubs. These effects of maternal rank on offspring growth and early weaning are most apparent when resource availability is very low, or when mothers must spend a considerable time away from the den on foraging trips (Hofer & East 1993c; Holekamp et al. 1996). The age at which females first bear young is strongly correlated with maternal rank, with daughters of the alpha female experiencing their first parturition at around 2.5 years of age, and daughters of the lowest-ranking females first giving birth at 5–6 years of age (Holekamp et al. 1996; Hofer & East 2003). Although rank does not affect litter size in hyenas, interlitter intervals are much shorter in dominant than in subordinate females, and dominants are more frequently able to support pregnancy and lactation concurrently, so the annual rate of cub production is substantially higher among dominant than subordinate females (Holekamp et al. 1996). Maternal rank affects the likelihood that cubs will survive to reproductive maturity, and it also has a pronounced effect on longevity among adult females (Watts 2007). Because high-ranking females start breeding earlier, live longer, and produce more surviving cubs per unit time, we have observed as much as a fivefold difference in lifetime reproductive success between the highest and lowest-ranking females in our *Crocuta* study populations in Kenya (Holekamp & Smale 2000). Thus these maternal effects have enormously important fitness consequences (Watts 2007; see also McAdam, this volume).

Finally, one last mechanism mediating maternal rank effects on offspring phenotype in hyenas involves pre- or postnatal hormone exposure. It is not yet known whether any maternal effects of social rank on behavioral development in hyenas are mediated by long-term alteration of the HPA axis. There does not appear to be a simple linear relationship between social status and basal glucocorticoid concentrations in adult hyenas (Goymann et al. 2001; 2003; Dloniak 2004; Dloniak et al. 2006), but rank effects on endocrine responses to acute stressors have not been investigated in this species. We have recently obtained preliminary results (Holekamp et al., unpublished data) suggesting that circulating concentrations of insulin-like growth factor (IGF-1) in young hyenas vary with maternal rank, with the cubs of the highest-ranking mothers having the highest concentrations. IGF-1 is known to play an important role in such processes as growth, immune stimulation, and wound healing in other mammals (e.g., Jones & Clemmons 1995), and IGF-1 concentrations have been found to vary with social rank among adult members of some primate species (e.g., Sapolsky & Spenser 1997). In spotted hyenas, rank-related variation in IGF-1 concentrations may be strongly affected by priority of access to food, as caloric restriction and decreased protein intake can both depress IGF-1 concentrations in other mammals (e.g., Prewitt et al. 1982; Isly et al. 1983).

Whereas glucocorticoids and IGF-1 typically must be present in sufficient concentrations to immediately "activate" particular behavioral or physiological processes, "organizational" hormone effects typically occur during early development, and permanently alter the behavior of exposed individuals, even in the total absence of the hormone later in life when the behavior occurs. In many mammals, the development of behaviors such as aggression and sexual play are strongly affected by organizational effects of certain hormones during fetal development, particularly androgens (e.g., Phoenix et al. 1959). If prenatal exposure of offspring to hormones varies with maternal rank during pregnancy, we might expect the organization of behavior to vary along these lines as well. We recently tested this hypothesis in wild spotted hyenas in Kenya (Dloniak et al. 2006), and found that high-ranking female hyenas have higher concentrations of androgens during the second half of

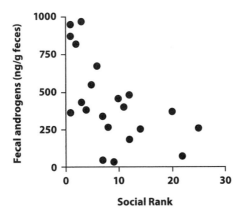

Figure 11-2. The relationship between fecal androgens and social rank in pregnant female spotted hyenas during the second half of gestation. The highest social rank possible is 1. Figure originally published in Dloniak et al. (2006).

gestation than do low-ranking females (Figure 11-2). In addition, both male and female cubs born to the females that exhibited high androgen concentrations late in pregnancy engaged in more dyadic aggressive behavior with non-littermates and more play-mounting behavior than did cubs born to low-ranking females (Figure 11-3). We were able to rule out littermates as the source of differential androgen concentrations to which hyena fetuses are exposed (see Vandenbergh, this volume) because only one fetus usually occupies each uterine horn in this species, and because litter composition has no effect on measured androgen concentrations (Dloniak et al. 2006). Thus, our study demonstrated that androgen profiles in female hyenas during late pregnancy appear to be affected by the social environment, and that rank-related variation in maternal androgens at this time can affect offspring phenotype. Although similar mechanisms for the transfer of status-related traits from mothers to offspring via prenatal hormone exposure has been well-documented in birds and other oviparous animals (e.g., Schwabl 1993), it has not yet been investigated in other mammals. However, we encourage additional studies in order to determine its prevalence and importance in mammalian behavioral development, ecology, and evolution.

CONCLUSIONS

A wide array of traits expressed in female carnivores affect the phenotypic development of their offspring independent of offspring genotype. In some

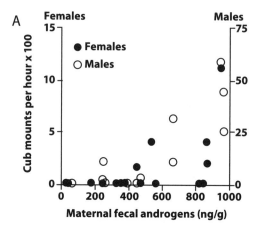

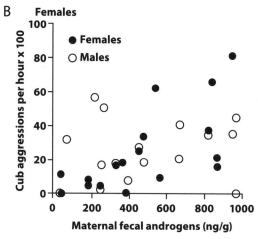

Figure 11-3. The relationship between maternal androgens measured during the second half of gestation and rates of mounting (A) and aggression (B) exhibited by male and female hyena cubs aged 2-6 months. Figure originally published in Dloniak et al. (2006).

species, offspring phenotype is affected by the parity, age, size, condition, or demeanor of the mother. In various cooperatively breeding carnivores, offspring phenotype and survival are affected by the number of allo-mothers or helpers participating concurrently in parental care. Maternal effects in fissiped carnivores play a particularly important role in helping offspring adjust to prevailing local ecological conditions. Mothers influence many different aspects of offspring interactions with prey, and they do this mainly by creating opportunities either for socially facilitated behavior directed at prey or for associative learning in relation to prey. Mothers also regularly create opportunities to learn about travel routes, territorial boundaries, potential dangers, and other key features of the local environment, but these maternal effects have been less intensively studied than those associated with foraging. Among carnivores whose societies are structured by linear dominance hierarchies, mothers may also help their offspring adjust to occupancy of a particular rank position in their group's hierarchy. Maternal rank effects are particularly important among spotted hyenas, where they affect virtually every aspect of an individual's existence for much of its life, as also occurs in many primates. Maternal rank in hyenas influences behavior, patterns of growth, and life history traits. Maternal effects in hyenas and other carnivores are mediated by learning, the social and physical environments created by the mother, nutrition, and endocrine physiology. Given their widespread occurrence, maternal effects on offspring phenotype are likely to represent important considerations in relation to the behavioral development, social structure, and population ecology of fissiped carnivores, and thus perhaps even to their conservation biology.

ACKNOWLEDGMENTS

This chapter was produced with partial support of funds to KEH from U.S. National Science Foundation grants IBN0343381 and IOB0618022, and to SMD from The Lakeside Foundation.

REFERENCES

Baerends-van Roon, J. M. & Baerends, G. P. 1979. *The Morphogenesis of the Behaviour of the Domestic Cat with Special Emphasis on the Development of Prey Catching.* Amsterdam: North Holland.

Bakken, M. 1995. Sex ratio variation and maternal investment in relation to social environment among farmed silver-fox vixens (*Vulpes vulpes*) of high competition capacity. *Journal of Animal Breeding and Genetics*, 112, 463–468.

Bekoff, M., Daniels, T. J. & Gittleman, J. L. 1984. Life history patterns and the comparative socioecology of carnivores. *Annual Review of Ecology and Systematics*, 15, 191–232.

Belsky, J., Steinberg, L. & Draper, P. 1991. Childhood experience, interpersonal development, and reproductive strategy: an evolutionary theory of socialization. *Child Development*, 62, 647–670.

Bernardo, J. 1996. The particular maternal effect of propagule size, especially egg size: patterns, models, quality of evidence and interpretations. *American Zoologist*, 36, 216–236.

Bininda-Emonds, O. R. P. & Gittleman, J. L. 2000. Are pinnipeds functionally different from fissiped carnivores? the importance of phylogenetic comparative analyses. *Evolution*, 54, 1011–1023.

Boydston, E. E., Kapheim, K. M., Szykman, M. & Holekamp, K. E. 2003. Individual variation in space utilization by female spotted hyenas (*Crocuta crocuta*). *Journal of Mammalogy*, 84, 1006–1018.

Boydston, E. E., Kapheim, K. M., Van Horn, R. C., Smale, L. & Holekamp, K. E. 2005. Sexually dimorphic patterns of space use throughout ontogeny in the spotted hyaena (*Crocuta crocuta*). *Journal of Zoology*, 267, 271–281.

Boydston, E. E., Kapheim, K. M. & Holekamp, K. E. 2006. Patterns of den occupation by the spotted hyena (*Crocuta crocuta*). *African Journal of Ecology*, 44, 77–86.

Bowen, W. D. 2009. Maternal effects on offspring size and development in pinnipeds. In: *Maternal Effects in Mammals* (Ed. by D. Maestripieri & J. M. Mateo), pp. 104–132. Chicago: University of Chicago Press.

Byrne, R. W. 1994. The evolution of intelligence. In: *Behaviour and Evolution* (Ed. by P. J. B. Slater & T. R. Halliday), pp. 223–265. Cambridge: Cambridge University Press.

Cameron, N. M., Champagne, F. A., Parent, C., Fish, E. W., Ozaki-Kuroda, K. & Meaney, M. J. 2005. The programming of individual differences in defensive responses and reproductive strategies in the rat through variations in maternal care. *Neuroscience and Biobehavioral Reviews*, 29, 843–865.

Caro, T. M. 1980a. The effects of experience on the predatory patterns of cats. *Behavioral and Neural Biology*, 29, 1–28.

Caro, T. M. 1980b. Effects of the mother, object play and adult experience on predation in cats. *Behavioral and Neural Biology*, 29, 29–51.

Caro, T. M. 1980c. Predatory behaviour in domestic cat mothers. *Behaviour*, 74, 128–147.

Caro, T. M. 1994. *Cheetahs of the Serengeti Plains*. Chicago: University of Chicago Press.

Caro, T. M. & Hauser, M. D. 1992. Is there teaching in nonhuman animals? *Quarterly Review of Biology*, 67, 151–174.

Champagne, F. & Curley, J. P. 2009. The trans-generational influence of maternal care on offspring gene expression and behavior in rodents. In: *Maternal Effects in Mammals* (Ed. by D. Maestripieri & J. M. Mateo), pp. 182–202. Chicago: University of Chicago Press.

Chesler, P. 1969. Maternal influence in learning by observation in kittens. *Science*, 166, 901–903.

Clutton-Brock, T. H., Russell, A. F., Sharpe, L. L., Brotherton, P. N. M., McIlrath, G. M., White, S. & Cameron, E. Z. 2001. Effects of helpers on juvenile development and survival in meerkats. *Science*, 293, 2446–2449.

Corbett, L. K. 1995. *The Dingo in Australia and Asia*. Ithaca, NY: Cornell University Press.

Creel, S. & Creel, N. M. 1991. Energetics, reproductive suppression and obligate communal breeding in carnivores. *Behavioral Ecology and Sociobiology*, 28, 263–270.

Creel, S. & Waser, P. M. 1991. Failure of reproductive suppression in dwarf mongooses (*Helogale parvula*): accident or adaptation? *Behavioral Ecology*, 2, 7–15.

Creel, S., Creel, N., Wildt, D. E. & Monfort, S. L. 1992. Behavioural and endocrine mechanisms of reproductive suppression in Serengeti dwarf mongooses. *Animal Behaviour*, 43, 231–245.

Creel, S., Creel, N. M., Mills, M. G. L. & Monfort, S. L. 1997. Rank and reproduction in cooperatively breeding African wild dogs: behavioural and endocrine correlates. *Behavioral Ecology*, 8, 298–306.

Creel, S., Creel, N. M. & Monfort, S. L. 1998. Birth order, estrogens, and sex-ratio adaptation in African wild dogs (*Lycaon pictus*). *Animal Reproduction Science*, 53, 315–320.

Creel, S., Mills, M. G. L. & McNutt, J. W. 2004. African wild dogs. In: *The Biology and Conservation of Wild Canids* (Ed. by D. W. Macdonald & C. Sillero-Zubiri), pp. 337–350. Oxford: Oxford University Press.

Dloniak, S. M. 2004. Socioendocrinology of spotted hyenas: patterns of androgen and glucocorticoid excretion within a unique social system. PhD Dissertation. Michigan State University, East Lansing.

Dloniak, S. M., French, J. A. & Holekamp, K. E. 2006. Rank-related maternal effects of androgens on behaviour in wild spotted hyaenas. *Nature*, 439, 1190–1193.

Drea, C. M., Place, N. J., Weldele, M. L., Coscia, E. M., Licht, P. & Glickman, S. E. 2002.

Exposure to naturally circulating androgens during foetal life incurs direct reproductive costs in female spotted hyenas, but is prerequisite for male mating. *Proceedings of the Royal Society of London, Series B*, 269, 1981–1987.

East, M. L., Hofer, H. & Turk, A, 1989. Functions of birth dens in spotted hyaenas (*Crocuta crocuta*). *Journal of Zoology*, 219, 690–697.

Eaton, R. L. 1970. The predatory sequence, with emphasis on killing behavior and its ontogeny in the cheetah (*Acinonyx jubatus* Schreber). *Zeitschrift für Tierpsychologie*, 27, 492–504.

Engh, A. L., Esch, K., Smale, L. & Holekamp, K. E. 2000. Mechanisms of maternal rank "inheritance" in the spotted hyaena. *Animal Behaviour*, 60, 323–332.

Estes, R. D. 1991. *The Behavior Guide to African Mammals*. Berkeley: University of California Press.

Ewer, R. F. 1969. The "instinct to teach." *Nature*, 222, 698.

Ewer, R. F. 1973. *The Carnivores*. Ithaca, NY: Cornell University Press.

Fish, E. W., Shahrokh, D., Bagot, R., Caldji, C., Bredy, T., Szyf, M. & Meaney, M. J. 2004. Epigenetic programming of stress responses through variations in maternal care. *Annals of the New York Academy of Science*, 1036, 167–180.

Fox, M. W. 1972. Socio-ecological implications of individual differences in wolf litters: a developmental and evolutionary perspective. *Behaviour*, 41, 298–313.

Fox, M. W. 1978. *The Dog: Its Domestication and Behavior*. New York: Garland STPM Press.

Fox, M. W. & Stelzner, D. 1966. Behavioural effects of differential early experience in the dog. *Animal Behaviour*, 14, 273–281.

Frank, L. G. 1986. Social organisation of the spotted hyaena (*Crocuta crocuta*). II. Dominance and reproduction. *Animal Behaviour*, 35, 1510–1527.

Frank, L. G., Glickman, S. E. & Licht, P. 1991. Fatal sibling aggression, precocial development, and androgens in neonatal spotted hyaenas. *Science*, 252, 702–704.

Frank, L. G., Holekamp, K. E. & Smale, L. 1995. Dominance, demography, and reproductive success of female spotted hyenas. In: *Serengeti II: Conservation, Research, and Management* (Ed. by A. R. E. Sinclair & P. Arcese), pp. 364–384. Chicago: University of Chicago Press.

Galef, B. G., Jr. 2009. Maternal influences on offspring food preferences and feeding behaviors in mammals. In: *Maternal Effects in Mammals* (Ed. by D. Maestripieri & J. M. Mateo), pp. 159–181. Chicago: University of Chicago Press.

Georges, J.Y., Groscolas, R. Guinet, C. & Robin, J. P. 2001. Milking strategy in subantarctic fur seals *Arctocephalus tropicalis* breeding on Amsterdam Island: evidence from changes in milk composition. *Physiological and Biochemical Zoology*, 74, 548–559.

Giger, U. & Casal, M. L. 1997. Feline colostrum—friend or foe: maternal antibodies in queens and kittens. *Journal of Reproduction and Fertility Supplement*, 51, 313–316.

Gilbert, B. K. 1999. Opportunities for social learning in bears. In: *Mammalian Social Learning: Comparative and Ecological Perspectives* (Ed. by H. O. Box & K. R. Gibson), pp. 225–235. Cambridge: Cambridge University Press.

Gilchrist, J. S. Otali. E. & Mwanguhya, F. 2004. Why breed communally? factors affecting fecundity in a communal breeding mammal: the banded mongoose (*Mungos mungo*). *Behavioral Ecology and Sociobiology*, 57, 119–131.

Gittleman, J. L. 1986. Carnivore life history patterns: allometric, phylogenetic, and ecological associations. *American Naturalist*, 127, 744–771.

Golla, W., Hofer, H. & East, M. L. 1999. Within-litter sibling aggression in spotted hyaenas: effect of maternal nursing, sex and age. *Animal Behaviour*, 58, 715–726.

Goymann, W., East, M. L., Wachter, B., Höner, O. P., Mostl, E., Van't Hof, T. J. & Hofer, H. 2001. Social, state-dependent and environmental modulation of faecal corticosteroid levels in free-ranging female spotted hyaenas. *Proceedings of the Royal Society of London, Series B*, 268, 2453–2459.

Goymann, W., East, M. L., Wachter, B., Höner, O. P., Mostl, E., Van't Hof, T. J. & Hofer, H. 2003. Social status does not predict corticosteroid levels in post-dispersal male spotted hyenas. *Hormones and Behavior,* 43, 474–479.

Grindstaff, J. L., Brodie, E. D. & Ketterson, E. D. 2003. Immune function across generations: integrating mechanism and evolutionary process in maternal antibody transmission. *Proceedings of the Royal Society of London, Series B,* 270, 2309–2319.

Gubernick, D. J. 1981. Parent and infant attachment in mammals. In: *Parental Care in Mammals* (Ed. by D. J. Gubernick & P. H. Klopfer), pp. 243–305. New York: Plenum Press.

Hamilton, W. J., III, Tilson, R. L. & Frank, L. G. 1986. Sexual monomorphism in spotted hyaenas, *Crocuta crocuta. Ethology,* 71, 63–73.

Henschel, J. R. & Skinner, J. D. 1987. Social relationships and dispersal patterns in a clan of spotted hyaenas *Crocuta crocuta* in the Kruger National Park. *South African Journal of Zoology,* 22, 18–24.

Henry, J. D. 1985. The little foxes. *Natural History,* 94, 46–57.

Hofer, H. & East, M. L. 1993a. The commuting system of Serengeti spotted hyaenas: how a predator copes with migratory prey. I. Social organization. *Animal Behaviour,* 46, 547–557.

Hofer, H. & East, M. L. 1993b. The commuting system of Serengeti spotted hyaenas: how a predator copes with migratory prey. II. Intrusion pressure and commuters' space use. *Animal Behaviour,* 46, 559–574.

Hofer, H. & East, M. L. 1993c. The commuting system of Serengeti spotted hyaenas: how a predator copes with migratory prey. III. Attendance and maternal care. *Animal Behaviour,* 46, 575–589.

Hofer, H. & East, M. L. 1996. The components of parental care and their fitness consequences: a life history perspective. *Verhandlungen der Deutschen Gesellschaft für Zoologie,* 89, 149–164.

Hofer, H. & East, M. L. 2003. Behavioral processes and costs of co-existence in female spotted hyenas: a life history perspective. *Evolutionary Ecology,* 17, 315–331.

Holekamp, K. E. & Smale, L. 1990. Provisioning and food sharing by lactating spotted hyenas, *Crocuta crocuta* (Mammalia, Hyaenidae). *Ethology,* 86, 191–202.

Holekamp, K. E. & Smale, L. 1991. Dominance acquisition and mammalian social development: the "inheritance" of maternal rank. *American Zoologist,* 31, 306–317.

Holekamp, K. E. & Smale, L. 1993. Ontogeny of dominance in free-living spotted hyaenas: juvenile rank relations with other immature individuals. *Animal Behaviour,* 46, 451–466.

Holekamp, K. E. & Smale, L. 1995. Rapid change in offspring sex ratios after clan fission in the spotted hyena. *American Naturalist,* 145, 261–278.

Holekamp, K. E. & Smale, L. 2000. Feisty females and meek males: reproductive strategies in the spotted hyena. In: *Reproduction in Context* (Ed. by K. Wallen & J. E. Schneider), pp. 257–285. Cambridge, MA: MIT Press.

Holekamp, K. E., Ogutu, J. O., Dublin, H. T., Frank, L. G. & Smale, L. 1993. Fission of a spotted hyena clan: consequences of prolonged female absenteeism and causes of female emigration. *Ethology,* 93, 285–299.

Holekamp, K. E., Smale, L. & Szykman, M. 1996. Rank and reproduction in the female spotted hyaena. *Journal of Reproduction and Fertility,* 108, 229–237.

Holekamp, K. E., Smale, L., Berg, R. & Cooper, S. M. 1997a. Hunting rates and hunting success in the spotted hyena (*Crocuta crocuta*). *Journal of Zoology,* 242, 1–15.

Holekamp, K. E., Cooper, S. M., Katona, C. I., Berry, N. A., Frank, L. G. & Smale, L. 1997b. Patterns of association among female spotted hyenas (*Crocuta crocuta*). *Journal of Mammalogy,* 78, 55–64.

Höner, O. P., Wachter, B., East, M. L., Runyoro, V. A. & Hofer, H. 2005. The effect of prey abundance and foraging tactics on the population dynamics of a social carnivore, the spotted hyena. *Oikos*, 108, 544–554.

Isly, W., Underwood, L. & Clemmons, D. 1983. Dietary components that regulate serum somatomedin C concentrations in humans. *Journal of Clinical Investigation*, 71, 175–181.

Jones, J. & Clemmons, D. 1995. Insulin-like growth factors and their binding proteins: biological actions. *Endocrine Reviews*, 16, 3–34.

Kitchener, A. C. 1999. Watch with mother: a review of social learning in the Felidae. In: *Mammalian Social Learning: Comparative and Ecological Perspectives* (Ed. by H. O. Box & K. R. Gibson), pp. 236–258. Cambridge: Cambridge University Press.

Kolb, E. 2003. The significance and composition of the colostrum and milk of the bitch: a review. *Tierarztliche Umschau*, 58,125–131.

Kolowski, J. M., Katan, D., Theis, K. R. & Holekamp, K. E. 2007. Daily patterns of activity in the spotted hyena. *Journal of Mammalogy*, 88, 1017–1028.

Kruuk, H. 1972. *The Spotted Hyena*. Chicago: University of Chicago Press.

Kruuk, H. 1995. *Wild Otters: Predation and Populations*. Oxford: Oxford University Press.

Kruuk, H. & Turner, M. 1967. Comparative notes on predation by the lion, leopard, cheetah and wild dog in the Serengeti area, East Africa. *Mammalia*, 31, 1–27.

Kuo, Z. Y. 1930. The genesis of a cat's response to the rat. *Journal of Comparative Psychology*, 11, 1–35.

Kuo, Z. Y. 1938. Further study of the behaviour of the cat toward the rat. *Journal of Comparative Psychology*, 25, 1–8.

Laurenson, M. K. 1993. Early maternal behavior of wild cheetahs: implications for captive husbandry. *Zoo Biology*, 12, 31–43.

Laurenson, M. K., Wielebnowski, N. & Caro, T. M. 1995. Extrinsic factors and juvenile mortality in cheetahs. *Conservation Biology*, 9, 1329–1331.

Levine, S. 1967. Maternal and environmental influences on the adrenocortical response to stress in weanling rats. *Science*, 156, 258–260.

Leyhausen, P. 1979. *Cat Behaviour: The Predatory and Social Behaviour of Domestic and Wild Cats* (translated by B. A. Tonkin) New York: Garland Press.

Liers, E. E. 1951. Notes on the river otter (*Lutra canadiensis*). *Journal of Mammalogy*, 32, 1–9.

Maas, B. & Macdonald, D. W. 2004. Bat-eared foxes. In: *The Biology and Conservation of Wild Canids* (Ed. by D. W. Macdonald & C. Sillero-Zubiri), pp. 227–242. Oxford: Oxford University Press.

Macdonald, D. W. 1980. The red fox, *Vulpes vulpes*, as a predator upon earthworms, *Lumbricus terrestris*. *Zeitschrift für Tierpsychologie*, 52, 171–200.

Macdonald, D. W., Creel, S. & Mills, M. G. L. 2004. Society. In: *The Biology and Conservation of Wild Canids* (Ed. by D. W. Macdonald & C. Sillero-Zubiri), pp. 85–106. Oxford: Oxford University Press.

Maestripieri, D. 2005. Effects of early experience on female behavioural and reproductive development in rhesus macaques. *Proceedings of the Royal Society of London, Series B*, 272, 1243–1248.

Maestripieri, D. 2009. Maternal influences on offspring growth, reproduction, and behavior in primates. In: *Maternal Effects in Mammals* (Ed. by D. Maestripieri & J. M. Mateo), pp. 256–291. Chicago: University of Chicago Press.

Malcolm, J. R. & Marten, K. 1982. Natural selection and the communal rearing of pups in African wild dogs (*Lycaon pictus*). *Behavioral Ecology and Sociobiology*, 10, 1–13.

Mandigers, P. J. J., Ubbink, G. J., Vandenbroek, J. & Bouw. J. 1994. Relationship between litter size and other reproductive traits in the Dutch Kooiker dog. *Veterinary Quarterly*, 16, 229–232.

Marma, B. B. & Yunchis, U. U. 1986. Observations on the breeding, management, and physiology of the snow leopard at the Kaunas Zoo from 1962 to 1967. *International Zoo Yearbook*, 8, 66–74.

Mateo, J. M. 2009. Maternal influences on development, social relationships and survival behaviors. In: *Maternal Effects in Mammals* (Ed. by D. Maestripieri & J. M. Mateo), pp. 133–158. Chicago: University of Chicago Press.

McAdam, A. G. 2009. Maternal effects on evolutionary dynamics in wild small mammals. In: *Maternal Effects in Mammals* (Ed. by D. Maestripieri & J. M. Mateo), pp. 64–82. Chicago: University of Chicago Press.

Meier, G. W. 1961. Infantile handling and development in Siamese kittens. *Journal of Comparative and Physiological Psychology*, 54, 284–286.

Mills, M. G. L. 1990. *Kalahari Hyenas: The Behavioral Ecology of Two Species*. London: Unwin Hyman.

Moehlman, P. D. 1979. Jackal helpers and pup survival. *Nature*, 277, 382–383.

Moehlman, P. D. & Hofer, H. 1997. Cooperative breeding, reproductive suppression, and body mass in canids. In: *Cooperative Breeding in Mammals* (Ed. by N. G. Solomon & J. A. French), pp. 76–128. Cambridge: Cambridge University Press.

Mousseau, T. A. & Fox, C. W. (Eds.). 1998. *Maternal Effects as Adaptations*. Oxford: Oxford University Press.

Nel, J. A. J. 1999. Social learning in canids: an ecological perspective. In: *Mammalian Social Learning: Comparative and Ecological Perspectives* (Ed. by H. O. Box & K. R. Gibson), pp. 259–278. Cambridge: Cambridge University Press.

Nutter, F. B., Levine, J. F. & Stoskopf, M. K. 2004. Reproductive capacity of free-roaming domestic cats and kitten survival rate. *Journal of the American Veterinary Medicine Association*, 225, 1399–1402.

Omata, Y., Oikawa, H., Kanda, M., Mikazuki, K., Dilorenzo, C., Claveria, F. G., Takahashi, M., Igarashi, I., Saito, A. & Suzuki, N. 1994. Transfer of antibodies to kittens from mother cats chronically infected with *Toxoplasma gondii*. *Veterinary Parasitology*, 52, 211–218.

Owens, M. & Owens, D. 1979. Communal denning and clan associations in brown hyaenas (*Hyaena brunnea*) of the central Kalahari. *African Journal of Ecology*, 17, 35–44.

Packer, C., Herbst, L., Pusey, A. E., Bygott, J. D., Hanby, J. P., Cairns, S. J. & Borgerhoff-Mulder, M. 1988. Reproductive success of lions. In: *Reproductive Success* (Ed. by T. H. Clutton-Brock), pp. 363–383. Chicago: University of Chicago Press.

Palomares, F., Revilla, E., Calzada, J., Fernandez, N. & Delibes, M. 2005. Reproduction and pre-dispersal survival of Iberian lynx in a sub-population of the Doñana National Park. *Biological Conservation*, 122, 53–59.

Pangle, W. T. 2008. Threat-sensitive behavior and its ontogenetic development in top mammalian carnivores. PhD Dissertation. Michigan State University, East Lansing.

Pauw, A. 2000. Parental care in a polygynous group of bat-eared foxes, *Otocyon megalotis* (Carnivora: Canidae). *African Zoology*, 35, 139–145.

Pedersen, V. & Jeppesen, L. L. 1990. Effects of early handling on later behaviour and stress response in the silver fox (*Vulpes vulpes*). *Applied Animal Behaviour Science*, 26, 383–393.

Persson, J. 2005. Female wolverine (*Gulo gulo*) reproduction: reproductive costs and winter food availability. *Canadian Journal of Zoology*, 83, 1453–1459.

Phoenix, C. H., Goy, R. W., Gerall, A. A. & Young, W. C. 1959. Organizing action of prenatally administered testosterone proprionate on the tissues mediating mating behavior in the female guinea pig. *Endocrinology*, 65, 369–382.

Prewitt, T., D'Ercole, A., Switzer, B. & Van Wyk, J. 1982. Relationship of serum immunoreactivity somatomedin C to dietary protein intake and energy in growing rats. *Journal of Nutrition*, 112, 144–157.

Pu, R., Okada, S., Little, E. R., Xu, B., Stoffs, W. V. & Yamamoto, J. K. 1995. Protection of neonatal kittens against feline immunodeficiency virus-infection with passive maternal antiviral antibodies. *AIDS*, 9, 235–242.

Pusey, A. E. & Packer, C. 1992. Non-offspring nursing in social carnivores: minimizing the costs. *Behavioral Ecology*, 5, 362–374.

Ramsay, M. A. & Stirling, I. 1988. Reproductive biology and ecology of female polar bears (*Ursus maritimus*). *Journal of Zoology*, 214, 601–634.

Rasa, A. O. E. 1973. Prey capture, feeding techniques and their ontogeny in the African dwarf mongoose, *Helogale undulate rufula*. *Zeitschrift für Tierpsychologie*, 32, 449–488.

Rasa, A. O. E. 1986. Parental care in carnivores. In: *Parental Behaviour* (Ed. by W. Sluckin & M. Herbert), pp. 117–151. New York. Basil Blackwell Inc.

Read, A. F. & Harvey, P. H. 1989. Life history differences among the eutherian radiations. *Journal of Zoology*, 219, 329–353.

Rheingold. H. L. 1963. Maternal behaviour in the dog. In: *Maternal Behaviour in Mammals* (Ed. by H. L. Rheingold), pp. 169–202. New York: Wiley.

Roitt, I., Brostoff, J. & Male, D. 1996. *Immunology*, 4th ed. London: Mosby.

Rood, J. P. 1974. Banded mongoose males guard young. *Nature*, 248, 176.

Rood, J. P. 1990. Group size, survival, reproduction and routes to breeding in dwarf mongooses. *Animal Behaviour*, 39, 566–572.

Roulin, A. & Heeb, P. 1999. The immunological function of allosuckling. *Ecology Letters*, 2, 319–324.

Russell, A. F., Clutton-Brock, T. H., Brotherton, P. N. M., Sharpe, L. L., McIlrath, G. M., Dalerum, F. D. & Cameron, E. Z. 2002. Factors affecting pup growth and survival in co-operatively breeding meerkats *Suricata suricatta*. *Journal of Animal Ecology*, 71, 700–709.

Russell, A. F. Brotherton, P. N. M., McIlrath, G. M., Sharpe, L. L. & Clutton-Brock, T. H. 2003. Breeding success in cooperative meerkats: effects of helper number and maternal state. *Behavioral Ecology*, 14, 486–492.

Russell, A. F. Carlson, A. A., McIlrath, G. M., Jordan, N. R. & Clutton-Brock, T. H. 2004. Adaptive size modification by dominant female meerkats. *Evolution*, 58, 1600–1607.

Sapolsky, R. M. & Spenser, E. M. 1997. Insulin-like growth factor I is suppressed in socially subordinate male baboons. *American Journal of Physiology*, 273, R1346-R1351.

Scantlebury, M., Russell, A. F., McIlrath, G. M., Speakman, J. R. & Clutton-Brock, T. H. 2002. The energetics of lactation in cooperatively breeding meerkats *Suricata suricatta*. *Proceedings of the Royal Society of London, Series B*, 269, 2147–2153.

Schaller, G. B. 1972. *The Serengeti Lion*. Chicago: University of Chicago Press.

Schaller, G. B. 1976. *The Deer and the Tiger*. Chicago: University of Chicago Press.

Schenkel, R. 1966. Play, exploration, and territoriality in the wild lion. *Symposia of the Zoological Society of London*, 18, 11–22.

Schwabl, H. 1993. Yolk is a source of maternal testosterone for developing birds. *Proceedings of the National Academy of Sciences USA*, 90, 11446–11450.

Seitz, P. F. D. 1959. Infantile experience and adult behavior in animal subjects. II. Age of separation from the mother and adult behavior in the cat. *Psychonomic Medicine*, 21, 353–378.

Serpell, J. & Jagoe, J. A. 1995. Early experience and development of behavior. In: *The Domestic Dog* (Ed. by J. Serpell), pp. 79–102. Cambridge: Cambridge University Press.

Sillero-Zubiri, C., Marino, J., Gottelli, D. & Macdonald, D. W. 2004. Ethiopian wolves. In *The Biology and Conservation of Wild Canids* (Ed. by D. W. Macdonald & C. Sillero-Zubiri), pp. 311–322. Oxford: Oxford University Press.

Slabbert, J. M. & Rasa, A. O. E. 1993. The effect of early separation from the mother on pups in bonding to humans and pup health. *Journal of the South African Veterinary Association*, 64, 4–8.

Smale, L., Frank, L. G. & Holekamp, K. E. 1993. Ontogeny of dominance in free-living spotted hyaenas: juvenile rank relations with adult females and immigrant males. *Animal Behaviour,* 46, 467–477.

Smale, L., Nunes, S. & Holekamp, K. E. 1997. Sexually dimorphic dispersal in mammals: patterns, causes, and consequences. *Advances in the Study of Behavior,* 26, 180–250.

Smale, L., Holekamp, K. E. & White, P. A. 1999. Siblicide revisited in the spotted hyaena: does it conform to obligate or facultative models? *Animal Behaviour,* 58, 545–551.

Smith, J. E., Memenis, S. K. & Holekamp, K. E. 2007. Rank-related partner choice in the fission-fusion society of the spotted hyena (*Crocuta crocuta*). *Behavioral Ecology and Sociobiology,* 61, 753–765.

Spence, K. W. 1937. Experimental studies of learning and higher mental processes in infra-human primates. *Psychological Bulletin,* 34, 806–830.

Szykman, M., Engh, A. L., Van Horn, R. C. Funk, S. M., Scribner, K. T. & Holekamp, K. E. 2001. Association patterns among male and female spotted hyenas (*Crocuta crocuta*) reflect male mate choice. *Behavioral Ecology and Sociobiology,* 50, 231–238.

Szykman, M., Engh, A. L., Van Horn, R. C., Scribner, K. T, Smale, L. & Holekamp, K. E. 2003. Rare male aggression directed toward females in a female-dominated society: baiting behavior in the spotted hyena. *Aggressive Behavior,* 29, 457–474.

Tanner, J. B. 2007. Behavioral and morphological development in a female-dominated species, the spotted hyena *Crocuta crocuta*. PhD Dissertation. Michigan State University, East Lansing.

Thornton, A. & McAuliffe, K. 2006. Teaching in wild meerkats. *Science,* 313, 227–229.

Tilson, R. T. & Hamilton, W. J. 1984. Social dominance and feeding patterns of spotted hyaenas. *Animal Behaviour,* 32, 715–724.

Vandenbergh, J. G. 2009. Effects of intrauterine position in litter-bearing mammals. In: *Maternal Effects in Mammals* (Ed. by D. Maestripieri & J. M. Mateo), pp. 203–226. Chicago: University of Chicago Press.

Van Horn, R. C., McElhinney, T. L. & Holekamp, K. E. 2003. Age estimation and dispersal in the spotted hyena (*Crocuta crocuta*). *Journal of Mammalogy,* 84, 1019–1030.

Van Horn, R. C., Engh, A. L., Scribner, K. T., Funk, S. M. & Holekamp, K. E. 2004. Behavioral structuring of relatedness in the spotted hyena (*Crocuta crocuta*) suggests direct fitness benefits of clan-level cooperation. *Molecular Ecology,* 13, 449–458.

Venkataraman, A. B. 1998. Male-biased adult sex ratios and their significance for cooperative breeding in dhole, *Cuon alpinus,* packs. *Ethology,* 104, 671–684.

Wachter, B., Höner, O. P., East, M. L., Golla, W. & Hofer, H. 2002. Low aggression levels and unbiased sex ratios in a prey-rich environment: no evidence of siblicide in Ngorongoro spotted hyenas (*Crocuta crocuta*). *Behavioral Ecology and Sociobiology,* 52, 348–356.

Wahaj, S. A. & Holekamp, K. E. 2006. Functions of sibling aggression in the spotted hyaena (*Crocuta crocuta*). *Animal Behaviour,* 71, 1401–1409.

Wahaj, S. A., Guse, K. & Holekamp. K. E. 2001. Reconciliation in the spotted hyena (*Crocuta crocuta*). *Ethology,* 107, 1057–1074.

Watt, J. 1993. Ontogeny of hunting behaviour in otters (*Lutra lutra*) in a marine environment. *Symposia of the Zoological Society of London,* 65, 87–104.

Watts, H. E. 2007. Social and ecological influences on survival and reproduction in the spotted hyena, *Crocuta crocuta*. PhD Dissertation. Michigan State University, East Lansing.

White, P. A. 2005. Maternal rank is not correlated with cub survival in the spotted hyena, *Crocuta crocuta. Behavioral Ecology,* 16, 606–613.

Wilsson, E. 1984. The social interaction between mother and offspring during weaning in German shepherd dogs: individual differences between mothers and their effects on offspring. *Applied Animal Behaviour Science,* 13, 101–112.

Woodroffe, R. & Macdonald, D. W. 1995. Female/female competition in European badgers *Meles meles*: effects on breeding success. *Journal of Animal Ecology*, 64, 12–20.

Wyrwicka, W. 1978. Imitation of mother's inappropriate food preference in weanling kittens. *Pavlovian Journal of Biological Science*, 13, 55–72.

Wyrwicka, W. & Long, A. M. 1980. Observations on the initiation of eating of new food by weanling kittens. *Pavlovian Journal of Biological Science*, 15, 115–122.

12

Maternal Influences on Offspring Growth, Reproduction, and Behavior in Primates

DARIO MAESTRIPIERI

INTRODUCTION

Primates are characterized by slower life histories than other mammals of similar body size. Specifically, primates have a long period of slow growth, late age of reproductive maturation, low reproductive output in adulthood, long gestation length, and long life span (Martin & MacLarnon 1988; Charnov 1991; Charnov & Berrigan 1993; Kappeler et al. 2003). Female primates also produce relatively large infants and provide high levels of maternal investment (Martin & MacLarnon 1985; Lee 1987; Lee et al. 1991). Paternal care is limited to a few species, and, in these species, it mostly takes the form of offspring carrying and protection (Whitten 1987; van Schaik & Paul 1997). Primate mothers are fully responsible for nourishing their offspring and do so in the context of extensive and intimate contact with them through pregnancy and lactation. Moreover, in some primate species, offspring of one sex, generally females, remain in close proximity to their mothers for the rest of their lives and continue to receive investment from them (e.g., Fairbanks 2000). Therefore, primate mothers have ample opportunity to affect their offspring's phenotype and reproductive success.

Body growth and behavioral, physiological, and reproductive development in primates are highly plastic and sensitive to environmental influences. The extended period of slow postnatal growth may in itself be an adaptation to facilitate developmental plasticity in relation to the environment (Janson & van

Schaik 1993; Kappeler et al. 2003). Moreover, there is remarkable variation in growth rates, age at weaning, age of reproductive maturation, and behavioral development even among individuals living in the same environment, suggesting the possibility that plasticity per se can be adaptive under certain environmental conditions (e.g., Borries et al. 2001; Pereira & Leigh 2003). Overall, due to their life history traits, their patterns of maternal investment, and the plasticity of their development, primates are a mammalian order in which maternal effects are expected to be especially strong and pervasive.

A wide range of maternal phenotypic characteristics could affect an offspring's phenotype, survival, and reproductive success. Some of these characteristics include the mother's body condition at the time of conception or during pregnancy or lactation, the mother's dominance rank, her age and parity, and her behavior. Although individual differences in some of these maternal phenotypic characteristics (e.g., body condition or behavior) might have a genetic basis (but see Dunbar 1990), genetic maternal effects and their consequences for evolutionary dynamics have not been systematically investigated in nonhuman primates, in the way it has been done for some other mammals (e.g., Cheverud & Wolf, McAdam, and Wilson & Festa-Bianchet, this volume). Instead, the best-known examples of maternal effects in primates involve maternal phenotypic characteristics such as dominance rank, which are largely environmentally determined and socially inherited (e.g., Holekamp & Smale 1991; see below). Dominance rank, in turn, is generally a good predictor of body condition, as high-ranking mothers tend to be in good body condition, whereas low-ranking mothers tend to be in poor condition (e.g., Small 1981; Fairbanks & McGuire 1995; see also Wilson & Festa-Bianchet, this volume, for ungulates). Maternal age and parity/experience are generally unrelated to dominance rank (e.g., in many primate species, female rank is set early in life and does not increase steadily with age as in some other mammals such as elephants; Holekamp & Smale 1991; Archie et al. 2006), but they too can be strongly associated with body condition (see below).

Individual differences in females' behavior towards their offspring and their propensity to invest in them could, in part, be dependent on temperament or personality and be genetically inherited (e.g., Maestripieri 1993). The genetic bases of interindividual variation in maternal behavior, however, are largely unknown, whereas a large body of evidence already exists linking individual differences in maternal behavior to variation in environmental and experiential factors (Fairbanks 1996; Maestripieri 2001c). For example, maternal behavior and maternal investment can be affected by a female's body condition, dominance rank, age, and parity as well as by other experiential and environmental variables including early experience with her own mother, presence and number of other relatives and other social companions, and their age, sex, and behavior (Fairbanks 1996). If a female's behavior is shaped in an adaptive fashion either through experience early in life or in response to current environment, then any influence of the female's behavior on her offspring's phenotype could represent a mechanism for the nongenetic transmission of adaptation across generations. In other words, the offspring's phenotype could be affected by the mother's environmental experience (Mousseau & Fox 1998a, b).

Although genetic maternal effects may affect evolutionary dynamics in primates as much as they do in other taxa, environmental maternal effects may play a more important role in primates than in other taxa due to the characteristics of primate life histories and those of the environments in which they live. Nonhuman primates have long life spans and live in tropical environments, which tend to be highly ecologically stable over long periods of time (e.g., Richard 1985). For some species and individuals (e.g., females in primate species with female philopatry) the social environment can also be highly stable over long periods of time. Even if ecological and social conditions are unstable over long periods of time but there is stability from one generation to the next, it would be adaptive for mothers to transfer "information" about the environment to their offspring, not only through their genes, but through their behavior and physiology as well. If behavioral and physiological responses to the environment are maladaptive, however, these maladaptive phenotypes can also be transmitted across generations. Since these maladaptive phenotypes do not have a strong genetic basis, they may be relatively buffered from the action of natural selection. Therefore, primates are a taxon in which both adaptive and maladaptive maternal effects are expected to be prominent.

In this chapter, I review our current knowledge of maternal influences on offspring growth, reproduction, and behavior in nonhuman primates. This review will focus on Old World monkeys for which the most information is available (for apes, see Bjorklund et al., this volume). Maternal influences on offspring food preferences, tool use, or cognitive development in primates are reviewed and discussed elsewhere (Galef, Bjorklund et al., this volume) and therefore will not be considered in this chapter. Since the influence of maternal dominance rank on offspring phenotype and fitness is one of the best-documented maternal effects in primates, the chapter will begin with the description of the process by which dominance rank is acquired and transmitted across generations in cercopithecine monkeys. I will then re-view and discuss studies investigating the influence of maternal dominance and/or body condition on offspring sex ratios at birth. A great deal is known about maternal influences on offspring growth and reproductive matura-tion in primates. I will first review the effects of maternal rank and body condition on these variables, then the effects of maternal age and parity, and finally the effects of maternal behavior. The next section will address ma-ternal influences on offspring behavior and physiology, including offspring social and mating preferences, reactivity to the environment, and parental behavior. The chapter will conclude with a brief summary and discussion of future research directions.

DOMINANCE RANK: ITS ACQUISITION AND TRANSMISSION ACROSS GENERATIONS

Many primate studies have reported an association between maternal domi-nance rank and offspring sex ratios at birth as well as between maternal rank and offspring's survival, reproduction, and behavior. The influence of mater-nal rank on offspring fitness could be mediated by nutritional mechanisms involving offspring growth and physical maturation, or by other nonnutri-tional, physiological, or behavioral mechanisms. Maternal dominance rank can also affect the offspring's dominance rank, and the offspring's domi-nance rank, in turn, affects its survival and reproductive success. There is an extensive research literature on the relation between adult dominance rank and reproductive success in primates (see reviews by Fedigan 1983; Cowlishaw & Dunbar 1991). This chapter will therefore concentrate on direct effects of maternal dominance rank on the fitness of immature offspring. Before discussing such effects, however, it is important to describe how dominance rank is acquired and transmitted across generations.

In cercopithecine monkeys and other primates in which females are philopatric and establish clear dominance hierarchies within their group (e.g., most baboons and macaques), sons and daughters acquire their moth-ers' dominance rank early in life through the agonistic support they receive from mothers and through observations of their mother's behavior towards other individuals (see Holekamp & Smale 1991; Chapais 1992; Pereira 1995 for reviews). Offspring are aided by their mothers when they get involved in conflicts with other individuals. When mothers are higher ranking than the offspring's opponents and the opponents' mothers, the mother's interven-tion results in a positive outcome of the conflict. Therefore, offspring learn that they can consistently win fights with certain opponents, and these opponents begin to show submissive behavior to the offspring the way they do it to their mothers. In contrast, when mothers are lower ranking than their offspring's opponents and the opponents' mothers, both offspring and mothers are defeated. Therefore, offspring learn that they can be consistently defeated by certain opponents and, as a result, begin to show submissive behavior to them. This learning process is reinforced by the offspring's observations of interactions between their mothers and other individuals. Offspring observe that their mothers consistently attack and defeat some individuals but are consistently attacked and defeated by others. Through a combination of observational learning and shaping through punishment and reward, offspring eventually match their mothers' behavior towards other individuals. As a result, offspring acquire a dominance rank adjacent to that of their mothers, and in particular immediately below that of their mothers because offspring remain subordinate to their mothers. The nongenetic nature of rank transmission in cercopithecine monkeys has been confirmed by observations that immatures that are adopted at birth by unrelated females acquire the dominance rank of their adoptive mothers and not that of their biological mothers (e.g., Chapais 1992). Furthermore, the social mechanisms of rank acquisition and transmission have been elucidated with experimental manipulations of group composition and alteration of conflict outcomes (e.g., Chapais 1991).

In cercopithecine monkeys with female philopatry, all females belonging to the same matriline have adjacent dominance ranks (e.g., Chapais 1992). Therefore matrilines have dominance ranks too, so that in a group there are high-ranking, middle-ranking, and low-ranking matrilines. Individual ranks can be stable throughout a female's lifetime, and matriline ranks can be stable for generations. As long as males remain in their natal group, they acquire dominance ranks within their matrilines, just like females do (although their body size can also affect their dominance rank acquisition; e.g., Pereira 1995). When males leave their group at puberty and immigrate into a new group, however, their rank within the new group is established through fighting and is generally unrelated to their rank in their group of origin. Therefore, whereas for females rank-related maternal effects may last a lifetime, for males they generally end when males emigrate (Chapais 1992; but see van Noordwijk & van Schaik 2001, for exceptions).

The direct effects of maternal dominance rank on offspring fitness are expected to be stronger in some primate species and environments than in others. Socioecological theory predicts that rank-related maternal effects on offspring fitness should be strongest in female-bonded primate species in which female social relationships are classified as despotic and nepotistic (i.e., dominance relationships are strongly unidirectional and asymmetrical, and strongly affected by kinship; van Schaik 1989). Despotic and nepotistic female social relationships, in turn, should be associated with ecological conditions in which resources are highly clumped, leading to high within-group contest competition (van Schaik 1989; Sterck et al. 1997). Such within-group contest competition may be exacerbated in environments with high predation risk, resulting in large foraging groups (Sterck et al. 1997). The direct effects of maternal dominance rank on offspring fitness are also expected to be stronger in wild than in captive or free-ranging but food-provisioned primates. This is because one of the most important functions of rank is to regulate priority of access to food when resources are limited. When monkeys are food-provisioned, dominance rank may still have strong effects on social interactions but few or no effects on fitness because all individuals have enough resources to meet their requirements.

MATERNAL EFFECTS ON OFFSPRING SEX RATIO

Although the birth sex ratio in primate populations is maintained close to 1:1 as predicted by Fisher (1930), there may be deviations from this balance due to stochastic or adaptive processes (but see Krackow 2002). Adaptive maternal manipulations of offspring sex ratio at birth can potentially occur whenever the cost-benefit ratio of producing offspring is consistently different for sons and daughters. There are several well-known hypotheses for adaptive maternal manipulation of sex ratios, some of which emphasize the differential costs of producing and rearing sons and daughters while others emphasize differential benefits. Furthermore, some of these hypotheses emphasize the role of environmental factors such as demographic and ecological processes, whereas others identify the source of adaptive variation with particular maternal characteristics. The local resource competition (LRComp) and the local resource cooperation (LRCoop) hypotheses (e.g., Silk 1983) emphasize the role of environmental factors in adaptive variations in offspring sex ratios. The LRComp hypothesis states that whenever there is sex-biased offspring dispersal, resources are limited, and the offspring of the philopatric sex are likely to compete with their mothers for resources, then mothers are expected to produce more offspring of the dispersing sex. Conversely, the LRCoop hypothesis states that whenever there is sex-biased offspring dispersal, resources are limited, and the offspring of the philopatric sex cooperate with their mothers to gain access to resources, then mothers are expected to produce more offspring of the philopatric sex. These two hypotheses have been tested with data from a wide range of primate species and supported by some studies (e.g., Silk 1983) but not others (e.g., Silk et al. 2005). The offspring survival hypothesis is another hypothesis in which the source of adaptive variation in offspring sex ratios is environmental. According to this hypothesis, whenever offspring survival is consistently sex-biased, mothers are expected to produce more offspring of the sex that is more likely to survive (see Wasser & Norton 1993; Maestripieri 2001a, for primate data that support this hypothesis). All the above hypotheses predict adaptive variation in offspring sex ratios in the same direction for all females in a particular environment, i.e., there should be an overall offspring sex ratio bias in the population.

Hypotheses that emphasize particular maternal characteristics as the source of adaptive variation in offspring sex ratio predict that females that live in the same environment but differ in these characteristics might bias the sex ratio of the offspring in opposite directions. Therefore, if females with different characteristics are represented in equal number in the population and have a similar reproductive output, a population bias in offspring sex ratio at birth may be absent. The best known of these hypotheses is the Trivers-Willard (T-W) hypothesis (Trivers & Willard 1973), which emphasizes the role of maternal condition as an indicator of ability to provide parental investment. It predicts that mothers in good condition should produce more offspring of the sex with the higher variance in reproductive success, whereas mothers in poor condition should produce more offspring of the sex with lower variance in reproductive success. The T-W hypothesis depends on the following assumptions: maternal condition affects investment in offspring, and investment, in turn, affects offspring condition when offspring are weaned; offspring condition at weaning predicts physical condition in adulthood; and physical condition in adulthood affects the offspring of the sex with higher variance in reproductive success more than the offspring of the other sex. In most mammalian species, including the primate species in which the assumptions of the T-W hypothesis are met, males are the offspring with higher variance in reproductive success. Therefore, the T-W hypothesis predicts that mothers in good condition should produce more males while mothers in poor condition should produce more females.

The T-W hypothesis has been tested and supported in studies of invertebrates (e.g., West et al. 2000), birds (e.g., Gowaty 1993), and various mammalian groups including rodents (e.g., Krackow 1995), ungulates (e.g., Hewison

& Gaillard 1999), and primates (see below). Testing the T-W hypothesis in primates, however, has proved problematic, in part because the best-studied primate species tend to be species characterized by female philopatry and social inheritance of dominance rank, in which maternal condition and dominance rank tend to be correlated. The inheritance of female rank in these species allows for another possible mechanism involving the adaptive bias of offspring sex ratios, with predictions opposite to those of the T-W hypothesis. This inheritance of rank hypothesis states that since adult daughters acquire their mother's dominance rank but adult sons do not, and since high rank is more advantageous for reproductive success than low rank, females of high rank should produce more daughters than sons while females of low rank should produce more sons than daughters (e.g., Hiraiwa-Hasegawa 1993). Therefore, females of high rank and good body condition in many cercopithecine monkey species should produce more sons than daughters according to the T-W hypothesis but should produce more daughters than sons according to the inheritance of rank hypothesis, whereas the opposite is true for low-ranking females in poor body condition.

The possible association between maternal dominance rank and offspring sex ratio at birth has been investigated by a large number of studies and several excellent reviews of this literature already exist (e.g., Clutton-Brock & Iason 1986; Hiraiwa-Hasegawa 1993; Bercovitch 2002). Two recent meta-analyses of these studies have concluded that there is no overall significant effect of rank on offspring sex ratio (Brown & Silk 2002; Schino 2004). Furthermore, the magnitude of the sex ratio bias in relation to rank appears to decline as the sample size of the study increases, suggesting that the significant effects may be the product of stochastic variation in small samples (Brown & Silk 2002). Another possibility, however, is that adaptive variation in offspring sex ratio in relation to maternal rank may fluctuate over time or in relation to availability of resources. For example, van Schaik & Hrdy (1991) suggested that primate offspring sex ratios should vary adaptively in relation to maternal rank as predicted by the T-W hypothesis when resources are abundant and population growth is high but vary as predicted by the LRComp hypothesis when resources are scarce and population growth rate is low (see van Schaik & Hrdy 1991; Schino 2004 for supporting data; but see Packer et al. 2000; Silk et al. 2005, for critiques or negative results). Others have suggested that adaptive biases in offspring sex ratios in general should occur only, or be most prominent, in species with high sexual dimorphism because male variance in reproductive success would be particularly high in these species (see Schino 2004 for supporting primate data). The adaptive hypotheses of offspring sex ratio manipulation are inherently complex and attempts to test them in primates have been complicated by methodological issues such as variation in sample size, variation in operational definitions of maternal dominance rank (e.g., some studies used matriline rank while others used different measures of individual rank) or condition, and differences in the environment in which the data set originated (e.g., captive versus wild populations; populations with low versus high competition for resources).

The mechanisms through which maternal manipulations of offspring sex ratio occur, if they indeed occur, have not been investigated in primates. However, they may be similar to the mechanisms underlying adaptive sex ratio variation in other vertebrate species (Krackow 1995). Maternal manipulations of offspring sex ratio in relation to environmental conditions or maternal characteristics could result in significant evolutionary change. For example, by overproducing or underproducing female offspring, mothers could alter the distribution of traits linked to the X chromosome in the population, or affect the patterns of individual dispersal and philopatry in the future generations, or affect the patterns of cooperation or competition for resources or mating (e.g., Wade et al. 2003). Most research on adaptive maternal manipulations of offspring sex ratios in primates, however, has concentrated on attempting to demonstrate whether these manipulations occur, while little or no attention has yet been paid to their evolutionary consequences.

MATERNAL EFFECTS ON OFFSPRING GROWTH AND REPRODUCTIVE MATURATION

Effects of maternal dominance rank and body condition

Studies in a wide range of primate species have documented considerable interindividual variation in growth rates of immature individuals. Differences in growth rates can have important implications for fitness. Low body weight early in life can be a significant risk factor for survival. In primates and other mammals, there appears to be a weight threshold for weaning such that weaning tends to occur when infants reach about one third of adult body mass (Lee et al. 1991; Bowman & Lee 1995). Fast-growing individuals will reach weaning at an earlier age and be more likely to survive if they lose their mothers than slow-growing individuals. Finally, low body weight after weaning can be associated with lower reproductive success in adulthood. A number of primate studies have shown that the timing of reproductive maturation (i.e., the neuroendocrine changes associated with first ovulation in females and testis enlargement in males) is affected by food availability and depends on reaching a threshold of body weight/fat, especially in females (Mori 1979; Sugiyama & Ohsawa 1982; Loy 1988; Wilson 1989; Bercovitch & Berard 1993; Bercovitch et al. 1998; see Frisch & McArthur 1974; Frisch 1984, for humans). Thus, individuals that grow at faster rates can reach puberty at an earlier age than individuals that grow more slowly. The timing of reproductive maturation, in turn, can be an important determinant of reproductive success in primates, as individuals that begin to reproduce earlier are likely to have greater lifetime reproductive success than individuals that begin to reproduce later (Wilson et al. 1983; Strum 1991; but see Fedigan et al. 1986; humans: Käär et al. 1996). Early onset of reproduction, however, has fitness costs as well, and in some cases these costs may balance or outweigh the benefits (e.g., Wilson et al. 1983; Bercovitch & Berard 1993; Bercovitch et al. 1998).

Growth rates and adult body mass can also affect adult survival, fecundity, and fertility, as larger individuals are more likely to survive food shortages and more likely to have high mating success or produce large offspring that are more likely to survive (e.g., Smith & Smith 1988; Paul et al. 1992). Bercovitch et al. (2000), however, cautioned that these effects are modulated by demographic and ecological conditions as well as by social strategies. In particular, social skills may be so important for male mating success that the effects of social skills are likely to supersede the effects of maternal investment as determinants of reproductive success in many primate species (Bercovitch et al. 2000).

Growth rates and timing of reproductive maturation have a significant genetic component (e.g., timing of female reproductive maturation, in baboons: Williams-Blangero & Blangero 1995; in humans: Rowe 2002) but are also highly sensitive to the environment, and to the maternal environment in particular. A number of studies in a wide range of primate species have reported that the offspring of high-ranking mothers have faster growth rates in the first months or years of life, are weaned earlier, and begin reproducing at an earlier age than the offspring of low-ranking mothers (rhesus macaques, *Macaca mulatta*: Drickamer 1974; Wilson et al. 1983; Schwartz et al. 1985; 1988; Bercovitch & Goy 1990; Bercovitch 1993; Johnson & Kapsalis 1995; Dixson & Nevison 1997; Japanese macaques, *Macaca fuscata*: Mori 1979; Barbary macaques, *Macaca sylvanus*: Roberts 1978; Paul & Thommen 1984; longtail macaques, *Macaca fascicularis*: van Noordwijk & van Schaik 1999; baboons, *Papio* spp.: Altmann et al. 1988; Bercovitch & Strum 1993; Alberts & Altmann 1995; Johnson 2003; Altmann & Alberts 2005; mandrills, *Mandrillus sphinx*:

Setchell et al. 2001; 2002; 2006). There are also studies that reported no significant effects of maternal rank on these variables (e.g., Altmann & Alberts 1987), studies in which the effects were significant for offspring of one sex but not for the other (e.g., Bercovitch et al. 2000; Johnson 2003), and studies in which early maternal influences on offspring growth did not persist into adulthood (e.g., for male mandrills; Setchell et al. 2006). Some of this variability is accounted for by variation among studies in captivity versus wild, food-provisioning versus nonprovisioning, or high population density versus low density (e.g., Berman 1988; Altmann & Alberts 2005).

The association between maternal rank, offspring growth rates, and reproductive maturation could be the result of acceleration of growth and reproductive maturation in the offspring of high-ranking mothers or delay of growth and maturation in the offspring of low-ranking mothers, or both. The mechanisms through which high maternal rank may accelerate offspring growth rates and result in earlier weaning and reproductive maturation have not been thoroughly investigated. One possibility is that these mechanisms are nutritional. High-ranking mothers have priority of access to high-quality food and can eat larger quantities of food for longer periods of time without experiencing interruptions. Therefore, they are likely to be in better condition than low-ranking mothers and better able to transfer nutrients to their offspring during pregnancy and lactation (but see Setchell et al. 2002). Effects of maternal rank on offspring body mass and growth rate can continue into the postweaning period (e.g., Setchell et al. 2001), but clearly the mechanisms underlying these effects must be different. One possibility is that, after weaning, high-ranking mothers maintain proximity to their offspring, protect them, and alleviate the intensity of feeding competition for them, thus allowing the offspring to ingest higher quantities of higher-quality food (Johnson 2003; Altmann & Alberts 2005). The effects of maternal rank on offspring reproductive maturation could also be mediated by nonnutritional mechanisms. In female primates, first ovulation is driven by gradual increases in nocturnal pulsatility and amplitude of luteinizing hormone (LH). Wilson (1992a) speculates that the daughters of high-ranking females may have more undisturbed sleeping patterns, so that the mechanisms regulating the peripubertal increase in nocturnal LH release operate more efficiently in them.

Delay in the growth and reproductive maturation of the offspring of low-ranking mothers could also be the result of nutritional or nonnutritional mechanisms. Low-ranking mothers could be in poor body condition due to limited access to resources and thus be unable to transfer adequate nutrients to their offspring, thus retarding their growth and maturation. Although these effects are entirely plausible, they have not been unequivocally demonstrated in primates, in part because most studies of maternal rank and offspring development have been conducted with food-provisioned monkeys. In provisioned rhesus monkeys, Wilson (1992a) provided evidence suggesting that first ovulation is more likely to be accelerated in the daughters of high-ranking females than delayed in the daughters of low-ranking females. Bercovitch & Berard (1993) showed that under conditions of low social density, daughters of low-ranking females were able to accelerate the onset of reproduction (see also Setchell et al. 2006, for effects of low social density on acceleration of growth and maturation in male mandrills), suggesting that delays in growth and maturation in the offspring of low-ranking mothers may be due to social rather than nutritional factors. Low-ranking mothers and their offspring typically receive high rates of aggression and harassment from other individuals (e.g., Silk 1983; Gomendio 1990), and stress may interfere with growth and reproduction through specific neuroendocrine mechanisms (Wasser & Barash 1983; Sapolsky 1986). An extreme case of suppression of reproduction occurs in common marmosets (*Callithrix jacchus*), in which the subordinate young daughters of a dominant female are prevented from ovulating and conceiving through specific social and physiological mechanisms (Abbott et al. 1997).

Maternal rank and maternal body condition are likely to be highly correlated, and only a few studies have attempted to tease apart their effects on offspring growth and reproductive maturation. In rhesus macaques, maternal weight or body mass index independent of rank affects offspring weight (Bowman & Lee 1995; Johnson & Kapsalis 1995; see also DiGiacomo et al. 1978), whereas in mandrills greater maternal mass is associated with greater offspring length before weaning but not with greater offspring weight (Setchell et al. 2001). Bercovitch et al. (2000) reported that, in food-provisioned rhesus macaques, maternal condition independent of rank affected offspring weight and reproductive maturation in males but not in females, but that this influence on the offspring reproductive success was, overall, minimal. Specifically they reported that maternal pre-conception mass explained about 8% of the variance in male offspring weight at 1 year of age, male weight at 1 year accounted for about 71% of the variance in adult weight, and adult weight accounted for about 25% of the variance in the number of progeny produced by males. Therefore, maternal mass contributed about 1.5% of the variance in sons' reproductive output. Similarly, in mandrills, the sons of higher-ranking females were heavier than those of low-ranking females in adolescence, but this difference in body mass did not affect the sons' dominance rank, the development of their secondary sexual adornments, or the timing of their emigration from the group (Setchell & Dixson 2002; Setchell et al. 2006). Other studies of food-provisioned monkeys, however, showed that maternal dominance rank and nutritional condition, both independently and combined, affected the timing of female reproductive maturation such that young high-ranking females and females in good condition reached menarche at an earlier age (Schwartz et al. 1985; 1988). In males, heavier adolescents may dominate lighter age-mates and have higher testosterone levels as well (Lee & Johnson 1992; Pereira 1995).

Overall, the majority of primate studies conducted so far indicate that rank-related maternal effects on offspring growth and reproductive maturation are widespread and can have long-lasting consequences for offspring fitness. They affect offspring survivorship during the period of dependency, the age and size at which offspring enter the reproductive community, and possibly their survival and reproductive success in adulthood as well. It has been argued that these maternal effects on offspring growth and reproductive maturation might result in the high-ranking matrilines eventually replacing the low-ranking lineages in primate groups with a matrilineal structure such as baboon groups (Altmann & Alberts 2005). In reality, however, this is unlikely to occur due to other counterbalancing influences on fitness, some of which involve maternal effects as well.

Effects of Maternal Age and Parity

An infant born to a young and nulliparous female could be very different phenotypically from an infant born to the same female when she is an adult in her reproductive prime, as well as from an infant born when she is older and near the end of her reproductive life, even if these offspring all share the same father and live in the same ecological and social environment. There may be many different reasons for these phenotypic differences. One of them is that the mother could be in very different physical and health conditions in these three stages of her life, and this could impact the quality of the offspring she can produce and rear.

One consequence of the fact that primates have an extended period of postnatal growth is that they become capable of reproducing before their growth is complete and they have achieved full adult body size. Thus, in many primate species, young females are smaller than adults and still growing when they reproduce for the first time (e.g., Wilson 1989; Setchell et al. 2002). Therefore, they have fewer resources to allocate to offspring than fully grown adult females, and if they allocate too much to offspring, they may jeopardize their own growth and future reproductive potential. In many primate species, the offspring of primiparous mothers have reduced survivorship for a variety of reasons (e.g., Schino & Troisi 2005). Furthermore, when they survive, the offspring of primiparous mothers are likely to be small and grow relatively slowly (rhesus macaques: Small 1981; mandrills: Setchell et al. 2001). Several studies have indicated that young primiparous mothers may compensate for the production of smaller offspring by investing more in them and lengthening the following interbirth interval (Paul & Thommen 1984; Gomendio 1989a; 1989b; Wilson 1992a; Bercovitch & Berard 1993; but see Setchell et al. 2002). According to Gomendio (1989a) the suckling patterns of primiparous and multiparous mothers are different, and the longer interbirth intervals exhibited by primiparous mothers are the result of their allowing their infants to suck more frequently and for longer periods of time. Wilson (1992b), however, showed that longer interbirth intervals in primiparous mothers are not the result of differences in nursing patterns but of greater sensitivity to the nursing-induced inhibition of LH and ovarian steroid secretion. In other words, primiparous mothers are physiologically different from multiparous mothers and, because of that, more likely to experience long periods of lactational amenorrhea. Regardless of the occurrence of investment compensation and its underlying mechanisms, the effects of maternal young age and primiparity on offspring fitness can extend well beyond weaning (e.g., Setchell et al. 2001, for mandrills). Studies of mandrills also suggest that it may be maternal age per se, rather than small body mass associated with young age, that has long-term effects on offspring growth and reproductive maturation. Since dominance rank does not vary with age in female mandrills, nonnutritional maternal effects of age are not due to rank but most likely to behavior (see below). Behavior can also mediate maternal effects of age on offspring growth and development later in life, as maternal interactions with offspring tend to change in relation to increasing parity and experience (see below).

Maternal body condition and health can again impact offspring growth and reproductive maturation when mothers are near the end of their reproductive career. Although in the wild many primate females do not survive long enough to reproduce when they are old, in captive or free-ranging but food-provisioned monkeys females continue to reproduce in their late years. Since old age is accompanied by a decline in maternal body condition and health, it is likely that the offspring of old females may be smaller, grow at slower rates, or be generally less healthy than those of younger females (e.g., Small 1984; Fairbanks & McGuire 1995). The effects of maternal aging on offspring fitness, however, have not been systematically investigated in primates.

Effects of Maternal Behavior

The effects of maternal dominance rank, body condition, age, and parity on offspring growth and reproductive maturation could be mediated by maternal behavior. A large number of studies have shown that these variables can affect several maternal behaviors towards the offspring, and that these maternal behaviors, in turn, can affect offspring growth, reproductive maturation, and other aspects of the offspring phenotype. Variation in maternal behavior can also originate from sources other than differences in rank, condition, age, or parity.

Nutrition of offspring from birth through weaning occurs in the context of intimate contact between mother and infant (Figure 12-1).

Therefore, any behavior that affects the amount of mother-infant contact can also potentially affect offspring nutrition and growth. Primate mothers play an active role in the regulation of contact with their offspring through behaviors such as contact-making and contact-breaking, approaches and leaves, and restraining and rejection (Fairbanks 1996). Individual mothers differ dramatically from one another in their tendencies to maintain or tolerate contact with their infants, and these individual differences are generally highly consistent over time and across different infants. These differences in maternal behavior can have significant consequences for weaning. Studies of rhesus monkeys have shown that mothers who break contact with their infants and reject their attempts to make nipple contact at higher rates early in life are more likely to wean their infants early and conceive again in the following mating season, whereas mothers with low rejection rates are more likely to skip the mating season and reproduce every other year (Simpson et al. 1981; Gomendio 1989b; Johnson et al. 1993; Berman et al. 1993; see also Hauser & Fairbanks 1988, for wild vervet monkeys, *Chlorocebus aethiops*). In vervet monkeys, more protective mothers are likely to have longer interbirth intervals as well (Fairbanks 1996).

Although it is well established that rates of nipple contact and suckling can affect the length of the mother's lactational amenorrhea (Short 1983; Lee 1987; Stewart 1988; Gomendio 1989b), it is less clear whether they also translate into differential offspring growth patterns (Lee et al. 1991). This is because measures of suckling frequency and duration may not be accurate indicators of milk transfer and, therefore, of maternal investment (Cam-

Figure 12-1. Macaque mother and infant feeding in close proximity (Photo: Stephen Ross).

eron 1998). Moreover, among food-provisioned monkeys, individual differences in maternal behavior in relation to maternal dominance rank, age, or parity generally do not match the observed differences in offspring growth and reproductive maturation in relation to these maternal characteristics. For example, high-ranking mothers are generally reported as being less protective and more rejecting than low-ranking mothers, yet the offspring of high-ranking mothers grow faster and reproduce earlier than those of low-ranking mothers (Fairbanks 1996). Similarly, young and primiparous mothers are reported as being more protective and less rejecting than older and multiparous mothers (Fairbanks 1996), yet the offspring of young and primiparous mothers grow more slowly and reproduce later than those of older and multiparous mothers.

One possible explanation for these discrepancies is that the effects of maternal behavior on offspring growth and reproductive maturation are mediated by nutritional mechanisms but these mechanisms do not involve milk transfer. For example, it is possible that the main effect of maternal behavior is to encourage or delay the offspring's transition to a solid food diet and that it is the timing of this transition as well as the amount of independent feeding early in life that affects offspring growth and reproductive maturation (but see Altmann 1998 for lack of significant effects of early independent feeding on female reproductive maturation in baboons).

Another possibility is that maternal behavior affects offspring growth and reproductive maturation but that the primary mechanisms underlying these effects are nonnutritional. A study of rhesus macaques has shown that daughters reared by mothers with a harsh and inconsistent parenting style including high rates of maternal rejection develop interest in infants earlier in life and tend to conceive for the first time at an earlier age than other females, suggesting that these young females might be on a faster developmental and reproductive track (Maestripieri 2005a). Since these females were cross-fostered at birth and reared by unrelated mothers, these phenotypic traits did not appear to reflect genetic similarities between mothers and daughters. Instead, there was some evidence that these traits were associated with higher cortisol responses to novelty, suggesting that exposure to a particular maternal style early in life results in long-term neuroendocrine alterations that may affect reproductive maturation (see also Maestripieri et al. 2006a). These findings are consistent with those of human studies showing that girls who are exposed to early psychosocial stress reach menarche earlier, engage in sexual activity and conceive earlier, and develop interest in infants earlier in life than other girls (see Ellis 2004, for a recent review of this literature; Maestripieri et al. 2004). These effects have been hypothesized to be mediated by the quality of parenting (i.e., psychosocial stress is associated with harsh and inconsistent parenting style; Belsky et al. 1991), but this hypothesis and the possible neuroendocrine mechanisms underlying these effects have not yet been investigated in humans.

MATERNAL EFFECTS ON OFFSPRING BEHAVIOR AND PHYSIOLOGY

Maternal Influences on the Development of Offspring's Mating and Social Preferences

The mating and social systems of all primate species, just like those of other mammals, are determined at the proximate level by an interaction between the social preferences of individuals and the opportunities and constraints provided by their demographic and ecological environments. Mating systems are the result of sexual interactions between individuals within a group or population. These sexual interactions, in turn, occur because individuals develop a sexual attraction to members of the opposite sex and pursue particular social strategies to mate with them, which depend on the distribution of these individuals in space and time, competition or cooperation with other individuals to gain access to mates, and trade-offs between mating and other activities such as foraging or parenting.

Studies of imprinting in birds have shown that sexual preferences can be significantly affected by the early maternal environment so that individuals develop attractions to individuals that share certain characteristics with their mothers, even if their mothers belong to a different species. Thus, cross-fostered chicks that are reared by parents from another species develop sexual preferences for individuals of their foster parents' species rather than for their own conspecifics (Immelmann 1972). Since, with the exception of brood parasites, birds are normally reared by members of their own species, sexual imprinting has probably evolved to prevent cross-species matings. Sexual imprinting may also favor optimal outbreeding within a species, as cross-fostered individuals become particularly attracted to individuals that look slightly different from their mothers, e.g., their first cousins (Bateson 1982).

Kendrick et al. (1998) demonstrated that sexual imprinting can also occur in mammals, or at least in the males of some mammalian species. In their study, reciprocally cross-fostered sheep and goats developed, as adults, social and mating preferences for members of their maternal species rather than for conspecifics. For males, these effects were strong and stable over time, whereas for females they were weaker and reversible within 1–2 years. Cross-fostered males expressed preferences for their maternal species also in experimental tests in which they were exposed to pictures of faces of their own and their maternal species (Kendrick et al. 1998), a finding previously demonstrated in macaques as well (Fujita 1993; Fujita & Watanabe 1995). No clear demonstration of sexual imprinting exists yet for nonhuman primates, however, and the extent to which there may be maternal effects on mating preferences and mating strategies (aside from effects of maternal dominance rank on mating success) in primates has never been systematically investigated.

Social preferences for genetically related individuals, and especially for relatives of the same sex, play an important role in the origin and maintenance of primate social systems and may significantly affect individual social strategies as well (Gouzoules & Gouzoules 1987; Silk 2002). The development of these social preferences has been studied in some cercopithecine monkeys, and especially in rhesus macaques. This research has shown that the social relationships monkey infants develop with other individuals tend to mirror those of their mothers (see Berman 2004 for a recent review). For example, rhesus mothers associate preferentially with close female relatives and have antagonistic relationships with unrelated females from other matrilines. Similarly, rhesus infants spend more time and affiliate with members of their own matriline and tend to ignore or avoid unrelated individuals (Berman 1982a, b). Furthermore, in macaque species in which the relationships between adult females and close female relatives are strong, so are the relationships between infants and these individuals, whereas in species in which the relationships between adult females and close female relatives are weak, infants show only a moderate or weak preference for these individuals (Berman 2004). Although interspecific differences in the strength of kin bias could be genetically based, studies have suggested that the development of kin-biased social behavior is influenced by the maternal environment (see below).

In rhesus macaques, young infants spend most of their time in contact or close proximity with their mothers. Since their mothers associate and affiliate with their close kin, infants begin interacting mostly with these individuals simply because these individuals happen to be nearby and are available for interaction. Although young infants initially do not seem to have an active preference for any kin other than their mothers, their social interactions are kin-biased by virtue of the fact that their immediate social environment is also kin-biased. Not only do mothers provide their offspring with opportunities to interact with particular individuals; they also serve as models for these interactions. In other words, by observing their mothers being amicable towards close relatives and avoidant or aggressive towards nonrelatives, infants learn how to behave accordingly with these individuals. Thus, an important component of the formation of social relationships in young monkey infants involves being passively exposed to the mother's environment and the mother's behavior. Macaque mothers, however, are also actively engaged in shaping both the environment and the behavior of their infants. First, macaque mothers, and especially older and multiparous mothers, actively encourage their infants to walk independently and explore their environment (Maestripieri 1995a; 1996). Second, they discourage interactions between their infants and certain other individuals (Maestripieri 1994; 1995b; Berman & Kapsalis 1999). Primate mothers, however, do not actively encourage their infants to interact with particular individuals and form relationships with them (Maestripieri 1995c; Maestripieri et al. 2002), despite some suggestions to the contrary (de Waal 1990).

Through the joint action of multiple processes driven by mothers such as providing opportunities, modeling, and active shaping of behavior, monkey infants develop active social preferences for their close kin, and seek out these individuals and initiate interactions with them. Since kin and non-kin, especially adults, respond differently to the infant's behavior, learning through reinforcement and punishment in the course of social interactions with these individuals can contribute to the formation of infant social preferences. In rhesus macaques, the development of active kin bias occurs some time in the first year of life and, for females, will continue to strengthen for the rest of their lives (de Waal 1996; Nakamichi 1989; Berman 2004). The social developmental trajectory of males, instead, will begin to diverge from that of females after 1–2 years (Hinde & Spencer-Booth 1967; Nakamichi 1989). After this age, males will receive increasing amounts of aggression from their older female relatives and will become increasingly attracted to peers and adult males. Eventually, these processes will culminate in the peripheralization of the juvenile and subadult males in the group and their emigration.

It is important to emphasize that maternal influences on offspring social development probably act in conjunction with genetically inherited predispositions, and that the relative contribution of maternal versus genetic influences varies in relation to the type of relationships the offspring are forming. For example, whereas maternal influences on the development of offspring kin bias appear to be strong, at least in some primate species, the development of female attraction to infants, the development of play partner preferences, and the development of sex partner preferences seem to be relatively independent of maternal influences and more dependent on a combination of inherited propensities and the species-typical social environment (i.e., young individuals must develop in social groups with species-typical size and demographic composition to be able to fully express these propensities; Berman et al. 1997; Maestripieri & Ross 2004). The relative importance of maternal influences versus inherited propensities for the development of social preferences also seems to vary greatly among primate species. For example, whereas maternal influences on the development of matrilineal kin bias appear to be strong in nepotistic primate species with female philopatry and male dispersal, different types of kin bias and of maternal influences could occur in species in which males are philopatric and females disperse (e.g., chimpanzees or red colobus monkeys) or species in which both sexes disperse (e.g., gorillas; Berman 2004). Unfortunately, little is known about maternal influences on social development in these species. Little is also known about whether and how within-species variation in the maternal environment and in maternal behavior affects the offspring social and mating strategies in adulthood (but see Schino et al. 2004) and whether and how this behavioral variation translates into variation in survival or reproductive success. Therefore, although social behavior represents a potentially important avenue for maternal effects in nonhuman primates, these maternal effects have not yet been well characterized.

Maternal Behavior and Offspring's Reactivity to the Environment

Maternal behavior can affect not only the development of mating and social preferences in the offspring but also the offspring's reactivity to the environment in general. For example, naturally occurring interindividual variation in maternal behavior could result in long-term differences in offspring's responsiveness to novel or stressful stimuli, through physiological mechanisms similar to those demonstrated in rodents (Champagne & Curley, this volume). Long-term changes in physiological and behavioral responsiveness to stimuli could affect many possible behaviors related to survival and reproduction, and therefore, have adaptive or maladaptive consequences for fitness.

Studies of intraspecific variation in parenting styles of cercopithecine monkeys have shown that most variability occurs along the two orthogonal dimensions of maternal protectiveness and rejection (Tanaka 1989; Schino et al. 1995; Maestripieri 1998). The maternal protectiveness dimension of parenting style includes variation in the extent to which the mother physically restrains her infant, initiates proximity and contact, and cradles and grooms her infant (Figure 12-2). The maternal rejection dimension includes the extent to which the mother limits the timing and duration of contact, suckling, or carrying. Although maternal behavior changes as a function of infant age and the mother's own age and experience, individual differences in parenting styles tend to be consistent over time and across infants (Hinde & Spencer-Booth 1971; Fairbanks 1996). Several studies of macaques and vervet monkeys have examined variation in offspring independence from the mother and their tendency to explore the environment or respond to challenges at various ages in relation to exposure to variable levels of maternal protectiveness and maternal rejection experienced in early infancy.

An early study by Simpson (1985) shows that exposure to high levels of maternal rejection in the first few months of life was associated with re-

Figure 12-2. Protective behavior displayed by rhesus macaque mothers with their infants (Photos: Reuters).

duced infant's exploration at the end of the first year. Subsequent studies, however, show that infants reared by highly rejecting (or less responsive mothers) mothers generally develop independence at an earlier age (e.g., spend more time out of contact with their mothers, explore the environment more, and play more with their peers) than infants reared by mothers with low rejection levels (Simpson & Simpson 1985; Simpson et al. 1989; Simpson & Datta 1990; Bardi & Huffman 2005). In contrast, infants reared by more protective mothers appear to be delayed in the acquisition of their independence and are relatively fearful and cautious when faced with challenging situations (Fairbanks & McGuire 1988; 1993). Although these findings may be the result of inherited temperamental similarities between mothers and offspring, similar findings were also obtained in studies in which maternal protectiveness was experimentally enhanced through manipulations of the environment (Fairbanks & McGuire 1987; Vochteloo et al. 1993). Genetic correlations between maternal and infant behaviors regulating the maintenance of contact, however, are likely to exist and be the result of coadaptation or evolutionary mother-offspring conflict (Maestripieri 2004).

The issue of whether these possible behavioral maternal effects on off-spring reactivity to the environment also persist into later stages of development and adulthood has been addressed by a few studies of vervet monkeys and macaques. In vervet monkeys, juveniles who were exposed to greater maternal protectiveness in infancy had a higher latency to enter a new enclosure and to approach novel food containers (Fairbanks & McGuire 1988; 1993), whereas adolescent males reared by highly rejecting mothers were more willing to approach and challenge a strange adult male (Fairbanks 1996). Schino et al. (2001) found no significant association between variation in maternal protectiveness or rejection early in life and the offspring's behavior several years later in Japanese macaques. However, they did report a relationship between early maternal rejection and offspring responsiveness to stressful situations. Specifically, individuals that were rejected more by their mothers early in life were less likely to respond with submissive signals or with avoidance to an approach from another individual and exhibited lower rates of scratching in the 5-minute period following the receipt of aggression. Finally, Maestripieri et al. (2006b) showed that rhesus macaques that were rejected more by their mothers in the first 6 months of life engaged more in solitary play and greater avoidance of other individuals in the second year. In this study, the association between maternal behavior and offspring behaviors later in life was also reported in females that were cross-fostered at birth and reared by unrelated adult females, thus excluding the possibility of inherited temperamental similarities between mothers and offspring.

Developmental differences in reactivity to novel stimuli or responsiveness to other individuals are likely to be accompanied by differences in neurochemical and neuroendocrine substrates regulating emotional and social processes. Neuropeptides and hormones of the hypothalamic-pituitary-adrenal (HPA) axis such CRH, ACTH, and cortisol, along with the brain monoamine neurotransmitters norepinephrine, serotonin, and dopamine would be likely candidates, as these substances play an important role in the regulation of emotional and behavioral processes and their concentrations can be relatively stable over long periods of time. Very few and preliminary data are available on the relationship between variable maternal care and offspring hormonal profiles (e.g., Bardi et al. 2005), although some relevant data have been provided by recent human studies (e.g., Gunnar 2003; Hane & Fox 2006)

Maestripieri et al. (2006a) reported that offspring reared by mothers with higher levels of maternal rejection exhibited lower cerebrospinal fluid (CSF) levels of 5-HIAA, MHPG, and HVA in the first 3 years of life than offspring reared by mothers with lower levels of rejection. This difference was observed in both nonfostered and cross-fostered infants, suggesting that exposure to variable parenting style early in life interacts with genetically inherited propensities in determining CSF monoamine metabolite levels (Rogers et al. 2004). Furthermore, CSF MHPG levels in the second year of life were negatively correlated with solitary play and avoidance of other individuals, while CSF 5-HIAA levels were negatively correlated with scratching rates, suggesting that individuals with low CSF 5-HIAA had higher anxiety (Maestripieri et al. 1992). In contrast, variation in maternal protectiveness early in life did not predict later variation in CSF monoamine metabolite levels or offspring behavior (Maestripieri et al. 2006a).

The brain noradrenergic system has been associated with the regulation of arousal and an individual's fearful or aggressive responses to novel or threatening stimuli, whereas the serotonergic system is believed to play an important role in impulse control and in reducing the probability that risky, dangerous, or aggressive behaviors will be expressed in response to internal pressures or external stimuli (Higley 2003). Taken together, the studies by Maestripieri et al. (2006a, b) suggest that exposure to maternal rejection early in life may affect the development of different neural circuits underlying emotion regulation, ranging from fear to anxiety to impulse control. Other studies of rhesus and vervet monkeys have shown that low levels of CSF 5-HIAA and MHPG are associated, at least in males, with high impulsivity, risk-taking behavior, and propensity to engage in severe forms of aggression (see Higley 2003, for a review) and that young males with low levels of CSF 5-HIAA may be more likely to attain high dominance rank in adulthood (Fairbanks et al. 2004). Young males with low CSF 5-HIAA also appear to be more likely to emigrate from their natal group at a young age (Mehlman et al. 1995; Kaplan et al. 1995; but see Howell et al. 2007, for opposite results). However, low CSF 5-HIAA seems to be correlated with reduced survival and lower mating success, at least in male rhesus macaques (Mehlman et al. 1997; Howell et al. 2007). Thus Howell et al. (2007) argued that this trait has mostly deleterious consequences for male fitness but may be associated with phenotypic traits that are advantageous to females such as low levels of aggressiveness associated with the maintenance of dominance rank. Although the long-term and fitness consequences of particular neurochemical profiles are not yet well understood, long-term alterations in the offspring serotonergic system induced by varying levels of maternal rejection early in life may be one of the mechanisms through which maternal behavior can alter offspring's responsiveness to the environment and significantly affect offspring survival and reproduction.

Intergenerational Transmission of Maternal Behavior

Another example of behavioral maternal effects in primates involves the intergenerational transmission of maternal behavior from mothers to daughters. Multigenerational observational studies of maternal behavior in captive
vervet monkeys and semi-free-ranging rhesus macaques reported similarities in behavior between mothers and daughters and suggested that these
similarities may be the result of the daughters' early experience. Specifically,
in vervet monkeys, the amount of time young females spent in contact with
their offspring was predicted by the amount of time these females had spent
in contact with their own mothers in infancy (Fairbanks 1989), whereas for
rhesus macaques Berman (1990) reported a correlation between the maternal rejection rates of mothers and daughters. Although these studies could
not rule out unequivocally the possibility that intergenerational consistency
in maternal behavior may be the result of genetic similarities between mothers and daughters, in one case experience with the mother in infancy was
the best predictor of the offspring's behavior in adulthood (Fairbanks 1989;
see also Fairbanks 1996), while in the other, social learning as a juvenile
(through observations of interactions between the mother and younger siblings) seemed to play a more important role (Berman 1990). Since maternal
behavior can vary in relation to female dominance rank, and mothers and
daughters have very similar dominance ranks, it is also possible that similarities in dominance rank contribute to similarities in maternal behavior
between mothers and daughters (Fairbanks 1996).

The mechanisms underlying the intergenerational transmission of maternal behavior were further explored in a study of rhesus macaques in which
some females were cross-fostered at birth and reared by unrelated foster
mothers (Maestripieri et al. 2007). This study examined cross-generational
consistencies in composite measures of parenting style dimensions rather
than in specific maternal behaviors and found evidence for such consistency
in the rejection dimension but not in the protectiveness dimension of parenting style. Significant similarities in maternal rejection between mothers
and daughters were found for both nonfostered and cross-fostered rhesus
females, suggesting that the daughters' behavior was affected by exposure
to their mothers' rejection in their first 6 months of life. This study also provided evidence that behavioral maternal effects on the transmission of parenting style may also be mediated by physiological mechanisms. Specifically,
both nonfostered and cross-fostered rhesus females reared by mothers with
higher (above the median) rates of maternal rejection had significantly lower
CSF concentrations of the serotonin metabolite 5-HIAA in their first 3 years

of life than females reared by mothers with lower (below the median) rates of maternal rejection, and that low CSF 5-HIAA was associated with high rejection rates when the daughters produced and reared their first offspring (Maestripieri et al. 2006a; 2007). Related data analyses involving some of the same rhesus females also showed that abusive parenting is transmitted across generations, from mothers to daughters, through experiential mechanisms (Maestripieri 2005b; see also Maestripieri & Carroll 1998, and Maestripieri et al. 1997, for additional evidence), and that the females that were abused by their mothers in infancy and became abusive mothers themselves had lower CSF 5-HIAA than the abused females that did not become abusive mothers (Maestripieri et al. 2006a).

Individual differences in maternal rejection rates may represent adaptations to particular maternal characteristics (e.g., dominance rank, body condition, or age) or demographic and ecological circumstances (e.g., availability of food or social support from relatives). In cercopithecine monkeys such as rhesus macaques, mothers and daughters have very similar dominance ranks and share their environment as well. Therefore, the nongenomic transmission of maternal rejection may represent an example of nongenomic transmission of behavioral adaptations from mothers to daughters. Abusive parenting, however, is a maladaptive behavioral trait that occurs in about 5–10% of individuals in the population and is often associated with high rates of maternal rejection (Maestripieri & Carroll 1998). The finding that both maternal rejection and abusive parenting can be transmitted across generations through similar experiential and physiological mechanisms illustrates the notion that behavioral maternal effects can be adaptive or maladaptive and operate through very similar mechanisms.

CONCLUSIONS AND FUTURE DIRECTIONS

In many primate species including humans, a wide range of maternal phenotypic characteristics can have long-lasting consequences for offspring growth, reproduction, and behavior. Although individual differences in these maternal traits likely have a genetic component (e.g., Maestripieri 2003), the genetic bases of maternal characters are generally uninvestigated in primates, and therefore genetic maternal effects on evolutionary dynamics are poorly understood (see Kirkpatrick & Lande 1989). A promising avenue for future research would be to investigate the heritability of selected maternal physical, physiological, and behavioral traits using quantitative genetic methods and/or cross-fostering experiments (Roff 1998), obtain quantitative measures of genetic maternal effects involving these traits, and examine the impact of these maternal effects on evolutionary dynamics in wild primate populations. To this end, it is worth noting that although primate mothers can effectively discriminate their offspring from other young very early in life, cross-fostering experiments can be successfully performed in primates under particular conditions (Maestripieri 2001b; 2003; 2004).

Given the characteristics of primate life histories and those of the environments in which they live, nongenetic maternal effects play an important role in primates' adaptation to their environment. Nongenetic variation in maternal traits such as dominance rank, body condition, age, parity, and parenting style has long-term effects on offspring growth and behavior, and some of these effects have consequences for survival and reproduction as well. Some of these maternal traits have also been linked to adaptive variation of offspring sex ratios at birth. Primates are perhaps the mammalian taxon in which nongenetic maternal effects have been best documented and their underlying mechanisms best elucidated. Knowledge of the mechanisms underlying maternal effects in primates can inform research with other mammalian and nonmammalian taxa, in which such mechanisms are virtually unknown (but see Champagne & Curley, this volume). Studies of maternal influences on offspring phenotype in primates have mostly been conducted at the level of analysis of the individual and have attempted to address both the mechanisms underlying these effects and their consequences for fitness. Although more studies of mechanisms and individual fitness are needed in primates, future studies should also increasingly address the consequences of maternal effects at the population level and attempt to test the predictions of maternal effect evolution models derived from theory and work with other taxa. Finally, since primate studies of maternal influences on offspring growth, reproduction, and behavior have thus far concentrated on a few species such as macaques, baboons, and vervet monkeys, future studies should include other species of Old World monkeys as well as prosimians, New World monkeys, and apes.

REFERENCES

Abbott, D. H., Saltzman, W., Schultz-Darken, N. J. & Smith, T. E. 1997. Specific neuroendocrine mechanisms not involving generalized stress mediate social regulation of female reproduction in cooperatively breeding marmoset monkeys. *Annals of the New York Academy of Sciences*, 807, 219–238.

Alberts, S. C. & Altmann, J. 1995. Preparation and activation: determinants of age at reproductive maturity in male baboons. *Behavioral Ecology and Sociobiology*, 36, 397–406.

Altmann, J. & Alberts, S. C. 1987. Body-mass and growth-rates in a wild primate population. *Oecologia*, 72, 15–20.

Altmann, J. & Alberts, S. C. 2005. Growth rates in a wild primate population: ecological influences and maternal effects. *Behavioral Ecology and Sociobiology*, 57, 490–501.

Altmann, J., Altmann, S. A. & Hausfater, G. 1988. Determinants of reproductive success in savannah baboons (*Papio cynocephalus*). In: *Reproductive Success* (Ed. by T. H. Clutton-Brock), pp. 403–418. Chicago: University of Chicago Press.

Altmann, S. A. 1998. *Foraging for Survival: Yearling Baboons in Africa*. Chicago: University of Chicago Press.

Archie, E. A., Morrison, T. A., Foley, C. A. H., Moss, C. J. & Alberts, S. C. 2006. Dominance rank relationships among wild female African elephants, *Loxodonta africana*. *Animal Behaviour*, 71, 117–127.

Bardi, M. & Huffman, M. A. 2005. Maternal behavior and maternal stress are associated with infant behavioral development. *Developmental Psychobiology*, 48, 1–9.

Bardi, M., Bode, A. E. & Ramirez, S. M. 2005. Maternal care and development of stress responses in baboons. *American Journal of Primatology*, 66, 263–278.

Bateson, P. 1982. Preferences for cousins in Japanese quail. *Nature*, 295, 236–237.

Belsky, J., Steinberg, L. & Draper, P. 1991. Childhood experience, interpersonal development, and reproductive strategy: an evolutionary theory of socialization. *Child Development*, 62, 647–670.

Bercovitch, F. B. 1993. Dominance rank and reproductive maturation in male rhesus macaques. *Journal of Reproduction and Fertility*, 99, 113–120.

Bercovitch, F. B. 2002. Sex-biased parental investment in primates. *International Journal of Primatology*, 23, 905–921.

Bercovitch, F. B. & Berard, J. 1993. Life-history costs and consequences of rapid reproductive maturation in female rhesus macaques. *Behavioral Ecology and Sociobiology*, 32, 103–109.

Bercovitch, F. B. & Goy, R. W. 1990. The socioendocrinology of reproductive development and reproductive success in macaques. In: *Socioendocrinology of Primate Reproduction* (Ed. by T. E. Ziegler & F. B. Bercovitch), pp. 59–93. New York: Wiley-Liss.

Bercovitch, F. B. & Strum, S. C. 1993. Dominance rank, resource availability, and reproductive maturation in female savanna baboons. *Behavioral Ecology and Sociobiology*, 33, 313–318.

Bercovitch, F. B., Lebron, M. R, Martinez, S. & Kessler, M. J. 1998. Primigravidity, body weight, and costs of rearing first offspring in rhesus macaques. *American Journal of Primatology*, 46, 135–144.

Bercovitch, F. B., Widdig, A. & Nurnberg, P. 2000. Maternal investment in rhesus macaques (*Macaca mulatta*): reproductive costs and consequences of raising sons. *Behavioral Ecology and Sociobiology*, 48, 1–11.

Berman, C. M. 1982a. The ontogeny of social relationships with group companions among free-ranging infant rhesus monkeys. I. Social networks and differentiation. *Animal Behaviour*, 30, 149–162.

Berman, C. M. 1982b. The ontogeny of social relationships with group companions among free-ranging infant rhesus monkeys. II. Differentiation and attractiveness. *Animal Behaviour*, 30, 163–170.

Berman, C. M. 1988. Maternal condition and offspring sex ratio in a group of free-ranging rhesus monkeys: an eleven year study. *American Naturalist*, 131, 307–328.

Berman, C. M. 1990. Intergenerational transmission of maternal rejection rates among free-ranging rhesus monkeys. *Animal Behaviour*, 39, 329–337.

Berman, C. M. 2004. Developmental aspects of kin bias in behavior. In: *Kinship and Behavior in Primates* (Ed. by B. Chapais & C. M. Berman), pp. 317–346. Oxford: Oxford University Press.

Berman, C. M. & Kapsalis, E. 1999. Development of kin bias among rhesus monkeys: maternal transmission or individual learning? *Animal Behaviour*, 58, 883–894.

Berman, C. M., Rasmussen, K. L. R. & Suomi, S. J. 1993. Reproductive consequences of maternal care patterns during estrus among free-ranging rhesus monkeys. *Behavioral Ecology and Sociobiology*, 32, 391–399.

Berman, C. M., Rasmussen, K. L. R. & Suomi, S. J. 1997. Group size, infant development and social networks in free-ranging rhesus monkeys. *Animal Behaviour*, 53, 405–421.

Bjorklund, D. F., Grotuss, J. & Csinady, A. 2009. Maternal effects, social cognitive development, and the evolution of human intelligence. In: *Maternal Effects in Mammals* (Ed. by D. Maestripieri & J. M. Mateo), pp. 292–321. Chicago: University of Chicago Press.

Borries, C., Koenig, A. & Winkler, P. 2001. Variation of life history traits and mating patterns in female langur monkeys (*Semnopithecus entellus*). *Behavioral Ecology and Sociobiology*, 50, 391–402.

Bowman, J. E. & Lee, P. C. 1995. Growth and threshold weaning weights among captive rhesus macaques. *American Journal of Physical Anthropology*, 96, 159–175.

Brown, G. R. & Silk, J. B. 2002. Reconsidering the null hypothesis: is maternal rank associated with birth sex ratios in primate groups? *Proceedings of the National Academy of Sciences USA*, 99, 11252–11255.

Cameron, E. Z. 1998. Is suckling behaviour a useful predictor of milk intake? *Animal Behaviour*, 56, 521–532.

Champagne, F. & Curley, J. P. 2009. The trans-generational influence of maternal care on offspring gene expression and behavior in rodents. In: *Maternal Effects in Mammals* (Ed. by D. Maestripieri & J. M. Mateo), pp. 182–202. Chicago: University of Chicago Press.

Chapais, B. 1991. Matrilineal dominance in Japanese macaques: the contribution of an experimental approach. In: *The Monkeys of Arashiyama* (Ed. by L. M. Fedigan & P. J. Asquith), pp. 251–273. Albany: SUNY Press.

Chapais, B. 1992. The role of alliances in social inheritance of rank among female primates. In: *Coalitions and Alliances in Humans and Other Animals* (Ed. by A. Harcourt & F. B. M. d. Waal), pp. 29–59. New York: Oxford University Press.

Charnov, E. L. 1991. Evolution of life history variation among female mammals. *Proceedings of the National Academy of Sciences USA*, 88, 1134–1137.

Charnov, E. L. & Berrigan, D. 1993. Why do female primates have such long lifespans and so few babies? or life in the slow lane. *Evolutionary Anthropology*, 1, 191–194.

Cheverud, J. M. & Wolf, J. B. 2009. The genetics and evolutionary consequences of maternal effects. In: *Maternal Effects in Mammals* (Ed. by D. Maestripieri & J. M. Mateo), pp. 11–37. Chicago: University of Chicago Press.

Clutton-Brock, T. H. & Iason, G. R. 1986. Sex ratio variation in mammals. *Quarterly Review of Biology*, 61, 339–374.

Cowlishaw, G. & Dunbar, R. I. M. 1991. Dominance rank and reproductive success in male primates. *Animal Behaviour*, 41, 1045–1056.

de Waal, F. B. M. 1990. Do rhesus mothers suggest friends to their offspring? *Primates*, 31, 597–600.

de Waal, F. B. M. 1996. Macaque social culture: development and perpetuation of affiliative networks. *Journal of Comparative Psychology*, 110, 147–154.

DiGiacomo, R. F., Shaughnessy, P. W. & Tomlin, S. L. 1978. Fetal-placental weight relationships in the rhesus (*Macaca mulatta*). *Biology of Reproduction*, 17, 749–753.

Dixson, A. F. & Nevison, C. M. 1997. The socioendocrinology of adolescent development in male rhesus monkeys (*Macaca mulatta*). *Hormones and Behavior*, 31, 126–135.

Drickamer, L. C. 1974. A ten-year summary of reproductive data for free-ranging *Macaca mulatta*. *Folia Primatologica*, 21, 61–80.

Dunbar, R. I. M. 1990. Environmental determinants of intraspecific variation in body weight in baboons (*Papio* spp.). *Journal of Zoology*, 220, 157–169.

Ellis, B. J. 2004. Timing of pubertal maturation in girls: an integrated life history approach. *Psychological Bulletin*, 130, 920–958.

Fairbanks, L. A. 1989. Early experience and cross-generational continuity of mother-infant contact in vervet monkeys. *Developmental Psychobiology*, 22, 669–681.

Fairbanks, L. A. 2000. Maternal investment throughout the life span in Old World monkeys. In: *Old World Monkeys* (Ed. by P. F. Whitehead & C. J. Jolly), pp. 341–367. Cambridge: Cambridge University Press.

Fairbanks, L. A. 1996. Individual differences in maternal styles: causes and consequences for mothers and offspring. *Advances in the Study of Behavior*, 25, 579–611.

Fairbanks, L. A. & McGuire, M. T. 1987. Mother-infant relationships in vervet monkeys: response to new adult males. *International Journal of Primatology*, 8, 351–366.

Fairbanks, L. A. & McGuire, M. T. 1988. Long-term effects of early mothering behavior on responsiveness to the environment in vervet monkeys. *Developmental Psychobiology*, 21, 711–724.

Fairbanks, L. A. & McGuire, M. T. 1993. Maternal protectiveness and response to the unfamiliar in vervet monkeys. *American Journal of Primatology*, 30, 119–129.

Fairbanks, L. A. & McGuire, M. T. 1995. Maternal condition and the quality of maternal care in vervet monkeys. *Behaviour*, 132, 733–754.

Fairbanks, L. A., Jorgensen, M. J., Huff, A., Blau, K., Hung, Y. & Mann, J. J. 2004. Adolescent impulsivity predicts adult dominance attainment in male vervet monkeys. *American Journal of Primatology*, 64, 1–17.

Fedigan, L. M. 1983. Dominance rank and reproductive success in primates. *Yearbook of Physical Anthropology*, 26, 91–129.

Fedigan, L. M., Fedigan, L., Gouzoules, S., Gouzoules, H. & Koyama, N. 1986. Lifetime reproductive success in female Japanese macaques. *Folia Primatologica*, 47, 143–157.

Fisher, R. A. 1930. *The Genetical Theory of Natural Selection*. Oxford: Oxford University Press.

Frisch, R. E. 1984. Body fat, puberty and fertility. *Biological Reviews*, 59, 161–188.

Frisch, R. E. & McArthur, J. W. 1974. Menstrual cycles: fatness as a determinant of minimum weight for height necessary for their maintenance or onset. *Science*, 185, 949–951.

Fujita, K. 1993. Development of visual preference for closely related species by infant and juvenile macaques with restricted social experience. *Primates*, 34, 141–150.

Fujita, K. & Watanabe, K. 1995. Visual preference for closely related species by Sulawesi macaques. *American Journal of Primatology*, 37, 253–261.

Galef, B. G. Jr. 2009. Maternal influences on offspring food preferences and feeding behaviors in mammals. In *Maternal Effects in Mammals* (Ed. by D. Maestripieri & J. M. Mateo), pp. 159–181. Chicago: University of Chicago Press.

Gomendio, M. 1989a. Differences in fertility and suckling patterns between primiparous and multiparous rhesus monkey (*Macaca mulatta*) mothers. *Journal of Reproduction and Fertility*, 87, 529–542.

Gomendio, M. 1989b. Suckling behaviour and fertility in rhesus macaques (*Macaca mulatta*). *Journal of Zoology*, 217, 449–467.

Gomendio, M. 1990. The influence of maternal rank and infant sex on maternal investment trends in rhesus macaques: birth sex ratio, inter-birth intervals and suckling patterns. *Behavioral Ecology and Sociobiology*, 27, 365–375.

Gouzoules, S. & Gouzoules, H. 1987. Kinship. In: *Primate Societies* (Ed. by B. B. Smuts, D. L. Cheney, R. M. Seyfarth, R. W. Wrangham & T. T. Struhsaker), pp. 299–305. Chicago: University of Chicago Press.

Gowaty, P. 1993. Differential dispersal, local resource competition, and sex-ratio variation in birds. *American Naturalist*, 141, 263–280.

Gunnar, M. R. 2003. Integrating neuroscience and psychological approaches in the study of early experiences. *Annals of the New York Academy of Sciences*, 1008, 238–247.

Hane, A. A. & Fox, N. A. 2006. Ordinary variations in maternal caregiving influence human infants' stress reactivity. *Psychological Science*, 117, 550–556.

Hauser, M. D. & Fairbanks, L. A. 1988. Mother-offspring conflict in vervet monkeys: variation in response to ecological conditions. *Animal Behaviour*, 36, 802–813.

Hewison, A. J. M. & Gaillard, J. M. 1999. Successful sons or advantaged daughters? the Trivers-Willard model and sex-biased maternal investment in ungulates. *Trends in Ecology and Evolution*, 14, 229–234.

Higley, J. D. 2003. Aggression. In: *Primate Psychology* (Ed. by D. Maestripieri), pp. 17–40. Cambridge: Harvard University Press.

Hinde, R. A. & Spencer-Booth, Y. 1967. The behaviour of socially living rhesus monkeys in their first two and a half years. *Animal Behaviour*, 15, 169–196.

Hinde, R. A. & Spencer-Booth, Y. 1971. Towards understanding individual differences in rhesus mother-infant interaction. *Animal Behaviour*, 19, 165–173.

Hiraiwa-Hasegawa, M. 1993. Skewed birth sex ratios in primates: should high ranking mothers have daughters or sons? *Trends in Ecology and Evolution*, 8, 395–400.

Holekamp, K. E. & Smale, L. 1991. Dominance acquisition during mammalian social development: the "inheritance" of maternal rank. *American Zoologist*, 31, 306–317.

Howell, S., Westergaard, G., Hoos, B., Chavanne, T. J., Shoaf, S. E., Cleveland, A., Snoy, P. J., Suomi, S. J. & Higley, J. D. 2007. Serotonergic influences on life history outcomes in free-ranging male rhesus macaques. *American Journal of Primatology*, 69, 851–865.

Immelmann, K. 1972. The influence of early experience upon the development of social behaviour in estrildine finches. In: *Proceedings of the 15th International Ornithological Congress*, pp. 316–338. The Hague.

Janson, C. H. & van Schaik, C. P. 1993. Ecological risk aversion in juvenile primates: slow and steady wins the race. In: *Juvenile Primates: Life History, Development, and Behavior* (Ed. by M. E. Pereira & L. A. Fairbanks), pp. 57–76. Oxford: Oxford University Press.

Johnson, R. L. & Kapsalis, E. 1995. Determinants of postnatal weight in infant rhesus monkeys: implications for the study of inter-individual differences in neonatal growth. *American Journal of Physical Anthropology*, 98, 343–353.

Johnson, R. L., Berman, C. M. & Malik, I. 1993. An integrative model of the lactational and environmental control of mating in female rhesus monkeys. *Animal Behaviour*, 46, 63–78.

Johnson, S. E. 2003. Life history and the competitive environment: trajectories of growth, maturation, and reproductive output among chacma baboons. *American Journal of Physical Anthropology*, 120, 83–98.

Käär, P., Jokela, J., Helle, T. & Kojola, I. 1996. Direct and correlative phenotypic selection on life history traits in three pre-industrial human populations. *Proceedings of the Royal Society of London, Series B*, 263, 1475–1480.

Kaplan, J. R., Fontenot, M. B., Berard, J., Manuck, S. B. & Mann, J. J. 1995. Delayed dispersal and elevated monoamine activity in free-ranging rhesus monkeys. *American Journal of Primatology*, 35, 229–234.

Kappeler, P. M., Pereira, M. E. & van Schaik, C. P. 2003. Primate life histories and socioecology. In: *Primate Life Histories and Socioecology* (Ed. by P. M. Kappeler & M. E. Pereira), pp. 1–23. Chicago: University of Chicago Press.

Kirkpatrick, M. & Lande, R. 1989. The evolution of maternal characters. *Evolution*, 43, 485–503.

Kendrick, K. M., Hinton, M. R., Atkins, K., Haupt, M. A. & Skinner, J. D. 1998. Mothers determine sexual preferences. *Nature*, 395, 229–230.

Krackow, S. 1995. Potential mechanisms for sex-ratio adjustment in mammals and birds. *Biological Reviews*, 70, 225–241.

Krackow, S. 2002. Why parental sex ratio manipulation is rare in higher vertebrates. *Ethology*, 108, 1041–1056.

Lee, P. C. 1987. Nutrition, fertility, and maternal investment in primates. *Journal of Zoology*, 213, 409–422.

Lee, P. C. & Johnson, J. 1992. Sex differences in the acquisition of dominance status among primates. In: *Coalitions and Alliances in Humans and Other Animals* (Ed. by A. Harcourt & F. B. M. d. Waal), pp. 391–414. Oxford: Oxford University Press.

Lee, P. C., Majluf, P. & Gordon, I. J. 1991. Growth, weaning, and maternal investment from a comparative perspective. *Journal of Zoology*, 225, 99–114.

Loy, J. 1988. Effects of supplementary feeding on maturation and fertility in primate groups. In: *Ecology and Behavior of Food-Enhanced Primate Groups* (Ed. by J. E. Fa & C. H. Southwick), pp. 153–166. New York: Alan Liss.

Maestripieri, D. 1993. Maternal anxiety in rhesus macaques (*Macaca mulatta*). II. Emotional bases of individual differences in mothering style. *Ethology*, 95, 32–42.

Maestripieri, D. 1994. Costs and benefits of maternal aggression in lactating female rhesus macaques. *Primates*, 35, 443–453.

Maestripieri, D. 1995a. First steps in the macaque world: do rhesus mothers encourage their infants' independent locomotion? *Animal Behaviour*, 49, 1541–1549.

Maestripieri, D. 1995b. Assessment of danger to themselves and their infants by rhesus macaque (*Macaca mulatta*) mothers. *Journal of Comparative Psychology*, 109, 416–420.

Maestripieri, D. 1995c. Maternal encouragement in nonhuman primates and the question of animal teaching. *Human Nature*, 6, 361–378.

Maestripieri, D. 1996. Maternal encouragement of infant locomotion in pigtail macaques, *Macaca nemestrina*. *Animal Behaviour*, 51, 603–610.

Maestripieri, D. 1998. Social and demographic influences on mothering style in pigtail macaques. *Ethology*, 104, 379–385.

Maestripieri, D. 2001a. Female-biased maternal investment in rhesus macaques. *Folia Primatologica*, 72, 44–47.

Maestripieri, D. 2001b. Is there mother-infant bonding in primates? *Developmental Review*, 21, 93–120.

Maestripieri, D. 2001c. Intraspecific variability in parenting styles of rhesus macaques: the role of the social environment. *Ethology*, 107, 237–248.

Maestripieri, D. 2003. Similarities in affiliation and aggression between cross-fostered rhesus macaque females and their biological mothers. *Developmental Psychobiology*, 43, 321–327.

Maestripieri, D. 2004. Genetic aspects of mother-offspring conflict in rhesus macaques. *Behavioral Ecology and Sociobiology*, 55, 381–387.

Maestripieri, D. 2005a. Effects of early experience on female behavioural and reproductive development in rhesus macaques. *Proceedings of the Royal Society of London*, Series B, 272, 1243–1248.

Maestripieri, D. 2005b. Early experience affects the intergenerational transmission of infant abuse in rhesus monkeys. *Proceedings of the National Academy of Sciences USA*, 102, 9726–9729.

Maestripieri, D. & Carroll, K. A. 1998. Risk factors for infant abuse and neglect in group-living rhesus monkeys. *Psychological Science*, 9, 143–145.

Maestripieri, D. & Ross, S. R. 2004. Sex differences in play among western lowland gorilla (*Gorilla gorilla gorilla*) infants: implications for adult behavior and social structure. *American Journal of Physical Anthropology*, 123, 52–61.

Maestripieri, D., Schino, G., Aureli, F. & Troisi, A. 1992. A modest proposal: displacement activities as an indicator of emotions in primates. *Animal Behaviour*, 44, 967–979.

Maestripieri, D., Wallen, K. & Carroll, K. A. 1997. Infant abuse runs in families of group-living pigtail macaques. *Child Abuse and Neglect*, 21, 465–471.

Maestripieri, D., Ross, S. R. & Megna, N. L. 2002. Mother-infant interactions in western lowland gorillas (*Gorilla gorilla gorilla*): spatial relationships, communication, and opportunities for social learning. *Journal of Comparative Psychology*, 116, 219–227.

Maestripieri, D., Roney, J. R., DeBias, N., Durante, K. M. & Spaepen, G. M. 2004. Father absence, menarche, and interest in infants among adolescent girls. *Developmental Science*, 7, 560–566.

Maestripieri, D., Higley, J. D., Lindell, S. G., Newman, T. K., McCormack, K. M. & Sanchez, M. M. 2006a. Early maternal rejection affects the development of monoaminergic systems and adult abusive parenting in rhesus macaques. *Behavioral Neuroscience*, 120, 1017–1024.

Maestripieri, D., McCormack, K., Lindell, S. G., Higley, J. D. & Sanchez, M. M. 2006b. Influence of parenting style on the offspring's behavior and CSF monoamine metabolites levels in crossfostered and noncrossfostered female rhesus macaques. *Behavioural Brain Research*, 175, 90–95.

Maestripieri, D., Lindell, S. G. & Higley, J. D. 2007. Intergenerational transmission of maternal behavior in rhesus monkeys and its underlying mechanisms. *Developmental Psychobiology*, 49, 165–171.

Martin, R. D. & MacLarnon, A. M. 1985. Gestation period, neonatal size, and maternal investment in placental mammals. *Nature*, 313, 220–223.

Martin, R. D. & MacLarnon, A. M. 1988. Comparative studies of growth and reproduction. *Symposia of the Zoological Society of London*, 60, 39–80.

McAdam, A. G. 2009. Maternal effects on evolutionary dynamics in wild small mammals. In: *Maternal Effects in Mammals* (Ed. by D. Maestripieri & J. M. Mateo), pp. 64–82. Chicago: University of Chicago Press.

Mehlman, P. T., Higley, J. D., Faucher, I., Lilly, A. A., Taub, D. M., Vickers, J. H., Suomi, S. J. & Linnoila, M. 1995. Correlation of CSF 5-HIAA concentration with sociality and the timing of emigration in free-ranging primates. *American Journal of Psychiatry*, 152, 907–913.

Mehlman, P. T., Higley, J. D., Fernald, B. J., Sallee, F. R., Suomi, S. J. & Linnoila, M. 1997. CSF 5-HIAA, testosterone, and sociosexual behaviors in free-ranging male rhesus macaques in the mating season. *Psychiatry Research*, 72, 89–10.

Mori, A. 1979. Analysis of population changes by measurement of body weight in the Koshima troop of Japanese monkeys. *Primates*, 20, 371–398.

Mousseau, T. A. & Fox, C. W. 1998a. The adaptive significance of maternal effects. *Trends in Ecology and Evolution*, 13, 403–407.

Mousseau, T. A. & Fox, C. W. (Eds.). 1998b. *Maternal Effects as Adaptations*. Oxford: Oxford University Press.

Nakamichi, M. 1989. Sex differences in social development during the first 4 years in a free-ranging group of Japanese monkeys, *Macaca fuscata*. *Animal Behaviour*, 38, 737–748.

Packer, C., Collins, D. A. & Eberly, L. E. 2000. Problems with primate sex ratios. *Philosophical Transactions of the Royal Society of London B*, 355, 1627–1635.

Paul, A. & Thommen, D. 1984. Timing of birth, female reproductive success and infant sex ratio in semifreeranging Barbary macaques (*Macaca sylvanus*). *Folia Primatologica*, 42, 2–16.

Paul, A., Kuester, J. & Arnemann, J. 1992. Maternal rank affects reproductive success of male Barbary macaques (*Macaca sylvanus*): evidence from DNA fingerprinting. *Behavioral Ecology and Sociobiology*, 30, 337–341.

Pereira, M. E. 1995. Development and social dominance among group living primates. *American Journal of Primatology*, 37, 143–175.

Pereira, M. E. & Leigh, S. R. 2003. Modes of primate development. In: *Primate Life Histories and Socioecology* (Ed. by P. M. Kappeler & M. E. Pereira), pp. 149–176. Chicago: University of Chicago Press.

Richard, A. F. 1985. *Primates in Nature*. New York: W. H. Freeman & Company.

Roberts, M. S. 1978. The annual reproductive cycle of captive *Macaca sylvana*. *Folia Primatologica*, 29, 229–235.

Roff, D. A. 1998. The detection and measurement of maternal effects. In: *Maternal Effects as Adaptations* (Ed. by T. A. Mousseau & C. W. Fox), pp. 83–96. Oxford: Oxford University Press.

Rogers, J., Martin, L. J., Comuzzie, A. G., Mann, J. J., Manuck, S. B., Leland, M. & Kaplan, J. R. 2004. Genetics of monoamine metabolites in baboons: overlapping sets of genes influence levels of 5-hydroxyindolacetic acid, 3-hydroxy-4-methoxyphenylglycol, and homovanillic acid. *Biological Psychiatry*, 55, 739–744.

Rowe, D. C. 2002. On genetic variation in menarche and age at first sexual intercourse. A critique of the Belsky-Draper hypothesis. *Evolution and Human Behavior*, 23, 365–372.

Sapolsky, R. M. 1986. Stress, social status and reproductive physiology in free-living baboons. In: *Psychobiology of Reproductive Behavior: An Evolutionary Perspective* (Ed. by D. Crews), pp. 291–321. Englewood Cliff: Prentice Hall.

Schwartz, S., Wilson, M. E., Walker, M. L. & Collins, D. C. 1985. Social and growth correlates of puberty onset in female rhesus monkeys. *Nutrition and Behavior*, 2, 225–232.

Schwartz, S., Wilson, M. E., Walker, M. L. & Collins, D. C. 1988. Dietary influences on growth and sexual maturation in premenarcheal rhesus monkeys. *Hormones and Behavior*, 22, 231–251.

Schino, G. 2004. Birth sex ratio and social rank: consistency and variability within and between primate groups. *Behavioral Ecology*, 15, 850–856.

Schino, G. & Troisi, A. 2005. Neonatal abandonment in Japanese macaques. *American Journal of Physical Anthropology*, 126, 447–452.

Schino, G., D'Amato, F. R. & Troisi, A. 1995. Mother-infant relationships in Japanese macaques: sources of interindividual variation. *Animal Behaviour*, 49, 151–158.

Schino, G., Speranza, L. & Troisi, A. 2001. Early maternal rejection and later social anxiety in juvenile and adult Japanese macaques. *Developmental Psychobiology*, 38, 186–190.

Schino, G., Aureli, F., Ventura, R. & Troisi, A. 2004. A test of the cross-generational transmission of grooming preferences in macaques. *Ethology*, 110, 137–146.

Setchell, J. M. & Dixson, A. F. 2002. Developmental variables and dominance rank in adolescent male mandrills (*Mandrillus sphinx*). *American Journal of Primatology*, 56, 9–25.

Setchell, J. M., Lee, P. C., Wickings, E. J. & Dixson, A. F. 2001. Growth and ontogeny of sexual size dimorphism in the mandrill (*Mandrillus sphinx*). *American Journal of Physical Anthropology*, 115, 349–360.

Setchell, J. M., Lee, P. C., Wickings, E. J. & Dixson, A. F. 2002. Reproductive parameters and maternal investment in mandrills (*Mandrillus sphinx*). *International Journal of Primatology*, 23, 51–68.

Setchell, J. M., Wickings, E. J. & Knapp, L. A. 2006. Life history in male mandrills (*Mandrillus sphinx*): physical development, dominance rank, and group association. *American Journal of Physical Anthropology*, 131, 498–510.

Short, R. V. 1983. The biological bases for the contraceptive effects of breastfeeding. In: *Advances in International Maternal and Child Health* (Ed. by D. B. Jelliffe & E. F. B. Jelliffe), pp. 27–39 Oxford: Oxford University Press.

Silk, J. B. 1983. Local resource competition and facultative adjustment of sex ratios in relation to competitive abilities. *American Naturalist*, 121, 56–66.

Silk, J. B. 2002. Kin selection in primate groups. *International Journal of Primatology*, 23, 849–875.

Silk, J. B., Willoughby, E. & Brown, G. R. 2005. Maternal rank and local resource competition do not predict birth sex ratios in wild baboons. *Proceedings of the Royal Society of London, Series B*, 272, 859–864.

Simpson, A. E. & Simpson, M. J. A. 1985. Short-term consequences of different breeding histories for captive rhesus macaque mothers and their young. *Behavioral Ecology and Sociobiology*, 18, 83–89.

Simpson, M. J. A. 1985. Effects of early experience on the behaviour of yearling rhesus monkeys (*Macaca mulatta*) in the presence of a strange object: classification and correlation approaches. *Primates*, 26, 57–72.

Simpson, M. J. A. & Datta, S. B. 1990. Predicting infant enterprise from early relationships in rhesus macaques. *Behaviour*, 116, 42–63.

Simpson, M. J. A., Simpson, A. E., Hooley, J. & Zunz, M. 1981. Infant-related influences on birth intervals in rhesus monkeys. *Nature*, 290, 49–51.

Simpson, M. J. A., Gore, M. A., Janus, M. & Rayment, F. D. G. 1989. Prior experience of risk and individual differences in enterprise shown by rhesus monkey infants in the second half of their first year. *Primates*, 30, 493–509.

Small, M. F. 1981. Body fat, rank, and nutritional status in a captive group of rhesus macaques. *International Journal of Primatology*, 2, 91–95.

Small, M. F. 1984. Aging and reproductive success in female *Macaca mulatta*. In: *Female primates: Studies by Women Primatologists* (Ed. by M. F. Small), pp. 249–259. New York: Alan Liss.

Smith, D. G. & Smith, S. 1988. Parental rank and reproductive success of natal rhesus males. *Animal Behaviour*, 36, 554–562.

Sterck, E. H. M., Watts, D. P. & van Schaik, C. P. 1997. The evolution of female social relationships in nonhuman primates. *Behavioral Ecology and Sociobiology*, 41, 291–309.

Stewart, K. 1988. Suckling and lactational anoestrus in wild gorillas. *Journal of Reproduction and Fertility*, 83, 627–634.

Strum, S. C. 1991. Weight and age in wild olive baboons. *American Journal of Primatology*, 25, 219–237.

Sugiyama, Y. & Ohsawa, H. 1982. Population dynamics of Japanese monkeys with special reference to the effect of artificial feeding. *Folia Primatologica*, 39, 238–263.

Tanaka, I. 1989. Variability in the development of mother-infant relationships among free-ranging Japanese macaques. *Primates*, 30, 477–491.

Trivers, R. L. & Willard, D. 1973. Natural selection of parental ability to vary the sex ratio of offspring. *Science*, 179, 90–92.

van Noordwijk, M. A. & van Schaik, C. P. 1999. The effects of dominance rank and group size on female lifetime reproductive success in wild long-tailed macaques, *Macaca fascicularis*. *Primates*, 40, 105–130.

van Noordwijk, M. A. & van Schaik, C. P. 2001. Career moves: transfer and rank challenge decisions by male long-tailed macaques. *Behaviour*, 138, 359–395.

van Schaik, C. P. 1989. The ecology of social relationships amongst female primates. In: *Comparative Socioecology: The Behavioural Ecology of Humans and Other Mammals* (Ed. by V. Standen & R. Foley), pp. 195–218. Boston: Blackwell.

van Schaik, C. P. & Hrdy, S. B. 1991. Intensity of local resource competition shapes the relationship between maternal rank and sex ratios at birth in cercopithecine primates. *American Naturalist*, 138, 1555–1562.

van Schaik, C. P. & Paul, A. 1997. Male care in primates: does it ever reflect paternity? *Evolutionary Anthropology*, 5, 152–156.

Vochteloo, J. D., Timmermans, P. J. A., Duijghuisen, J. A. H. & Vossen, J. M. H. 1993. Effects of reducing the mother's radius of action on the development of mother-infant relationships in longtailed macaques. *Animal Behaviour*, 45, 603–612.

Wade, M. J., Shuster, S. M. & Demuth, J. P. 2003. Sexual selection favors female-biased sex

ratios: the balance between the opposing forces of sex-ratio selection and sexual selection. *American Naturalist* 162, 403–414.

Wasser, S. K. & Barash, D. P. 1983. Reproductive suppression among female mammals:
implications for biomedicine and sexual selection theory. *Quarterly Review of Biology,*
58, 513–538.

Wasser, S. K. & Norton, G. 1993. Baboons adjust secondary sex ratio in response to predictors of sex-specific offspring survival. *Behavioral Ecology and Sociobiology,* 32, 273–281.

West, S. A., Alle Herre, E. & Sheldon, B. C. 2000. The benefits of allocating sex. *Science,*
290, 288–290.

Whitten, P. L. 1987. Infants and adult males. In: *Primate Societies* (Ed. by B. B. Smuts, D. L.
Cheney, R. M. Seyfarth, R. W. Wrangham & T. T. Struhsaker), pp. 343–357. Chicago:
University of Chicago Press.

Williams-Blangero, S. & Blangero, J. 1995. Heritability of age at first birth in captive olive
baboons. *American Journal of Primatology,* 37, 233–239.

Wilson, A. J. & Festa-Bianchet, M. 2009. Maternal effects in wild ungulates. In: *Maternal
Effects in Mammals* (Ed. by D. Maestripieri & J. M. Mateo), pp. 83–103. Chicago: University of Chicago Press.

Wilson, M. E. 1989. Relationship between growth and puberty in the rhesus monkey. In:
Control of the Onset of Puberty. Vol. 3 (Ed. by H. A, Delemarre, T. M. Plant, G. P. van Rees
& J. Schoemaker), pp. 137–149. Amsterdam: Elsevier.

Wilson, M. E. 1992a. Factors determining the onset of puberty. In: *Handbook of Behavioral
Neurobiology.* Vol. 11. *Sexual Differentiation* (Ed. by A. A. Gerall, H. Moltz & I. L. Ward),
pp. 275–312. New York: Plenum Press.

Wilson, M. E. 1992b. Primiparous rhesus monkey mothers are more sensitive to the
nursing-induced inhibition of LH and ovarian steroid secretion. *Journal of Endocrinology,* 134, 493–503.

Wilson, M. E., Walker, M. L. & Gordon, T. P. 1983. Consequences of first pregnancy in
rhesus monkeys. *American Journal of Physical Anthropology,* 61, 103–110.

13

Maternal Effects, Social Cognitive Development, and the Evolution of Human Intelligence

DAVID F. BJORKLUND, JASON GROTUSS,
AND ADRIANA CSINADY

INTRODUCTION

The relation between ontogeny and phylogeny (i.e., development and evolution) has been debated by evolutionary biologists virtually since the publication of Darwin's *Origin of Species*. Different systems or parts of an organism can develop at different rates, and these rates may be accelerated or retarded relative to the developmental rates experienced by one's ancestors. Genetic mutations that produce seemingly small changes in the onset or offset of a developmental process (e.g., mutations in regulatory as opposed to structural genes) have cascading effects on ontogenetic timing and can produce substantial changes in phenotypes. If these phenotypic changes are selected for, they eventually produce phylogenetic changes. Genetic mutations, however, are only one possible avenue of modification in the developmental process. Other sources of developmental change can be attributed to interactions with various elements in the environment. For mammals, mothers are an important component of a young organism's environment and can exert a crucial influence on its development. Maternal influences on development, in turn, can have implications for evolutionary processes.

Mothers are of central importance to all mammals. Internal conception, gestation, and obligate nursing make it unavoidable that mothers will have a substantial influence on their offspring. Mothers' influences may be particularly potent early in their offspring's lives, when developmental and neural plasticity is most pronounced. Mothers serve as a filter or buffer by which their young experience the environment, and as a consequence contribute greater than 50% of heritable variance in some species (Moore 2003). As a result, correlations between the phenotypes of offspring and mothers in some species are often greater than with fathers (Moore 1995).

The focus of this chapter is on maternal effects in human cognitive evolution. In the first section, we briefly discuss what we see as major changes in the evolution of intelligence, specifically social intelligence, in *Homo sapiens*. We argue that the confluence of increased brain size, delayed maturation, and social complexity contributes to the changing nature of mental representation, and that this was most critical in the social realm. We next examine maternal effects in the ontogeny of some social-cognitive abilities in children. We then discuss epigenetic theories of inheritance and evolution, and the nongenetic trans-generational transmission of behaviors via maternal effects in nonhuman mammals. We follow this with a discussion of the possible influence that mothers had in affecting social-cognitive changes over the course of human evolution. The central concept of this chapter is that mothers are a central component to the developmental systems that have evolved since we last shared a common ancestor with chimpanzees 5 to 7 million years ago (Chen & Li 2001; Wood 2006) and that mothers have been changed as much by the process as their offspring.

THE EVOLUTION OF HUMAN INTELLIGENCE

Major physical differences between *Homo sapiens* and extant apes include bipedality, a long digestive track, and modified hands that permit greater manual dexterity. However, it is our large brain and the computing power that it provides that truly differentiate humans from other primates. Human beings are able to represent events in both the past and the future, conceive of what another person is thinking, and communicate about events that occurred in distant places and times. This form of explicit, symbolic, meta-representational cognition permits people to transmit nongenetic information across generations with impressive fidelity (e.g., Richerson & Boyd 2005), a defining characteristic of material culture.

Big Brains, Delayed Maturation, and Social Complexity

Although our species' technological accomplishments are the most obvious product of human intelligence, we, and others, have argued that the modern human mind has its roots in social cognition (e.g., Alexander 1989; Dunbar

1995; Flinn et al. 2005). More specifically, we have proposed that human intelligence evolved as a result of the confluence of an enlarged brain, an extended juvenile period, and life in socially complex groups (Bjorklund & Bering 2003; Bjorklund et al. 2005; Bjorklund & Rosenberg 2005).

The large brain of humans is afforded by a prolonged juvenile period in which brain development is extended into adolescence and young adulthood (e.g., Gould 1977). The pace of human brain development begins prenatally and continues throughout the second year of postnatal life. Neural changes associated with brain development include comparatively prolonged synaptogenesis, glial cell growth, myelination of axons, and dendritic growth in the cortex. The result is great expansion of the human brain, especially in the neocortex (see Langer 2000).

The slow rate of human somatic growth not only affords increased brain growth, but provides more time for children to learn the skills necessary to become a competent adult. Although an extended juvenile period was likely beneficial for acquiring technological skills such as those involved in hunting, extractive foraging, and tool manufacture and use (e.g., Kaplan et al. 2000), children also need to master social relationships. The long road to reproductive maturity is necessitated by the need to navigate a social terrain and to successfully deal with conspecifics (e.g., Dunbar 1995; Flinn et al. 2005).

The diversity of human living conditions has made it impossible for people to inherit a set of relatively fixed cognitive operations to apply in well-defined situations. Rather, human social complexity and diversity necessitated a flexible intelligence and a prolonged period of time to learn (Bjorklund 1997). Moreover, social relationships can change rapidly, requiring quick modification of strategies. As Geary & Huffman (2002, p. 675), point out, the important social variable is not merely aggregating in large groups as in herding animals, "but rather dynamics that involve developing and maintaining long-term relationships with conspecifics and competition that involves, for instance, social deception (e.g., furtive mating)."

Complex social dynamics are associated with larger neocortical volumes among primates (e.g., Dunbar 1995; Reader & Laland 2002). Species with a greater ratio of neocortex to brain stem show more tool use, social learning, and innovation (discovering novel solutions to environmental or social problems) than species with smaller neocortex-to-brain stem ratios (Reader & Laland 2002). The relationship between a species' brain size and social complexity is further enhanced by the length of the juvenile period. For example, Joffe (1997) examined the size of the nonvisual neocortex, length of the juvenile period, and social complexity (i.e., size of typical social group) among 27 primate species and reported that larger brains, longer juvenile periods, and larger social groups were positively correlated and likely coevolved.

Inhibitory Control

What type of intelligence must develop in order to deal with these new social pressures? An initial step in evolving a human-like form of social intelligence includes an increased ability to inhibit prepotent responding. Increased social complexity requires greater voluntary inhibitory control of sexual and aggressive behaviors in order to function harmoniously and cooperate successfully with conspecifics (see Bjorklund & Harnishfeger 1995; Bjorklund & Kipp 2002; Bjorklund et al. 2005). Theoretically, neural circuits initially involved in the control of emotional and appetitive behaviors could be coopted for other purposes, such as playing a critical role modifying social behavior. Over time, inhibitory mechanisms could become increasingly under cortical (and thus intentional) control, resulting in the evolution of the cognitive architecture in modern humans.

Bjorklund & Kipp (1996) have proposed that, in some contexts, there was a greater pressure on females to develop inhibitory abilities than males. Following the tenets of parental investment theory (Trivers 1972), females invest more in offspring than males. In mammals, conception and gestation occur within the female body, so the mother is the sole source of nourishment for her infant. Although there are some species of South American monkeys in which the father also contributes parental care (see Allman 1999), for primates in general, and for humans historically and cross-culturally (e.g., Whiting & Whiting 1975; Eibl-Eibesfeldt 1989), females do the bulk of the child care after birth. In humans, and likely for our recent ancestors, nursing and/or attentive parenting continues for 3 or 4 years before a child is even remotely able to fend for itself. Because of this sex differential in parental investment, there are far greater consequences for engaging in sex for women than for men (pregnancy and potential years caring for a dependent child).

Women may also have needed greater political skill to keep sexual interests in other men hidden from a mate. The major cause of male violence against their mates is suspected female infidelity (see Daly & Wilson 1988; Fisher 1992). Infidelity also frequently leads to divorce, which, historically and in contemporary societies, is more emotionally and financially detrimental to women and their offspring than to men (Fisher 1992).

Consistent with Bjorklund and Kipp's theorizing, there is limited evidence that women are better able to inhibit sexual arousal than are men (Cerny

1978). Evidence indicates that women are better able to control the expression of their emotions than are men, despite the fact that females are more emotionally expressive than males (Buck 1982). One technique that has been used to assess sex differences in expressive control of emotions involves asking people to show positive emotion following a negative experience, or vice versa. For instance, participants may be given a foul-tasting drink and asked to pretend that it tastes good. Females from the age of 3 years are better able to control their emotional expressions (that is, fool a judge watching their reactions) than are males (e.g., Feldman & White 1980; Saarni 1984).

Bjorklund & Kipp (1996) proposed that ancestral females also needed greater inhibitory skills to deal with the difficulties presented by infants and young children (see also Stevenson & Williams 2000; Bjorklund & Kipp 2002). Caring for infants often requires the inhibition of aggressive responses and delaying one's own gratification. Again, research performed mostly with children has consistently shown a female advantage on tasks that involve resisting temptation (Slaby & Parke 1971; Kochanska et al. 1996) and delaying gratification (Kochanska et al. 1996), exactly the pattern that one would predict if pressures associated with taking care of young children were greater on hominid females than males.

Social Learning

Social learning refers to "situations in which one individual comes to behave similarly to others" (Boesch & Tomasello 1998, p. 598). Comparative psychologists differentiate between different types of social learning, based on presumed underlying mechanisms (see Tomasello & Call 1997; Tomasello 1999; see also Galef, this volume). For example, social learning can be achieved via *local enhancement*, in which an individual observes the actions of others at a particular location (for instance, some chimpanzees crack nuts at a location that has plentiful stones), moves to that location, and, in a process of trial and error, discovers a useful behavior (for instance, cracking nuts with stones, although using other techniques than the ones observed). A more sophisticated form of social learning involves *mimicry*, in which an observer copies aspects of a model's behavior without understanding the goal of those behaviors, and *emulation*, in which an observer understands the general goal of a model, but does not reproduce specific behaviors in attempts to attain that goal. For example, one child may observe another sifting sand through her fingers to search for seashells. That child may then start searching through sand with his hands, but tossing handfuls of sand that separates sand from the shells. These forms of social learning are contrasted with what Tomasello and his colleagues have referred to as true imitation, in which the observer understands the goal, or intention, of the model and reproduces important aspects of the modeled behavior to achieve a similar goal (Tomasello et al. 1993a; Tomasello 1999). Most sophisticated yet is teaching, which requires that the teacher understand what the learner knows and his or her goals, and also that the student appreciate the teacher's goal (Caro & Hauser 1992; Tomasello et al. 1993a). That is, teachers must view their students as intentional agents, and vice versa. *Homo sapiens'* remarkable ability to transmit nongenetic information across generations is made possible by our ability to reflect upon our own thoughts, feelings, and behaviors and those of others, and to use this insight to acquire, and sometimes explicitly teach, information.

The perspective-taking abilities required for some forms of social learning are also needed (although perhaps not to the same degree) in the expression of empathy, an emotion of critical importance to a highly social animal as *Homo sapiens*. Empathy, defined as the ability to identify with and understand another person's feelings, develops in the second year of life in humans, presumably based on enhanced representational abilities that emerge at this time (e.g., Lewis 2000). Humans' success as a species may be attributed in large part to the social cohesion afforded by the ability to empathize. This is reflected by Hrdy (1999, p. 392), who wrote: "What makes us humans rather than just apes is the capacity to combine intelligence with articulate empathy." Empathy develops over the course of ontogeny as a result of the social interaction between infants and their caretakers.

The Evolution of Consciousness

Perhaps the jewel in the crown of human intelligence is consciousness, or self-awareness. We use Bering & Bjorklund's (2007) definition of consciousness as "that naturally occurring cognitive representational capacity permitting explicit and reflective accounts of the—mostly causative—contents of mind, contents harbored by the psychological frame of the self and, as a consequence, also the psychological frames of others" (p. 596). Conscious awareness essentially reflects an awareness of self—the ability to experience one's own feelings, desires, and behavior—and correspondingly the realization that other individuals have the same insight.

Humphrey (1976) was the first psychologist to suggest the significance of self-awareness in human cognitive evolution, recognizing that self-consciousness allows one to interpret and predict the feelings and behaviors of others, a valuable tool for a social animal. The term used to describe such phenomena is *theory of mind*, based on the assumption that people possess a "theory" of how their and other people's minds work. According to Wellman (1990), people's theories are based on *belief-desire reasoning*. People understand that one's behavior is a function of what one knows or believes, and what one wants or desires, and that different people can have different, and often conflicting, beliefs and desires. Although rudiments of theory of mind can be found in toddlers (see discussion to follow), belief-desire reasoning, as assessed by the false-belief task (understanding that someone can have a false belief, for example, believing that a cereal box contains cereal when it really contains pencils) is not reliably observed in children until about 4 years of age (e.g., Wellman et al. 2001).

MATERNAL EFFECTS ON CHILDREN'S SOCIAL COGNITION

The purported impact of mothers on their children is almost mythical within psychology. Psychoanalytic theory put the spotlight on mothers as the constructors of their children's personalities, including their future pathologies (Freud 1938). Bettelheim (1972) believed that autism was caused by "refrigerator mothers," whose unconscious distain for their unwanted children caused them to be cold and distant and their infants to retreat into an inner world of their own making. The social, emotional, and intellectual retardation of infants and children reared in stultifying institutions was attributed to "maternal deprivation" (e.g., Spitz 1945), and the academic underachievement of children born into poverty was attributed primarily to mothers who provided insufficient intellectual stimulation for their offspring (e.g., Hunt 1961). Of course, mothers' behaviors related to weaning or toilet training are not the keys to their children's subsequent mental health, autism's cause (although still largely unknown) is not due to lack of maternal warmth, and there are many interacting factors that are responsible for children's intellectual functioning beyond competent "mothering." Nonetheless, early theorists understood implicitly that mothers were important for their children's psychological development, even if they sometimes overstated mothers' influence and ignored the influence of other significant people in children's lives (such as their fathers), genetics, and the child itself.

Contemporary research continues to illustrate the importance of mothers in child development, albeit in a more tempered way than in decades past. In fact, child developmental psychologists, while still studying the effects of mothers on their children's behavior and development, have effectively eliminated "mothering" as a concept and replaced it with "parenting," based on the recognition that fathers, too, can and do play an important role in children's ontogeny. In the 2006 edition of the *Handbook of Child Psychology* (Damon & Lerner 2006), there are over 130 lines in the indexes to the four volumes for "parents" or "parenting" and no listings for "mothering," "mothers," or "maternal effects" (except for the Maternal Maltreatment Classification Interview (MMCI)). This is more than political correctness, for fathers, grandparents, and other parent-substitutes often interact with, care for, and thus affect children in ways comparable to mothers. But just as earlier theorists gave mothers both more credit and more blame for their children's outcomes than they deserved, this modern perspective tends to sell mothers a little short, we believe. Throughout the history of our species, mothers have presumably had the bulk of responsibility for child care. Furthermore, this sex difference in the care of offspring has been observed in all cultures studied (e.g., Whiting & Whiting 1975; Eibl-Eibesfeldt 1989), and continues in Western societies in which women work outside the home (Hetherington et al. 1999). Even in an era when many men are increasing the amount of time they spend caring for their children, other men are abandoning their children and many women are choosing to raise their children without a husband, as reflected by the fourfold increase in homes headed by single females since 1960 (see Cabrera et al. 2000). The bottom line, we believe, is that although most types of postweaning experience can be provided as readily by fathers as by mothers, this has not been the case historically nor is it even the case in contemporary society. Among humans, as it is in nearly all other mammal species (Clutton-Brock 1991), it is the mothers who have born the brunt of child care responsibilities and who have had the greatest postnatal influence on their children's development (to say nothing of prenatal influence), and we believe that the phrase *maternal effects* accurately describes such influences, particularly when conceptualized across cultures and historical time.

Mothers, of course, have effects on their children before birth. Early development in some species is directed not by the embryo's own genes but by genes and other biological factors found in the cytoplasm of the egg (see West-Eberhard 2003; Wade et al., this volume). In mammals, mothers provide nutrition to the fetus through the placenta so that experiences of mothers that have biochemical consequences can, in turn, affect the developing child. The effects on prenatal development of fetal malnutrition (e.g., Lukas & Campbell 2000) and exposure to teratogens including drugs, alcohol, lead, and cigarette smoke (e.g., Baghurst et al. 1992; Olds et al. 1994), are well established. Maternal emotional state during pregnancy has been found to be related to children's subsequent development. For example, high degrees of maternal anxiety are related to the incidence of ADHD in children at 8 and 9 years of age (Van den Bergh & Marcoen 2004), whereas moderate levels of psychological stress have been shown to be related to enhanced mental development at age 2 (DiPietro et al. 2006). Intelligence, as measured by IQ, has been shown to be related to degree of fluctuating body asymmetry (Furlow et al. 1997), which is believed to be caused by perturbations in prenatal development. In fact, the genetic contribution to individual differences in IQ in behavior genetics model is reduced when the prenatal environment is considered (Devlin et al. 1997).

In the remainder of this section we examine briefly several areas of research that have demonstrated significant postnatal maternal effects on children's development. We start with topics indirectly related to social cognition: intelligence as measured by IQ, stress reactivity, and attachment. We then examine maternal effects in several areas of social cognition: joint attention, social learning, theory of mind, language, and autobiographical memory.

Intelligence as Measured by IQ

Stemming from research documenting that factors such as socioeconomic status (SES) and parental level of education relate to children's IQ and academic achievements, psychologists began to look for specific maternal behaviors (or maternal teaching styles) associated with children's intellectual outcomes. This early research reported a relationship between SES and teaching styles, with middle-class mothers providing more "advanced cognitive environments" for their children than mothers from lower-SES homes (e.g., Hess & Shipman 1965). In subsequent work, aspects of the home environment, and particularly mothers' interactions with their children (e.g., vocalizing to child, responding to child's vocalization, use of punishment or restrictions on child's behavior, mother's involvement with child) were carefully examined and related to child IQ and school performance. In general, moderate correlations (typically between 0.30 and 0.60) have been reported between quality of the home environment and children's IQs (e.g., Bradley & Caldwell 1976; Espy et al. 2001). Other research indicates that children at biological risk (for example, prematurity, low birth weight) are especially susceptible to nonresponsive behaviors by their mothers (e.g., Caughy 1996; Landry et al. 1997). For example, Landry and her colleagues (1997) studied "very low birth weight" infants (less than 1600 grams at birth), some of whom were described as being at high medical risk and others at low medical risk. They reported that mothers who tried to maintain their children's attention or ongoing behavior had children with higher language and cognitive abilities at 36 months of age than mothers who used such strategies less often. In addition, mothers who were restrictive, who used physical or verbal attempts to stop what their children were doing or saying, had children with lower cognitive and language scores. The effects of restrictiveness were especially pronounced for the high-risk children.

Stress reactivity

Early experience has been shown to influence mammals' reactivity to stress (see Francis et al. 1999; Ellis et al. 2006; Fenoglio et al. 2006; Champagne & Curley, Maestripieri, this volume). In rats, mothers who engage in high amounts of licking/grooming (LG) and arched-back nursing (ABN), in which the dam arches her back and splays her legs outward during nursing, have offspring who show less fear of novelty and accompanying differences in hypothalamic-pituitary-adrenal (HPA) responses, including gene expression associated with corticotropin-releasing hormone receptors (e.g., Caldji et al. 1998; Francis et al. 1999; see Meaney 2001; Champagne & Curley, this volume, for a review). Similar effects presumably occur in humans. Children who experience elevated stressors early in life, as reflected by living in foster care (Dozier et al. 2006), stultifying orphanages (Gunnar et al. 2001), or other types of family trauma (e.g., Flinn & England 2003; Flinn 2006), show higher levels of cortisol (a stress-related hormone) and poorer health than children without such early stress experiences. That these effects are mediated at least in part by mothers is illustrated by higher cortisol levels and more sick days for children who experienced stressors prenatally than for control children (see Flinn 2006).

The effects of early stressors may be particularly harmful for "high-risk" children (e.g., those with medical problems during childhood, including prematurity). Bugental and her colleagues (2006) found elevated levels of cortisol for young adults with a history of medical problems who reported having poor emotional relationships with their parents; no such association was found for adults who reported positive emotional relationships with their parents. Bugental et al. also reported that 18-month-old children who had been born prematurely had elevated levels of cortisol if their mothers showed depressive symptoms, whereas there was no relation between cortisol levels and maternal depression for toddlers who had been born full term. Bugental and her colleagues concluded that children at medical risk are especially susceptible to elevated stress reactions associated with poor parenting.

One must be cautious in interpreting these correlational studies, of course, and the early stressors that infants experience cannot all be attributed to their mothers. For example, father absence and living with distant relatives are associated with elevated cortisol levels (Flinn & England 2003). Nonetheless, experimental research with laboratory animals (e.g., Francis et al. 1999) clearly demonstrates that maternal behaviors can affect stress reactivity in their offspring, permanently altering gene expression governing behavioral and neurohormonal stress responses (see Champagne & Curley, this volume). And because mammal mothers, including human mothers, are the primary caregivers for their infants—filtering environmental experience for their young offspring—we believe that many of the neurohormonal responses that have long-term consequences for the child can be appropriately described as maternal effects.

Attachment

A much-studied topic for which maternal effects in humans are well documented is attachment. Individual differences in maternal behavior are related to subsequent quality of attachment, which in turn is related to cognitive, social, and emotional functioning later in life (see Belsky 1999; Thompson 2006). Attachment quality is typically defined in terms of a four-part classification scheme: secure, insecure-resistant, insecure-avoidant, and disorganized (e.g., Ainsworth et al. 1978; Main & Solomon 1986). The most studied correlate of attachment quality is parental sensitivity to infants' signals of physical and social need. Ainsworth and her colleagues (1978) demonstrated that mothers of securely attached infants were more responsive to their babies' emotional signals, encouraged them to explore, and enjoyed close contact with them. These women responded appropriately and reliably to their babies' cues, and their infants were able to predict reasonably well what to expect in various situations. Mothers of infants displaying insecure or disorganized attachment have generally been found to be less responsive to their infants (see Belsky 1999). Other maternal factors, such as *interactional synchrony*, the degree to which mothers and infants establish a fluid, dyadic, and coordinated relationship, have also been found to be related to quality of attachment (e.g., Isabella et al. 1989; see De Wolff & van IJzendoorn 1997 for a review).

Behaviors associated with attachment relationships influence neurohormonal functioning, which may be the link to subsequent psychological functioning. For example, endogenous opioids are associated with reward or pleasure, and are recognized by receptors in the brain, primarily in the limbic system. There is a reduction in opioids in young mammals, including humans, in response to distress due to either changing internal states (such as hunger) or separation from the mother. Mothers' own opioid levels decrease when they hear their infants' distress cries, which prompts them to care for their infants, resulting in increased opioid levels for both mothers and infants. Simultaneously, levels of oxytocin are released in both mothers and infants when the mothers comfort their babies, promoting positive feelings and recognition of significant social partners, which, when coupled with the increased production of rewarding opioids, sets the stage for the formation of a durable infant-mother attachment (see Bugental 2000; Chisholm et al. 2005).

Opioid and oxytocin levels influence and are influenced by the HPA axis, the same neuroendocrine system that affects response to stress. Attachment-related behaviors in infancy may organize HPA functioning later in life, affecting people's subsequent psychosocial functioning. Children who establish insecure attachments to their mothers and who experience a stressful and unpredictable environment, although possibly hampering future physical and mental health, may be compensated by more immediate benefits of having a neurohormonal system that is primed to deal with chronic conflict, and preparation for dealing with an equally problematic adult life (the best predictor of future environments often is the present environment). Boyce & Ellis (2005) proposed the concept of conditional adaptations in which individual differences in early experience can produce different patterns of development that may prove to be adaptive for later environments. Early infant-mother attachment may adapt children's HPA axis for adjusting not only to their current environment but also for probable future ones.

Joint Attention

At the core of social cognition is the ability to understand the perspective of another individual. Without it, more advanced forms of social learning, such as true imitation and teaching (e.g., Tomasello et al. 1993a) and theory of mind (e.g., Wellman 1990), would not be possible. Perspective-taking ability does not emerge fully formed in the adult but has a gradual developmental history beginning in infancy with joint attention (Carpenter et al. 1998; Tomasello 1999). In joint attention, social partners focus on a common reference. It is a form of triadic interaction, in which two people are attending to a common third object, each one being aware of what the other is seeing. This is something that mothers (and fathers) of infants do readily, pointing to or gazing at objects while catching their infant's attention, drawing the baby into a social relationship that extends beyond the dyad. Infants do not start life engaging in these types of interactions, although they appear to be oriented to social stimuli from the first days of life. From birth, infants orient to the human face and quickly learn to seek their mothers' faces (Feldman & Eidelman 2004). A few months after birth, infants recognize self-produced, biological motion and adjust their gaze to look at the same object another is looking at (Tomasello et al. 2005). It is not until about 9 months that infants begin to gaze in the direction that adults are pointing or looking, engage in protractive interaction with an adult and an object, point or hold up objects to another person, and imitate an adult's action (see Carpenter et al. 1998; Tomasello 1999). These abilities increase over the next 6 months and reflect, according to Tomasello (1999), infants' under-standing of other people as intentional agents. Between 12 and 18 months of age, infants use others' eye gaze to achieve joint attention (Brooks & Meltzoff 2002), and by 18 to 24 months of age they use it along with other directional cues, such as pointing and head orientation, for word learn-ing and social referencing (Poulin-Dubois & Forbes 2002). The cognitive abilities underlying joint attention include the social partners knowing that they are attending to a common object, monitoring each other's attention, and coordinating their efforts.

The development of joint attention appears highly canalized, with the various forms of joint attention developing in concert and following a reg-ular sequence between about 9 and 15 months of age. However, joint at-tention clearly requires a supportive social partner, and in most cases that partner is the child's mother. Although the timetable for the development of joint-attention skills is similar across cultures (e.g., Adamson & Bakeman 1991), there are cultural differences (e.g., Chavajay & Rogoff 1999), as well as individual differences in how mothers (and presumably fathers) interact with their infants and toddlers in joint-attention situations (e.g., Tomasello & Farrar 1986; Goldsmith & Rogoff 1997). These individual differences in turn are predictive of children's attention to objects (e.g., Chavajay & Rogoff 1999), language development (e.g., Tomasello & Farrar 1986), and play (Big-elow et al. 2004). For example, Carpenter and her colleagues (1998) examined various aspects of social interaction between mothers and their infants in a 6-month longitudinal study beginning when the infants were 9 months old. They reported that the amount of time children spent in joint engagement with their mothers and the extent to which mothers used language that contained some reference to an object the infant was holding predicted infants' subsequent communication skills.

Social Learning

True imitation is preceded developmentally by other forms of social learning, which are supported by social interactions, initially with mothers. Neonatal imitation, in which a newborn matches the gestures (e.g., tongue protrusion, mouth opening, and finger movements) of a model is now well established (e.g., Meltzoff & Moore 1977; Nagy & Molnar 2004). Newborns are not learning anything new during these episodes, for the copied actions are already within their behavioral repertoire. Byrne (2005) proposed that such matching behavior can best be described as social mirroring, in which one member of a dyad copies the behavior of the other as a demonstration of being "in tune" with one another, or as an expression of a form of empathy or mutual identification. This fosters and consolidates social interaction between the interactants. That neonatal imitation serves such a function is illustrated by research demonstrating that newborns both imitate an adult and also provoke imitation from an adult, with different patterns of heart-rate changes accompanying imitation (decreasing heart rate) and provocations (increasing heart rate; Nagy & Molnar 2004; Nagy 2006). This is consistent with arguments that neonatal imitation serves a different function from the imitation seen later in infancy, namely to foster social interaction or communication between the young infant and its mother (Bjorklund 1987; Legerstee 1991).

The interactions between infants and their mothers serve as the basis for subsequent social learning, with infants beginning to imitate sounds and gestures by 8 months (Piaget 1962) or earlier (Collie & Hayne 1999). Infants seem to recognize and enjoy bouts of mother-infant imitation, as illustrated by research reporting that 3.5-month-old infants vocalized and smiled more during and immediately after maternal imitative than nonimitative behavior (Field et al. 1985). Mothers imitate infants as much as or more than infants imitate them. Piaget (1962) defines the phenomenon of mutual imitation as an infant copying the behavior of an adult who is imitating the infant, which he claimed is seen early in infancy. Over the course of infancy, mothers and infants continue this reciprocal imitative relationship. For example, in a longitudinal study, Masur & Rodemaker (1999) observed interactions of mothers and their infants during free play or bath time when the infants were 10, 13, 17, and 21 months old. Mothers or infants displayed imitation at a rate of about one episode per minute; mothers imitated infants more than infants imitated mothers; and the incidence of imitation increased with age, as did the average duration of an imitative episode. Unlike neonatal imitation, in which infants are not learning anything new when they match the behavior of a model, beginning at least in the second half of the first year of life infants are learning new things (many being vocabulary words; see Masur & Eichorst 2002). For example, based on diary reports provided by parents, Barr & Hayne (2003) reported that 12-, 15-, and 18-month-olds learn, on average, one or two new behaviors a day simply by watching.

Imitation may play an important role in the subsequent development of theory of mind (Meltzoff 1995; 2005; Fenstermacher & Saudino 2006). Through reciprocal imitative bouts, infants learn that other beings are "like me," a prerequisite for theory of mind. Meltzoff (2005) proposed that the perspective-taking abilities that underlie both true imitation and theory of mind develop as children learn, through copying the behavior of others, the connection between their own imitative actions and their underlying mental states, and the insight that other people who behave similarly to the child may also have similar mental states (i.e., be "like me").

Theory of Mind

Although theory of mind follows a universal developmental trajectory and timetable, there are individual differences in the rate of acquisition and sophistication of children's theory of mind. Understanding others' mental states has been found to be related to factors in the child's environment, including attachments, parenting styles, and parent-child communication (Carpendale & Lewis 2004). For example, several researchers have reported a connection between a child's level of language development and subsequent performance on false-belief tasks, with amount and content of mothers' language to their children predicting theory-of-mind development (Ruffman et al. 2002; de Rosnay et al. 2004). However, it is not just linguistic ability or the amount of interaction a child has (Cassidy et al. 2005), but the quality and type of interaction that predicts theory-of-mind reasoning. For example, in their research on linguistic acquisitional style, Meins & Fernyhough (1999) reported that mothers' use of nonstandard words in their children's vocabularies, their meaningful interpretations of their children's early vocalizations, and their propensity to focus on their children's mental attributes were positively related with children's performance on a false-belief task at age 5. In their work on language and theory-of-mind understanding, Ruffman et al. (2002) concluded that the mothers' language is related to children's language and theory-of-mind development and that mothers' mental-state utterances facilitate children's subsequent false-belief performance (see also de Rosnay et al. 2004).

It is not just the interaction itself, but the quality of the interaction that impacts the child's understanding of the social environment. The mother's behavior and the manner in which she communicates with the child will affect how the child comes to understand and interpret the behavior of others. According to Ruffman and his colleagues (2006), maternal mental-state talk and maternal warmth affect children's theory of mind, as well as child cooperation. If mothers are a primary source of cooperative behavior in children, it would seem that maternal behavior has influenced the cognitive capacity of children to understand the intentions, desires, and beliefs of other individuals in their environment. This, in turn, would increase cooperative and pro-social behavior.

Language

Differences in how mothers use language when talking to their children not only influence children's developing theory of mind, but also their language development. Parents, and particularly mothers, carefully present language to children in a way that fosters not only language acquisition but also other aspects of social cognition (e.g., Bruner 1983; Carpenter et al. 1998). According to Carpenter et al. (1998, p. 126), "children's initial skills of linguistic communication are a natural outgrowth of their emerging understanding of other persons as intentional agents."

When speaking with infants, mothers (and other people as well) tend to use infant-directed speech, which usually involves high-pitched tones and exaggerated modulations (see Hoff 2005). Infants are more attentive to adults when they use infant-directed as opposed to adult-directed speech (e.g., Cooper & Aslin 1994), are better able to discriminate words spoken in infant-directed as opposed to adult-directed speech (e.g., Moore et al. 1997), and although there are individual and cultural differences in the use of infant-directed speech, some aspects of it appear to be universal (Fernald 1992; Kuhl et al. 1997). The primary reason mothers seem to use infant-directed speech is not to "teach" language to their babies, but to regulate their infants' emotions and behaviors and to convey their own feelings to their infants (e.g., Fernald 1992). For example, Fernald (1992) identified different patterns of infant-directed speech that mothers, speaking various languages, use to seek the infant's attention, provide comfort to the infant, convey approval, and express prohibition. She proposed that the origins of language can be found in mothers' attempts to regulate their infants' emotion (see also Locke 1994; Trainor et al. 2000).

Autobiographical Memory

Autobiographical memory, people's abilities to recall personally experienced events, develops in children over the preschool years. Autobiographical memory is socially constructed, in that it develops during episodes in which children and their parents remember events together (see Nelson 1996; Gauvian 2001). Parents, especially mothers, guide children in recalling information from the past, and in the process learn about remembering, how people in their culture form narratives, and what is important to remember and tell others about events.

How well young children recall past events is related to the conversational styles of their mothers during shared memory. For example, children of "high-elaborative" mothers, who provide support for their children's recollections, confirm their point of view, ask open-ended questions (e.g., "What did we see at the zoo today?"), and introduce new information about the event, recall more information about past events than children of "low-elaborative" mothers (e.g., Reese et al. 1993; Cleveland & Reese 2005). There are cultural differences in how often parents talk to their children about the past, and these differences have consequences for some aspects of memory development. For example, American mothers talk to their 3-year-old children about past events about three times more frequently than do Korean mothers (Mullen & Yi 1995), which corresponds to differences in how often children in these cultures talk about the past (American children do it more often, Han et al. 1998), and of their earliest childhood memory (Americans report earlier memories, Mullen 1994).

No one would seriously question that mothers have an important impact on their children's lives. Their interactions with their children facilitate social-cognitive development and shape individual differences in how they come to deal with their social world. However, for many of the topics discussed in this section, fathers, grandparents, siblings, or staff members of an orphanage could have the same effect on children's development as mothers, and in many cases they do. Human infants enter the world with a biological predisposition for social interaction, and it is likely that any responsive social partner would be sufficient for the species-typical pattern of social cognitive development to emerge. However, in most situations in contemporary society, in all cultures, and over the course of human history, it is not just "any" social partner who serves this role; it is the child's mother. Mothers, we believe, have been predisposed by evolution to be especially receptive to their infants' signals of social need. Juvenile females in a number of primate species, including humans, engage in and show greater interest in infant and parenting than juvenile males (see Maestripieri & Pelka 2002). Girls also show a preference for neoteny (i.e., juvenile facial features as described by Lorenz) earlier than boys (Fullard & Reiling 1976). Thus, although fathers, older siblings, and well-educated institution staff members can serve as effective mother-substitutes, we believe that over the course of our species' history, mothers were the most important influence on young children's social-cognitive development.

In the next section we discuss epigenetic inheritance and evolution, explaining how neural plasticity and early experience could influence evolutionary change. In the final section we discuss the possibility that maternal effects not only play an important role in the ontogeny of social cognition, but may also have played a significant role in their phylogeny.

EPIGENETIC INHERITANCE AND EVOLUTION

It seems clear that there are significant maternal effects on human cognitive development. However, for phylogenetic changes to be realized, there must be some way in which maternal effects induce trans-generational changes. Following the Modern Synthesis, evolutionary changes occur when random mutations produce adaptive phenotypes. However, recently, proponents of epigenetic theories of evolution have argued that it is the developing organism's response to environmental change that is the creative force of evolution; developmental plasticity generates new phenotypes upon which natural selection operates (e.g., Gottlieb 1992; 2002; West-Eberhard 2003; Bjorklund 2006). New phenotypes are created through a combination of genetic and environmental influences. In other words, any mechanism, including developmental experience that results in phenotypic variation, creates the grist upon which natural selection acts.

The idea that changes in ontogeny can influence phylogeny (see Bateson 1988; Gottlieb 1992; 2002; Bjorklund & Pellegrini 2002), although out of mainstream biology for most of the last century for fears of accusations of Lamarckism, can be traced back to James Mark Baldwin (1896; 1902). (Similar ideas were suggested about the same time by Conway Lloyd Morgan and H. F. Osborn, but it is Baldwin's name that became associated with the theory; see Depew 2003.) Baldwin proposed that if members of a population had sufficient "ontogenetic plasticity" to survive substantial variations in environmental conditions, they would be more apt to become primogenitors for new lines of descendents (see Gottlieb 1992).

Evidence for a Baldwin-type effect was provided by the British biologist Conrad Waddington (1975), who demonstrated in a series of experiments with fruit flies that features induced by extreme environments (e.g., lack of veins in the wings due to heat shock) would be expressed in future generations of selectively bred flies in the absence of the conditions that initially induced the change. For example, when Waddington raised fruit flies on a high-salt medium, some developed larger anal papillae that helped them excrete salt from their bodies. He selectively bred flies with larger anal papillae, and after 21 generations some flies developed the larger anal papillae in the absence of the high-salt medium. Waddington referred to this phenomenon as genetic assimilation, which he defined as "the conversion of an acquired character into an inherited one; or better, as a shift towards a greater importance of heredity in the degree to which the character is acquired or inherited" (p. 61). (For other examples and replications of Waddington's work, see Waddington 1975 and Gibson & Hogness 1996).

Since Waddington's experiments, there have been numerous demonstrations of genetic assimilation (or epigenetic inheritance) in a variety of species (see Jablonka & Lamb 1995; West-Eberhard 2003 for reviews), including mammals. Some studies, in fact, have shown modification of behavior in infant animals as a result of maternal behavior, which is then transmitted to subsequent generations. For instance, Denenberg & Rosenberg (1967) showed that female rats that had been handled as infants (removed daily from the home cage and placed in a tin can with shavings for 3 minutes) had grandoffspring that were more active and weighed less than the grandoffspring of nonhandled females, at least when the second generation of rats was allowed to explore their environments. Ressler (1966) demonstrated that C57BL and BALB strains of mice performed better on operant conditioning tests when they were raised by mothers from the BALB as opposed to the C57BL strain, and that this effect was maintained into the next generation. Ressler did not identify the mechanism responsible for the trans-generational effect, but he speculated that some aspect of maternal behavior may be responsible for this effect. As we noted in an earlier section, the amount of licking/grooming (LG) and arched-back nursing (ABN) by mother rats is associated with their offspring's subsequent reactions to stress (see Meaney 2001). These effects have been shown to persist for at least two generations, with the foster offspring of high LG/ABN displaying high levels of LG/ABN themselves and having offspring who show low levels of stress in novel situations, irrespective of their genetic background (Francis et al. 1999).

One need not postulate a Lamarckian-type mechanism of "inheritance of acquired characteristics" to explain these phenomena. The most likely explanation is that organisms possess substantial unexpressed genetic variability. New or extreme environments alter gene expression (e.g., some latent genes become activated, some active genes become deactivated, genes are expressed earlier, later, or for different durations, see Gottlieb 1992), which results in new morphology (as in Waddington's fruit flies) or novel behaviors (as in foster-reared rats and mice, see Meaney 2001). If the environment that induced phenotypic change remains stable, individuals who are able to respond adaptively to the novel environment (i.e., who possess genes permitting the phenotypic change) will have a greater chance of successfully reproducing and preserving their phenotype into the next generation. If these new phenotypes are better adapted to their local environment than other phenotypes, they will increase in frequency as will the genes associated with these phenotypes. These individuals may fill or construct new niches and expose themselves to new selection pressures, prompting further ontogenetic adaptations, and, eventually, phylogenetic changes (Odling-Smee 1988; Laland et al. 2000). Ecological factors that bring about phenotypic change can influence all members of a population, resulting in relatively high frequencies of individuals with altered phenotypes, assuming individuals have the genetic potential to respond. This is in contrast to phenotypic change induced by random mutation, which would be expected to affect few individuals within a population at any one time. Thus, some have proposed that the initial stages of phylogenetic changes are more likely to occur as a result of environmental induction, which would be followed by genetic changes (see Gottlieb 1992; 2002). Although the role of the Baldwin effect in evolution remains controversial, many evolutionary biologists argue that the phenomenon is not inconsistent with the fundamental principles of the Modern Synthesis (see Weber & Depew 2003).

MATERNAL EFFECTS IN THE EVOLUTION OF HUMAN SOCIAL COGNITION

There can be many precipitating environmental events for epigenetic evolution, but one important source, we believe, is maternal effects. Mammal mothers are often the sole source of food, shelter, and social interaction for a young animal at a time in its life when cognitive and behavioral plasticity are at their greatest. Modifications in maternal behavior can therefore have a profound effect of the development of their offspring, bringing about the phenotypic changes proposed by epigenetic evolutionary theorists. This may have been particularly true for human cognitive evolution.

Evolutionary biologists since Darwin (1871) have believed there is a continuity of mental function in evolution, so that the cognition of a species such as *Homo sapiens* should share many features with species with whom they recently shared a common ancestor. Two questions that must be addressed with respect to possible maternal effects on the evolution of social cognition in humans, are: (a) what were the cognitive abilities of our ancestors? and (b) did these ancestors have the cognitive and behavioral plasticity necessary for epigenetic evolution via maternal effects? These questions can best be addressed by examining social cognitive abilities in primates, our best extant models for what our common ancestor with chimpanzees may have been like. Although it is beyond the scope of this chapter to review this literature, chimpanzees and other great apes display substantial social-cognitive abilities. Chimpanzees (e.g., Whiten et al. 1999) and orangutans (van Schaik et al. 2003) transmit nongenetic information across generations, including forms of greeting, grooming, and foraging, a characteristic of maternal culture. Chimpanzees engage in forms of social learning, including local enhancement and emulation (e.g., Call et al. 2004; Horner & Whiten 2005), although there is less evidence that chimpanzees engage in true imitation (see Tomasello & Call 1997; Whiten et al. 2004). An exception to this seems to be for enculturated (human-reared) apes, that have demonstrated true imitation both immediately after viewing a model's behavior (e.g., Tomasello et al. 1993b) and following a significant delay (e.g., Tomasello et al. 1993b; Bering et al. 2000; Bjorklund et al. 2002).

Were the social-cognitive abilities possessed by contemporary chimpanzees also possessed by humans' and apes' common ancestor, they would have served as a solid foundation for the emergence of the suite of social-cognitive skills that characterize modern people. Moreover, chimpanzees, and thus most likely our ape ancestors, also possess the cognitive plasticity to modify these social-cognitive skills under some conditions. Apes provided with a species-atypical rearing environment from shortly after birth, in which they are treated much as human children are treated, develop some limited cognitive abilities (e.g., referential pointing, true imitation, possibly teaching and cognitive empathy) more similar to those of human children than to mother-reared chimpanzees. As we have argued previously (e.g., Bjorklund & Pellegrini 2002; Bjorklund & Rosenberg 2005; Bjorklund 2006), these abilities, both in enculturated chimpanzees and human children, emerge via an epigenetic process in which the inherited cognitive and behavioral systems of the young animal are sensitive to the ecology in which it is raised (see also Leavens et al. 2005).

Such an epigenetic process, based on the social-cognitive abilities similar to those of extant chimpanzees, could have served as a mechanism for epigenetic inheritance. In fact, because of prolonged and intimate contact between mammal mothers and their offspring and the ubiquity of maternal effects in mammalian development, we believe that maternal effects are the most likely source of epigenetic inheritance in the line that led to modern humans. Changes in maternal behavior among only a handful of mothers could have produced offspring with, initially, marginally greater social-learning and communication skills. Individuals with these enhanced abilities would have been at a selective advantage, may have bred with other individuals with similar social-cognitive abilities, and treated their own offspring in such a way as to promote the continuance of these skills within the population.

Such an account of epigenetic inheritance via maternal effects in human social-cognitive evolution must remain speculative, of course. Moreover, the claim here is that such effects were not *the* cause of human intellectual evolution but one of a likely suite of causes occurring over the past 4 to 7 million years. We argue that an understanding of the ontogeny of social-cognitive abilities in both humans and our close genetic relatives, as well as a focus on the role that mothers play in the emergence of such abilities, provides greater insight into both the ontogeny and phylogeny of our species' seemingly unique abilities.

REFERENCES

Adamson, L. B. & Bakeman, R. 1991. The development of shared attention. In: *Annals of Child Development* (Ed. by R. Vasta), pp. 1–41. London: Kingsley.

Ainsworth, M. D. S., Blehar, M. C., Waters, E. & Wall, S. 1978. *Patterns of Attachment: A Psychological Study of the Strange Situation.* Hillsdale, NJ: Erlbaum.

Alexander, R. D. 1989. Evolution of the human psyche. In: *The Human Revolution: Behavioural and Biological Perspectives on the Origins of Modern Humans* (Ed. by P. Mellers & C. Stringer), pp. 455–513. Princeton, NJ: Princeton University Press.

Allman, J. M. 1999. *Evolving Brains.* New York: Scientific American Library.

Baghurst, P. A., McMichael, A. J., Wigg, N. R., Vimpani, G. V., Robertson, E. F., Roberts, R. G. & Tong, S.-L. 1992. Environmental exposure to lead and children's intelligence at the age of seven years. *New England Journal of Medicine,* 327, 1279–1284.

Baldwin, J. M. 1896. A new factor in evolution. *American Naturalist,* 30, 441–451.

Baldwin, J. M. 1902. *Development and Evolution.* New York: McMillan.

Barr, R. & Hayne, H. 2003. It's not what you know, it's who you know: siblings facilitate imitation during infancy. *International Journal of Early Years Education*, 11, 7–21.

Bateson, P. P. G. 1988. The active role of behaviour in evolution. In: *Process and Metaphors in Evolution* (Ed. by M.-W. Ho & S. Fox), pp. 191–207. Chichester: Wiley.

Belsky, J. 1999. Interactional and contextual determinants of attachment security. In: *Handbook of Attachment* (Ed. by J. Cassidy & P. Shaver), pp. 249–264. New York: Guilford.

Bering, J. M. & Bjorklund, D. F. 2007. The serpent's gift: evolutionary psychology and consciousness. In: *Cambridge Handbook of Consciousness* (Ed. by P. D. Zelazo, M. Moscovitch & E. Thompson), pp. 595–627. New York: Cambridge University Press.

Bering, J. M., Bjorklund, D. F. & Ragan, P. 2000. Deferred imitation of object-related actions in human-reared juvenile chimpanzees and orangutans. *Developmental Psychobiology*, 36, 218–232.

Bettelheim, B. 1972. *Empty Fortress*. New York: Free Press.

Bigelow, A. E., MacClean, K. & Proctor, J. 2004. The role of joint attention in the development of infants' play with objects. *Developmental Science*, 7, 518–526.

Bjorklund, D. F. 1987. A note on neonatal imitation. *Developmental Review*, 7, 86–92.

Bjorklund, D. F. 1997. The role of immaturity in human development. *Psychological Bulletin*, 122, 153–169.

Bjorklund, D. F. 2006. Mother knows best: epigenetic inheritance, maternal effects, and the evolution of human intelligence. *Developmental Review*, 26, 213–242.

Bjorklund, D. F. & Bering, J. M. 2003. Big brains, slow development, and social complexity: the developmental and evolutionary origins of social cognition. In: *The Social Brain: Evolutionary Aspects of Development and Pathology* (Ed. by M. Brüne, H. Ribbert & W. Schiefenhövel), pp. 133–151. New York: Wiley.

Bjorklund, D. F. & Harnishfeger, K. K. 1995. The role of inhibition mechanisms in the evolution of human cognition and behavior. In: *New Perspectives on Interference and Inhibition in Cognition* (Ed. by F. N. Dempster & C. J. Brainerd), pp. 141–173. New York: Academic Press.

Bjorklund, D. F. & Kipp, K. 1996. Parental investment theory and gender differences in the evolution of inhibition mechanisms. *Psychological Bulletin*, 120, 163–188.

Bjorklund, D. F. & Kipp, K. 2002. Social cognition, inhibition, and theory of mind: the evolution of human intelligence. In: *The Evolution of Intelligence* (Ed. by R. J. Sternberg & J. C. Kaufman), pp. 27–53. Mahwah, NJ: Erlbaum.

Bjorklund, D. F. & Pellegrini, A. D. 2002. *The Origins of Human Nature: Evolutionary Developmental Psychology*. Washington, DC: American Psychological Association.

Bjorklund, D. F. & Rosenberg, J. S. 2005. The role of developmental plasticity in the evolution of human cognition. In: *Origins of the Social Mind: Evolutionary Psychology and Child Development* (Ed. by B. J. Ellis & D. F. Bjorklund), pp. 45–75. New York: Guilford.

Bjorklund, D. F., Yunger, J. L., Bering, J. M. & Ragan, P. 2002. The generalization of deferred imitation in enculturated chimpanzees (*Pan troglodytes*). *Animal Cognition*, 5, 49–58.

Bjorklund, D. F., Cormier, C. & Rosenberg, J. S. 2005. The evolution of theory of mind: big brains, social complexity, and inhibition. In: *Young Children's Cognitive Development: Interrelationships Among Executive Functioning, Working Memory, Verbal Ability and Theory of Mind* (Ed. by W. Schneider, R. Schumann-Hengsteler & B. Sodian), pp. 147–174. Mahwah, NJ: Erlbaum.

Boesch, C. & Tomasello, M. 1998. Chimpanzee and human culture. *Current Anthropology*, 39, 591–604.

Boyce, W. T. & Ellis, B. J. 2005. Biological sensitivity to context. I. An evolutionary-developmental theory of the origins and functions of stress reactivity. *Development and Psychopathology*, 17, 271–301.

Bradley, R. H. & Caldwell, B. M. 1976. The relation of infants' home environment to mental test performance at fifty-four months: a follow-up study. *Child Development*, 47, 1172–1174.

Brooks, R. & Meltzoff, A. N. 2002. The importance of eyes: how infants interpret adult looking behavior. *Developmental Psychology*, 38, 958–966.

Bruner, J. S. 1983. *Child's Talk: Learning to Use Language*. New York: Norton.

Buck, R. 1982. Spontaneous and symbolic nonverbal behavior and the ontogeny of communication. In: *Development of Nonverbal Behavior in Children* (Ed. by R. S. Feldman), pp. 29–62. New York: Springer-Verlag.

Bugental, D. B. 2000. Acquisition of the algorithms of social life: a domain based approach. *Psychological Bulletin*, 126, 187–219.

Bugental, D. B., Beaulieu, D. B., O'Brien, E., Schwartz, A., Cayan, L., Fowler, E., Ellerson, P., Godinho, T. & Kokotay, S. 2006. Physiological and psychological resilience in the face of early adversity. Unpublished manuscript.

Byrne, R. W. 2005. Social cognition: imitation, imitation, imitation. *Current Biology*, 15, 498–499.

Cabrera, N. J., Tamis-LeMonda, C. S., Bradley, R. H., Hofferth, S. & Lamb, M. E. 2000. Fatherhood in the twenty-first century. *Child Development*, 71, 127–136.

Caldji, C., Tannenbaum, B., Sharma, S., Francis, D., Plotsky, P. M. & Meaney, M. J. 1998. Maternal care during infancy regulates the development of neural systems mediating the expression of fearfulness in the rat. *Proceedings of the National Academy of Sciences USA*, 95, 5335–5340.

Call, J., Carpenter, M. Tomasello, M. 2004. Copying results and copying actions in the process of social learning: chimpanzees (*Pan troglodytes*) and human children (*Homo sapiens*). *Animal Cognition*, 8, 151–163.

Caro, T. M. & Hauser, M. D. 1992. Is there teaching in nonhuman animals? *Quarterly Review of Biology*, 67, 151–174.

Carpendale, J. I. M. & Lewis, C. 2004. Constructing an understanding of mind: the development of children's social understanding within social interaction. *Behavioral and Brain Sciences*, 27, 79–151.

Carpenter, M., Nagell, K. & Tomasello, M. 1998. Social cognition, joint attention, and communicative competence from 9 to 15 months of age. *Monographs of the Society for Research in Child Development*, 63 (4, Serial No. 255).

Cassidy, K. W., Fineberg, D. S., Brown, K. & Perkins, A. 2005. Theory of mind may be contagious, but you don't catch it from your twin. *Child Development*, 76, 97–107.

Caughy, M. O. 1996. Health and environmental effects on the academic readiness of school-age children. *Developmental Psychology*, 32, 515–522.

Cerny, J. A. 1978. Biofeedback and the voluntary control of sexual arousal in women. *Behavior Therapy*, 9, 847–855.

Champagne, F. & Curley, J. P. 2009. The trans-generational influence of maternal care on offspring gene expression and behavior in rodents. In: *Maternal Effects in Mammals* (Ed. by D. Maestripieri & J. M. Mateo), pp. 182–202. Chicago: University of Chicago Press.

Chavajay, P. & Rogoff, B. 1999. Cultural variation in management of attention by children and their caregivers. *Developmental Psychology*, 35, 1079–1090.

Chen, F. C. & Li, W. H. 2001. Genomic divergences between humans and other hominoids and the effective population size of the common ancestor of humans and chimpanzees. *American Journal of Human Genetics*, 68, 444–456.

Chisholm, J. S., Burbank, V. K., Coall, D. A. & Gemmit, F. 2005. Early stress: perspectives from developmental evolutionary ecology. In: *Origins of the Social Mind: Evolutionary Psychology and Child Development* (Ed. by B. J. Ellis & D. F. Bjorklund), pp. 76–107. New York: Guilford.

Cleveland, E. S. & Reese, E. 2005. Maternal structure and autonomy support in conversations about the past: contributions to children's autobiographical memory. *Developmental Psychology*, 41, 376–388.

Clutton-Brock, T. H. 1991. *The Evolution of Parental Care*. Princeton, NJ: Princeton University Press.

Collie, R. & Hayne, R. 1999. Deferred imitation by 6- and 9-month-old infants: more evidence for declarative memory. *Developmental Psychobiology*, 35, 83–90.

Cooper, R. P. & Aslin, R. N. 1994. Developmental differences in infant attention to the spectral properties of infant-directed speech. *Child Development*, 65, 1663–1677.

Daly, M. & Wilson, M. 1988. *Homicide*. New York: Aldine.

Damon, W. & Lerner, R. M. (Eds.). 2006. *Handbook of Child Psychology*, 6th ed. New York: Wiley.

Darwin, C. 1871. *The Descent of Man, and Selection in Relation to Sex*. London: John Murray.

Denenberg, V. H. & Rosenberg, K. M. 1967. Non-genetic transmission of information. *Nature*, 216, 549–550.

Depew, D. J. 2003. Baldwin and his many effects. In: *Evolution and Learning: The Baldwin Effect Reconsidered* (Ed. by B. H. Weber & D. J. Depew), pp. 3–31. Cambridge, MA: MIT Press.

de Rosnay, M., Pons, F., Harris, P. L. & Morrell, J. 2004. A lag between understanding false belief and emotion attribution in young children: relationships with linguistic ability and mothers' mental state language. *British Journal of Developmental Psychology*, 22, 197–218.

Devlin, B., Daniels, M. & Roeder, K. 1997. The heritability of IQ. *Nature*, 388, 468–471.

De Wolff, M. & van IJzendoorn, M. 1997. Sensitivity and attachment: a meta-analysis of parental antecedents of infant attachment. *Child Development*, 68, 571–591.

DiPietro, J. A., Noval, M. F. S. X., Costigan, K. A., Atella, L. D. & Reusing, S. P. 2006. Maternal psychological distress during pregnancy in relation to child development at age two. *Child Development*, 77, 573–587.

Dozier, M., Peloso, E., Gordon, M. K., Manni, M., Gunnar, M. R., Stovall-McClough, K. C. & Levine, S. 2006. Foster children's diurnal production of cortisol: an exploratory study. *Child Maltreatment*, 11, 189–197.

Dunbar, R. I. M. 1995. Neocortex size and group size in primates: a test of the hypothesis. *Journal of Human Evolution*, 28, 287–296.

Eibl-Eibesfeldt, I. 1989. *Human Ethology*. New York: Aldine & Gruyter.

Ellis, B. J., Jackson, J. J. & Boyce, W. T. 2006. The stress response systems: universality and adaptive individual differences. *Developmental Review*, 26, 175–212.

Espy, K. A., Molfese, V. J. & DiLalla, L. F. 2001. Effects of environmental measures on intelligence in young children: growth curve modeling of longitudinal data. *Merrill-Palmer Quarterly*, 47, 42–73.

Feldman, R. & Eidelman, A. I. 2004. Parent-infant synchrony and the social-emotional development of triplets. *Developmental Psychology*, 40, 1133–1147.

Feldman, R. S. & White, J. B. 1980. Detecting deception in children. *Journal of Communication*, 30, 121–128.

Fenoglio, K. A., Yuncai Chen, Y. & Baram, T. Z. 2006. Neuroplasticity of the hypothalamic-pituitary-adrenal axis early in life requires recurrent recruitment of stress-regulating brain regions. *Journal of Neuroscience*, 26, 2434–2442.

Fenstermacher, S. K. & Saudino, K. J. 2006. Understanding individual differences in young children's imitative behavior. *Developmental Review*, 26, 346–364.

Fernald, A. 1992. Human maternal vocalizations to infants as biologically relevant signals: an evolutionary perspective. In: *The Adapted Mind: Evolutionary Psychology and the Gen-*

eration of Culture (Ed. by J. H. Barkow, L. Cosmides & J. Tooby), pp. 391–428. New York: Oxford University Press.

Field, T., Guy, L. & Umbel, V. 1985. Infants' responses to mothers' imitative behaviors. *Infant Mental Health Journal,* 6, 40–44.

Fisher, H. E. 1992. *Anatomy of Love: The Natural History of Monogamy, Adultery, and Divorce.* New York: Norton.

Flinn, M. V. 2006. Evolution and ontogeny of stress response to social challenges in the human child. *Developmental Review,* 26, 138–174.

Flinn, M. V. & England, B. G. 2003. Childhood stress: endocrine and immune responses to psychosocial events. In: *Social and Cultural Lives of Immune Systems* (Ed. by J. M. Wilce), pp. 107–147. London: Routledge Press.

Flinn, M. V., Geary, D. C. & Ward, C. V. 2005. Ecological dominance, social competition, and coalitionary arms races: why humans evolved extraordinary intelligence. *Evolution and Human Behavior,* 26, 10–46.

Francis, D. D., Diorio, J., Liu, D. & Meaney, M. J. 1999. Nongenomic transmission across generations of maternal behavior and stress responses in the rat. *Science,* 286, 1155–1158.

Freud, S. 1938/1964. An outline of psychoanalysis. In: *The Standard Edition of the Complete Psychological Works of Sigmund Freud.* Vol. 23 (Ed. and Translated by J. Strachey), pp. 141–207. London: Hogarth Press.

Fullard, W. & Reiling, A. M. 1976. An investigation of Lorenz's "babyness." *Child Development,* 47, 1191–1193.

Furlow, F. B., Armijo-Prewitt, T., Gangstead, S. W. & Thornhill, R. 1997. Fluctuating asymmetry and psychometric intelligence. *Proceedings of the Royal Society of London, Series B,* 264, 823–829.

Galef, B. G., Jr. 2009. Maternal influences on offspring food preferences and feeding behaviors in mammals. In: *Maternal Effects in Mammals* (Ed. by D. Maestripieri & J. M. Mateo). Chicago: University of Chicago Press.

Gauvian, M. 2001. *The Social Context of Cognitive Development.* New York: Guilford.

Geary, D. C. & Huffman, K. J. 2002. Brain and cognitive evolution: forms of modularity and functions of mind. *Psychological Bulletin,* 128, 667–698.

Gibson, G. & Hogness, D. S. 1996. Effect of polymorphism in the *Drosophila* regulatory gene *Ultrabithorax* on homeotic stability. *Science,* 271, 200–203.

Goldsmith, D. & Rogoff, B. 1997. Mothers' and toddlers' coordinated joint focus of attention: variations with maternal dsyphoric symptoms. *Developmental Psychology,* 33, 113–119.

Gottlieb, G. 1992. *Individual Development and Evolution: The Genesis of Novel Behavior.* New York: Oxford University Press.

Gottlieb, G. 2002. Developmental-behavioral initiation of evolutionary change. *Psychological Review,* 109, 211–218.

Gould, S. J. 1977. *Ontogeny and Phylogeny.* Cambridge, MA: Harvard University Press.

Gunnar, M. R., Morison, S. J., Chisholm, K. & Schuder, M. 2001. Salivary cortisol levels in children adopted from Romanian orphanages. *Development and Psychopathology,* 13, 611–628.

Han, J. J., Leichtman, M. D. & Want, Q. 1998. Autobiographical memory in Korean, Chinese, and American children. *Developmental Psychology,* 34, 701–713.

Hess, R. D. & Shipman, V. C. 1965. Early experience and the socialization of cognitive modes in children. *Child Development,* 36, 869–886.

Hetherington, E. M., Henderson, S. H. & Reiss, D. 1999. Adolescent siblings in stepfamilies: family functioning and adolescent adjustment. *Monographs of the Society for Research in Child Development,* 64 (4, Serial No. 259).

Hoff, E. 2005. *Language Development,* 3rd ed. Belmont, CA: Wadsworth.

Horner, V. & Whiten, A. 2005. Causal knowledge and imitation/emulation switching in chimpanzees (*Pan troglodytes*) and children (*Homo sapiens*). *Animal Cognition*, 8, 164–181.

Hrdy, S. B. 1999. *Mother Nature: Maternal Instincts and How They Shape the Human Species*. New York: Ballantine Books.

Humphrey, N. K. 1976. The social function of intellect. In: *Growing Points in Ethology* (Ed. by P. P. G. Bateson & R. Hinde), pp. 303–317. Cambridge: Cambridge University Press.

Hunt, J. M. 1961. *Intelligence and Experience*. New York: Ronald Press.

Isabella, R. A., Belsky, J. & von Eye, A. 1989. The origins of infant-mother attachment: an examination of interactional synchrony during the infant's first year. *Developmental Psychology*, 25, 12–21.

Jablonka, E. & Lamb, M. 1995. *Epigenetic Inheritance and Evolution: The Lamarckian Dimension*. New York: Oxford University Press.

Joffe, T. H. 1997. Social pressures have selected for an extended juvenile period in primates. *Journal of Human Evolution*, 32, 593–605.

Kaplan, H., Hill, K., Lancaster, J. & Hurtado, A. M. 2000. A theory of human life history evolution: diet, intelligence, and longevity. *Evolutionary Anthropology*, 9, 156–185.

Kochanska, G., Murry, K., Jacques, T. Y., Koenig, A. L. & Vandegeest, K. A. 1996. Inhibitory control in young children and its role in emerging internalization. *Child Development*, 67, 490–507.

Kuhl, P. K., Andruski, J. E., Christovich, I. A., Christovich, L. A., Kozhevnikova, E. V., Ryskina, V. L., Stolyarova, E. I., Sundberg, U. & Lacerda, F. 1997. Cross-language analysis of phonetic units in language addressed to infants. *Science*, 277, 684–686.

Laland, K. N., Odling-Smee, J. & Freldman, M. W. 2000. Niche construction, biological evolution and cultural change. *Behavioral and Brain Sciences*, 23, 131–175.

Landry, S. H., Smith, K. E., Miller-Loncar, C. L. & Swank, P. R. 1997. Predicting cognitive-language and social growth curves from early maternal behaviors in children at varying degrees of biological risk. *Developmental Psychology*, 33, 1040–1053.

Langer, J. 2000. The heterochronic evolution of primate cognitive development. In: *Biology, Brains, and Behavior: The Evolution of Human Development* (Ed. by S. T. Parker, J. Langer & M. L. McKinney), pp. 215–235. Santa Fe, NM: School of American Research Press.

Leavens, D. A., Hopkins, W. D. & Bard, K. A. 2005. Understanding the point of chimpanzee pointing: epigenesis and ecological validity. *Current Directions in Psychological Science*, 14, 185–189.

Legerstee, M. 1991. The role of person and object in eliciting early imitation. *Journal of Experimental Child Psychology*, 51, 423–433.

Lewis, M. 2000. Self-conscious emotions: embarrassment, pride, shame, and guilt. In: *Handbook of Emotions*, 2nd ed. (Ed. by M. Lewis & J. Haviland), pp. 623–636. New York: Guilford.

Locke, J. L. 1994. Phases in the child's development of language. *American Scientist*, 82, 436–445.

Lukas, W. D. & Campbell, B. C. 2000. Evolutionary and ecological aspects of early brain malnutrition in humans. *Human Nature*, 11, 1–26.

Maestripieri, D. 2009. Maternal influences on offspring growth, reproduction, and behavior in primates. In: *Maternal Effects in Mammals* (Ed. by D. Maestripieri & J. M. Mateo). Chicago: University of Chicago Press.

Maestripieri, D. & Pelka, S. 2002. Sex differences in interest in infants across the lifespan: a biological adaptation for parenting? *Human Nature*, 13, 327–344.

Main, M. & Solomon, J. 1986. Discovery of a disorganized/disoriented attachment pattern. In: *Affective Development in Infancy* (Ed. by T. B. Brazelton & M. N. Youngman), pp. 95–124. Norwood, NJ: Ablex.

Masur, E. F. & Eichorst, D. L. 2002. Infants' spontaneous imitation of novel versus familiar words: relations to observational and maternal report measures o their lexicons. *Merrill-Palmer Quarterly*, 48, 405–426.

Masur, E. F. & Rodemaker, J. E. 1999. Mothers' and infants' spontaneous vocal, verbal and action imitation during the second year. *Merrill-Palmer Quarterly*, 45, 392–412.

Meaney, M. J. 2001. Maternal care, gene expression, and the transmission of individual differences in stress reactivity across generations. *Annual Review of Neuroscience*, 24, 1161–1192.

Meins, E. & Fernyhough, C. 1999. Linguistic acquisitional style and mentalising development: the role of maternal mind-mindedness. *Cognitive Development*, 14, 63–80.

Meltzoff, A. N. 1995. Understanding the intentions of others: re-enactment of intended acts by 18-month-old children. *Developmental Psychology*, 31, 838–850.

Meltzoff, A. N. 2005. Imitation and other minds: the "Like Me" hypothesis. In: *Perspectives on Imitation: From Cognitive Neuroscience to Social Science* (Ed. by S. Hurley & N. Chapter), pp. 55–77. Cambridge, MA: MIT Press.

Meltzoff, A. & Moore, M. K. 1977. Imitation of facial and manual gestures by human neonates. *Science*, 198, 75–78.

Moore, C. L. 1995. Maternal contributions to mammalian reproductive development and the divergence of males and females. In: *Advances in the Study of Behavior*. Vol. 24 (Ed. by P. J. B. Slater, J. S. Rosenblatt, C. T. Snowdon & M. Milinski), pp. 47–118. New York: Academic Press.

Moore, C. L. 2003. Evolution, development, and the individual acquisition of traits: what we've learned since Baldwin. In: *Evolution and Learning: The Baldwin Effect Reconsidered* (Ed. by B. H. Weber & D. J. Depew), pp. 115–139. Cambridge, MA: MIT Press.

Moore, D. S., Spence, M. J. & Katz, G. S. 1997. Six-month-olds' categorization of natural infant-directed utterances. *Developmental Psychology*, 33, 980–989.

Mullen, M. K. 1994. Earliest recollections of childhood: a demographic analysis. *Cognition*, 52, 55–79.

Mullen, M. K. & Yi, S. 1995. The cultural context of talk about the past: implications for the development of autobiographical memory. *Cognitive Development*, 10, 407–419.

Nagy, E. 2006. From imitation to conversation: the first dialogues with human neonates, *Infant and Child Development*, 15, 223–232.

Nagy, E. & Molnar, P. 2004. Homo imitans or homo provocans? human imprinting model of neonatal imitation. *Infant Behavior and Development*, 27, 54–63.

Nelson, K. 1996. *Language in Cognitive Development: The Emergence of the Mediated Mind*. New York: Cambridge University Press.

Odling-Smee, F. J. 1988. Niche constructing phenotypes. In: *The Role of Behavior in Evolution* (Ed. by H. C. Plotkin), pp. 73–132. Cambridge, MA: MIT Press.

Olds, D. L., Henderson, C. R. & Tatelbaum, R. 1994. Intellectual impairment in children of women who smoke cigarettes during pregnancy. *Pediatrics*, 93, 221–227.

Piaget, J. 1962. *Play, Dreams and Imitation in Childhood*. New York: Norton.

Poulin-Dubois, D. & Forbes, J. N. 2002. Toddlers' attention to intentions-in-action in learning novel action words. *Developmental Psychology*, 38, 104–114.

Reader, S. M. & Laland, K. N. 2002. Social intelligence, innovation, and enhanced brain size in primates. *Proceedings of the National Academy of Sciences USA*, 99, 4436–4441.

Reese, E., Haden, C. & Fivush, R. 1993. Mother-child conversations about the past: relationships of style and memory over time. *Cognitive Development*, 8, 403–430.

Ressler, R. H. 1966. Inherited environmental influences on the operant behavior of mice. *Journal of Comparative and Physiological Psychology*, 61, 264–267.

Richerson, P. J. & Boyd, R. 2005. *Not by Genes Alone: How Culture Transformed Human Evolution*. Chicago: University of Chicago Press.

Ruffman, T., Slade, L. & Crowe, E. 2002. The relation between children's and mother's mental state language and theory-of-mind understanding. *Child Development, 73,* 734–751.

Ruffman, T., Slade, L., Devitt, K. & Crowe, E. 2006. What mothers say and what they do: the relation between parenting, theory of mind, language and conflict/cooperation. *British Journal of Developmental Psychology, 24,* 105–124.

Saarni, C. 1984. An observational study of children's attempts to monitor their expressive behavior. *Child Development, 55,* 1504–1513.

Slaby, R. G. & Parke, R. D. 1971. Effects of resistance to deviation of observing a model's affective reaction to response consequence. *Developmental Psychology, 5,* 40–47.

Spitz, R. 1945. Hospitalism: an inquiry into the genesis of psychiatric conditions in early childhood. *Psychoanalytic Study of the Child, 1,* 53–74.

Stevenson, J. C. & Williams, D. C. 2000. Parental investment, self-control, and sex differences in the expression of ADHD. *Human Nature, 11,* 405–422.

Thompson, R. 2006. The development of the person: social understanding, relationships, conscience, self. In: *Handbook of Child Psychology,* 6th ed. (Ed. by W. Damon & R. M. Lerner), pp. 24–98. New York: Wiley.

Tomasello, M. 1999. *The Cultural Origins of Human Cognition.* Cambridge, MA: Harvard University Press.

Tomasello, M. & Call, J. 1997. *Primate Cognition.* New York: Oxford University Press.

Tomasello, M. & Farrar, J. J. 1986. Joint attention and early language. *Child Development, 57,* 1454–1463.

Tomasello, M., Kruger, A. C. & Ratner, H. H. 1993a. Cultural learning. *Behavioral and Brain Sciences, 16,* 495–511.

Tomasello, M., Savage-Rumbaugh, S. & Kruger, A. C. 1993b. Imitative learning of actions on objects by children, chimpanzees, and enculturated chimpanzees. *Child Development, 64,* 1688–1705.

Tomasello, M., Carpenter, M., Call, J., Behne, T. & Moll, H. 2005. Understanding and sharing intentions: the origins of cultural cognition. *Behavioral and Brain Sciences, 28,* 675–692.

Trainor, L. J., Austin, C. M. & Desjardins, R. N. 2000. Is infant-directed speech prosody a result of the vocal expression of emotion? *Psychological Science, 11,* 188–195.

Trivers, R. 1972. Parental investment and sexual selection. In: *Sexual Selection and the Descent of Man* (Ed. by B. Campbell), pp. 136–179. New York: Aldine de Gruyter.

Van den Bergh, B. R. H. & Marcoen, A. 2004. High antenatal maternal anxiety is related to ADHD symptoms, externalizing problems, and anxiety in 8- and 9-year-olds. *Child Development, 75,* 1085–1097.

van Schaik, C. P., Ancrenaz, M., Borgen, G., Galdikas, B., Knott, C. D., Singleton, I., Suzuki, A., Utami, S. S. & Merrill, M. 2003. Orangutan cultures and the evolution of material culture. *Science, 299,* 102–105.

Waddington, C. H. 1975. *The Evolution of an Evolutionist.* Ithaca, NY: Cornell University Press.

Wade, M. J., Priest, N. K. & Cruickshank, T. E. 2009. A theoretical overview of genetic maternal effects: evolutionary predictions and empirical tests with mammalian data. In: *Maternal Effects in Mammals* (Ed. by D. Maestripieri & J. M. Mateo), pp. 38–63. Chicago: University of Chicago Press.

Weber, B. H. & Depew, D. J. (Eds.). 2003. *Evolution and Learning: The Baldwin Effect Reconsidered.* Cambridge. MA: MIT Press.

Wellman, H. M. 1990. *The Child's Theory of Mind.* Cambridge, MA: MIT Press.

Wellman, H. M., Cross, D. & Watson, J. 2001. Meta-analysis of theory-of-mind development: the truth about false belief. *Child Development, 72,* 655–684.

West-Eberhard, M. J. 2003. *Developmental Plasticity and Evolution.* New York: Oxford University Press.

Whiten, A., Goodall, J., McGrew, W. C., Nishida, T., Reynolds, V., Sugiyama, Y., Tutin, C. E. G., Wrangham, R. W. & Boesch, C. 1999. Cultures in chimpanzees. *Nature*, 399, 682–685.

Whiten, A., Horner, V., Litchfield, C. A. & Marshall-Pescini, S. 2004. How do apes ape? *Learning & Behavior*, 32, 36–52.

Whiting, B. B. & Whiting, J. W. 1975. *Children of Six Cultures: A Psycho-cultural Analysis.* Cambridge, MA: Harvard University Press.

Wood, B. 2006. A precious little bundle. *Nature*, 443, 278–280.

Maternal Effects in Mammals:
Conclusions and Future Directions

JILL M. MATEO AND DARIO MAESTRIPIERI

This concluding chapter provides a synthesis of the major themes of this volume, discusses recent research on maternal effects in mammals in relation to theoretical advances in this area, integrates the studies done in disparate groups of mammals, and provides possible future directions for research. Maternal effects occur when a mother's phenotype influences her offspring's phenotype independent of the genes it inherits from its mother. They can arise before or after birth, or both, and can be mediated by the mother's nutrition, physiology, behavior, social status, physical environment, or some combination of these variables. The range of offspring phenotypes influenced by maternal effects is potentially unlimited, but as described below most empirical research to date has largely focused on offspring birth weight, sex ratio, growth rate, and survival. However, growing interest in maternal effects promises to yield new insights from both laboratory and field studies and from additional nonmodel species.

Cheverud and Wolf's chapter describes maternal effects as the influence of the mother's environment on the development of her offspring's phenotype, separate from the effects of the offspring's genes and environment on its own traits. Maternal environmental effects arise as a result of the environment experienced by the mother, and maternal genetic effects ultimately map back to the maternal genome. These two maternal effects are contrasted with "direct effects," in which the environment experienced by an individual or the individual's genotype directly affects its own phenotype. The authors point out that in maternal-effects models, maternal reproductive success is complete with her production of offspring, and that offspring survival affects its own fitness, not its mother's fitness. However, the model allows the mother to affect her offspring's survival and fitness through the environment she provides. As noted in all chapters, it is this environment that is the single most important factor contributing to differences among phenotypes and fitness in newborns and weanlings. Thus, the joint selection of maternal effects and offspring phenotypes plays a critical role in mammalian evolution. Cheverud and Wolf illustrate the interesting and unusual architecture of genetic maternal effects with a single-locus model and a two-locus model. In the single- locus model, they assume that a locus in the mother affects the development of her offspring with selection operating on the offspring phenotype. In the two-locus model, they assume that the effect of the "maternal" locus differs depending on the offspring genotype. This interaction is referred to as maternal-offspring epistasis and plays a role in a number of evolutionary processes.

How can one empirically test maternal-effect models? First, as described in McAdam's, Bowen's, and Wilson and Festa-Bianchet's chapters, long-term field studies with detailed pedigrees can be used to examine the effects of maternal condition on offspring birth weight and survival, among other traits. Second, Cheverud and Wolf as well as Wade, Priest, and Cruickshank detail the use of cross-fostering in laboratory studies to quantify maternal effects, wherein halves of litters (or single offspring) are transferred between mothers on the day of birth. (As discussed in Vandenbergh's chapter, the prenatal uterine environment has a very large hormonal effect, which is confounded with direct genetic effects in this design. However, some of these hormonal effects are mediated by siblings rather than the mother. Embryo transfer would reduce these confounds, but current technology limits broad applicability of this approach. Furthermore, genetic and environmental sources of variation in the context provided by the mother cannot be distinguished unless full sisters are included as mothers in the fostering design.) Typically, these cross-fostering studies show that the effects of the maternal environment are the major contributor to variance in offspring body weight and growth rate from birth to weaning. Data from the agricultural literature show that maternal effects are responsible for nearly half of the phenotypic variance in size among animals before weaning. After weaning, when offspring are physically separated from their mothers, maternal effects decline over time, reaching a modest value of about 10% for many adult traits in rodents. A third, molecular approach uses quantitative trait loci (QTL) with multigenerational crosses of inbred strains to differentiate direct and maternal effects (see also the chapters by Wade et al. and Wilson & Festa-Bianchet).

Wade et al.'s chapter continues the overview of maternal-effect theories and presents an empirical test with publicly available sequence data. They use mammalian gene sequences to test two theoretical predictions about the relaxed selective constraint (RSC) theory. For offspring phenotypes with an intermediate fitness optimum, such as body size, they show why the dual genetic control of maternal provisioning and selection among maternal lineages often leads to negative genetic correlations between coadapted maternal and zygotic traits. Further, they explain why such genes are expected to diversify among taxa even more rapidly than genes with only maternal effects and why they should tend to have higher levels of standing diversity within species. They test the RSC theory by examining patterns of polymorphism and divergence in two maternal-effect genes in humans and mice, the maternal-effect gene *Mater* and its zygotically expressed homolog *RHD*. They conclude by calling for more genetic studies in conjunction with expression arrays and systematic randomized sequencing regimes to understand the evolutionary processes and significance of genetic maternal effects.

McAdam and his colleagues have been observing red squirrels (*Tamiasciurus hudsonicus*) in the southwest Yukon of Canada as part of a long-term field project aimed at understanding the ecological relationship between food abundance and red-squirrel evolution. Their pedigree data allow them to quantify genetic maternal effects on offspring growth based on the heritabilities and contributions of two maternal traits, litter size and parturition date. They find that these genetic maternal effects result in a more than threefold increase in the potential for evolutionary change than would have been predicted based on direct genetic effects alone. In his chapter, McAdam argues that testing the actual importance of maternal effects to evolutionary dynamics requires measures of the strength of selection and the response to selection as well. Maternal effects will either accelerate or delay the response to selection depending on whether they act positively or negatively on offspring traits. In red squirrels, there are large maternal effects that act positively on offspring growth, so one would expect a much greater response to selection than would be predicted based on heritability alone. McAdam and his colleagues found correlations between responses to selection and the strength of both current and previous selection, and suggest this provides evidence of an evolutionary time-lag in which current responses also depend on previous selection. They show that the observed dynamics of red-squirrel growth rates support both of the fundamental predictions of models of maternal-effect evolution. Their findings are among the few quantifications of the contribution of maternal effects to the potential for evolution in a wild vertebrate. McAdam concludes with an interesting revisit of the Chitty hypothesis, which suggests that the large fluctuations in population densities of small mammals can be explained by cycles of low aggression and high reproductive output favored at low population densities, followed by high aggression and reduced reproductive output favored at high densities. The few empirical field tests of the Chitty hypothesis have produced mixed results, but McAdam proposes that traits associated with density fluctuations might possess a much greater potential for evolution in wild mammals than has been revealed by simple measurements of heritability, and could be better explained by maternal effects.

Wilson and Festa-Bianchet focus on the implications of maternal effects for the ecology and evolution of wild ungulates. This taxonomic group is well suited for long-term studies because ungulates are terrestrial and diurnal and can be (relatively) easily followed. Long-term studies have found that, in general, offspring birth weight and survival initially increase with mother's age and parity; however, as is the case with other mammalian groups, senescence-based maternal effects often reduce infant weight or survival. The relative importance of maternal effects varies from year to year according to environmental conditions, such as breeding after severe winters, and this makes the effects of senescence on fitness more difficult to detect if fewer older females survive to the next breeding season. Other extrinsic environmental variables have been documented to affect maternal condition and "performance" such as temperature, snowfall, and plant biomass. By tracking known individuals over multiple reproductive events, Wilson and Festa-Bianchet and other ungulate researchers can determine whether and to what extent early experience influences a female's ability to respond to changing environments. They also examine whether maternal effects on offspring phenotype decline over ontogeny or exert a permanent influence. For a long-lived animal, if maternal effects are mostly exerted during prenatal or lactation periods, then after weaning the mother's phenotype might be less important. In domestic ungulates and perhaps some wild species, maternal effects on offspring body weight decrease with offspring age, yet from an evolutionary perspective maternal effect on body size might be most important in the first year, when mortality is highest, so there may be no benefit of persistent effects. Although the actual influence of maternal effects on fitness has rarely been measured in ungulates, maternal effects are commonly associated with traits correlated with fitness, such as offspring birth date, birth weight, and early growth.

Bowen presents a broad review of maternal effects in pinnipeds, the aquatic carnivores including walruses, seals, and sea lions. Like the ungulates described above, these pinnipeds are long-lived and large-bodied and bear a single offspring per reproductive event. Unlike most of the mammals described in this volume, mothers invest little to no energy in nest building or predator defense, and maternal effects are most prominent during the lactation period. In some pinniped species, mothers fast on land or ice, feeding their young until weaned, whereas others cycle between foraging at sea and suckling on land. Only walruses swim with their nursing pups. The location and timing of births are affected by mothers' phenotypes and in turn affect offspring condition and survival. For example, in the well-studied elephant seals (*Mirounga* spp.), females with intermediate levels of maternal experience breed early and have more surviving offspring than inexperienced or highly experienced females. Across pinnipeds, however, there are notables differences in the roles of maternal age, weight, parity, timing of birth, and maternal postpartum mass (MPPM) on offspring phenotypes. The effects of maternal behavior on offspring traits are less well understood because of the time mothers spend at sea, but more efficient foragers are known to have higher mass and thus heavier offspring. The best predictor of pup weaning mass is total milk energy intake over the entire lactation period, which can vary in length depending on the mother's condition. Pups of larger females are not only heavier at weaning but are relatively fatter than lighter pups, and this is presumably one of the factors resulting in improved survival of larger pups. Long-term pinniped research has also revealed maternal effects mediated by El Niño events. Overall, pinniped offspring are affected by maternal strategies about the timing and location of birth, mother's foraging ability, and her ability to defend and maintain contact with her dependent offspring. Larger, older, and in some cases more experienced females give birth to larger pups, which then grow faster during lactation and are heavier at weaning. Furthermore, in a number of species offspring size at weaning predicts their survival for one or more years and survival to the age of first reproduction. These results have implications for life-history theory and for predictions about the trade-off between growth and reproductive expenditure over the life of a female, since a critical assumption is that the amount of resources allocated to offspring affects fitness. More long-term comparative analyses are called for, as well as cross-fostering studies where appropriate, although Bowen notes the difficulty of such studies with pinnipeds given their slow growth rates, single-pup litters, and in some species, endangered or threatened status.

Mateo's chapter summarizes various parental effects on offspring behavioral development, including the influence of mothers' physiological, social, and abiotic environments on their young. A mother's hormonal status during gestation and lactation, including gonadal and adrenal hormones, can influence the anatomy, physiology, and behavior of her offspring. These effects can persist throughout the lifetime of the offspring, and can even persist across generations. For example, in litter-bearing rodents, young gestating between two males can be masculinized by exposure to fetal androgens (detailed in Vandenbergh's chapter). Daughters in this location tend to give birth to male-biased litters themselves, thus having daughters that are more likely to gestate between two males. These proximate effects can have positive or negative fitness consequences if the local breeding population favors one sex over another. Maternal glucocorticoids can have effects on offspring morphology, development, and survivability, although these effects have not been as well studied in mammals as in other taxonomic groups (but see chapters by Champagne & Curley and Holekamp & Dloniak). As discussed in several chapters, timing of maternal reproduction in seasonal breeders can have a significant impact on offspring condition and survival. Furthermore, in seasonal breeders perinatal melatonin from mothers can adaptively prime their young for somatic and reproductive growth appropriate for the time of year in which they are born, accelerating puberty and mating if conditions are favorable, or delaying sexual maturation until breeding conditions are appropriate. In many species, the early social environment, including the mother and siblings but other conspecifics as well, can affect adult dominance relationships, mate-choice preferences, and kin-selected behaviors. Comparative studies on closely related species, or populations of a species living in different ecological habitats, allow examination of differences in the strength of maternal effects as a function of social structure, food availability, annual patterns, or length of maternal dependence. Finally, maternal effects may be especially important in shaping offspring behavior with survival consequences, such as foraging strategies, antipredator behavior, dispersal, and habitat choices. Although not all mammalian groups are well suited for quantification of the maternal effects described in this chapter, they are worth bearing in mind when considering the interface of ecology and evolution on behavioral development.

Milk provisioning may be a defining feature of mammalian life, but, as summarized by Galef, mothers can shape the food preferences expressed by their offspring well after weaning. Young animals have sensory biases that cause them to find some flavors attractive and others aversive. Subsequent trial-and-error learning influences which foods they eat and which they avoid. Finally, juveniles can use the feeding behavior of adults to guide the development of their own food preferences, and thus acquire an appropriate diet without relying on either unreliable sensory-affective responses to flavors or the potentially risky process of learning by trial and error. Avoidance of dangerous foods can thus spread through a population rapidly and can be transmitted to future generations. Regardless of the process of diet development by offspring, mothers play an important role because of their close association with young pre- and postnatally. Fetuses can learn about flavors ingested by pregnant females, but there is little evidence that this affects the subsequent food choices of offspring (except in domesticated rabbits, *Oryctolagus cuniculus*). Flavors in mother's milk can also lead to an increased preference for that flavor in lab animals; in contrast, diet cues in feces are not very salient unless a diet is predominated by a pungent item (e.g., rabbit diets rich in juniper). Young can learn what their mother eats by licking her mouth, rubbing her muzzle, browsing on dropped morsels, or stealing her food. Galef also describes how food preferences can be learned through demonstrator-observer effects and cultural transmission, and details several examples of transmission of food-extraction techniques among conspecifics. Galef urges caution in overinterpreting the existing literature, for example pointing out that despite careful research there is no clear evidence that predator mothers teach their young to hunt by bringing prey to the den or nest, and that results of food-learning studies in captivity may artificially increase the amount of time young spend in contact with mothers, and thus minimize what would normally be trial-and-error learning in the wild.

Most of the literature on the development of food preferences focuses on rats in the wild and in the laboratory, but there are increasing anecdotal and quantitative data from other species, such as meerkats (*Suricata suricatta*), primates, canids, and other predators. Galef argues that although the mechanisms involved in maternal influences on the food choices of offspring are often unknown, the functions of such influence are probably similar, increasing the probability that young wean to an adequate diet of safe foods before exhausting their internal reserves and without ingesting dangerous toxins. Although the potential benefits of these maternal effects seem obvious, as yet there is little empirical evidence that they result in optimal diets in free-living young.

Champagne and Curley focus on the transmission of maternal effects across generations. In particular, they discuss the implications of naturally occurring variation in maternal care as well as disruptions in maternal care on the development of offspring phenotypes. The manner in which mothers rear their young after birth can affect offspring neuroendocrine systems (including the stress axis), cognition, and social and reproductive behavior. Importantly, these effects include modification of daughters' maternal behaviors when they become mothers, representing a nongenetic or Lamarckian transmission of traits across generations. Laboratory research on the effects of variation in maternal care often involves a maternal-separation paradigm, in which the mother is separated from her pups for prolonged periods or, in rare cases, is permanently removed and pups are reared in isolation. Short periods of separation can stimulate maternal care upon reunion with positive effects on pups, whereas prolonged separation has negative effects, typically measured in terms of stress reactivity, glucocorticoid receptor (GR) levels, cognitive performance, and social behavior. A more ecologically relevant technique has been used with some rodent and primate species, in which mothers have to work at varying levels for food. In some species the difficulty in foraging does not have a large effect on maternal care, but rather the predictability of the required effort is most salient. In an often-cited example of trans-generational transmission of maternal care, mothers in the Long-Evans strain of rats (*Rattus norvegicus*) show high variability in the rates of arched-back nursing and licking and grooming (LG) of pups, but this variation is individually stable. Adult offspring of high-LG mothers are more neophilic and have lower acute stress responses and more stress-hormone-related receptors in the hippocampus and amygdala than offspring of low LG mothers. Cross-fostering newborns between high- and low-LG mothers shows that these effects are mediated behaviorally rather than genetically or through prenatal hormones; that is, they are determined directly by styles of maternal care. These maternal effects involve epigenetic modifications to DNA that are associated with the quality of mother-infant interactions occurring during the postnatal period. Differential methylation of a GR promoter prevents the binding of factors necessary for increased expression of the receptor; these differences arise in young of high- and low-LG mothers during the postpartum period and are maintained into adulthood, thus accounting for the alteration in maternal care of daughters. Further, low levels of maternal LG are associated with high levels of estrogen-receptor (ER) methylation, whereas high levels of LG are associated with low levels of ER methylation among female offspring. Although transmission of maternal-care styles from generation to generation is well documented in some rodents and primates (see also Maestripieri's chapter), additional research is needed in other taxonomic groups to determine just how robust a phenomenon it is among animals. In addition, little is known about possible trans-generational effects on sons, for example involving cognitive or social traits, or about the possible role of biparental care in the transmission of traits. Although the mechanisms of this maternal effect are becoming well elucidated, the ecological and evolutionary functions of such transmission remain unclear.

Vandenbergh discusses a powerful and persistent maternal effect observed in some polytocous or litter-bearing species, including rodents, swine, and possibly humans. Because female fetuses do not produce androgens, exposure to the steroids from adjacent males in the same uterine horn can have a masculinizing effect on them. Male fetuses are affected by their own androgens as well as those from neighboring brothers. The effects due to transfer of androgens among developing fetuses are known as intrauterine position (IUP) effects. In both sexes, prenatal androgen exposure influences the development of sexually dimorphic neural structures such as the sexually dimorphic nucleus of the preoptic area in the hypothalamus, hippocampus, and corpus callosum, although the degree to which IUP affects these structures varies across species. Females that gestate between two males (2M females) can have later vaginal opening and first estrus (measures of reproductive development) and fewer litters, and thus have lower lifetime reproductive success. In Mongolian gerbils (*Meriones unguiculatus*), but not mice (*Mus musculus*), males that gestate between two males are more attractive to females, achieve more inseminations, and sire more pups than those gestating between two females (0M males). Behaviorally, 2M female mice in seminatural environments are more aggressive and have larger home ranges but are less attractive to males and have lower fertility. IUP effects can persist across generations, with 2M females giving birth to male-biased litters. Using cesarean births, it has been shown for some species that anogenital distance (AGD) in rodents reliably correlates with IUP and, presumably, prenatal androgen exposure. Although it requires further validation, AGD might be used with free-living animals to estimate their prenatal exposure to androgens without invasive experimentation. Vandenbergh also notes the potential relevance of IUP effects for studies on toxicological agents and endocrine disruptors.

Holekamp and Dloniak's chapter complements Bowen's, focusing on maternal effects in fissiped carnivores (with toes separated to the base, in contrast to the pinniped carnivores with webbed toes), including felines, canids, otters, bears, and mongooses. Similar to other mammalian groups, in fissipeds maternal age, weight, and reproductive experience can influence offspring birth weight and survival, and again these effects vary by species. Unlike most pinnipeds, fissipeds often rear their young in dens or nests, and a mother's choice of den site, based on her prior reproductive experiences, ability to compete for sites, or her current condition, can affect cub survival. Among highly social species, a female's dominance rank and competitive ability can confound the effects of age, weight, and parity. A key maternal effect in spotted hyenas (*Crocuta crocuta*) is the nongenetic transmission of maternal rank to offspring. Offspring acquire their mother's rank in the clan, and because daughters are philopatric they maintain this rank for their life span. Sons disperse, but if their mother has a high rank they (and their sisters) are more likely to survive to adulthood and live longer, due to increased access to food and faster growth rates as cubs. Furthermore, a high-ranking female's sons are more successful at mating in their new clan and her daughters are more attractive as mating partners. In contrast, low-ranking females attend their dens less often because of increased foraging efforts, which has negative impacts on cub growth and survival. Holekamp and Dloniak present data on rank differences in steroid hormones and growth factors and how they affect hyena development. They hypothesize that other potential maternal effects are likely to play an important role in fissipeds. For example, in communally breeding species, offspring of first-time mothers might be more successful if their mothers had previously gained experience serving as helpers or allo-mothers than if their mothers did not. In addition, in communal groups the immunocompetence of offspring could be boosted by ingesting the colostrum of several different females. In domestic dogs, handling or mild stress results in neophilic pups that respond well to stressful situations and in learning contexts (see also Champagne & Curley's chapter). A mother's grooming and nursing style in the wild could have similar effects on her young, but as yet these effects have not been studied. Finally, Holekamp and Dloniak discuss how the length of dependence on mothers influences the potential for maternal effects (see also Mateo's chapter), as well as how diet differences (e.g., omnivore versus carnivore) affect growth rates and maternal dependency.

Maestripieri's chapter summarizes a wide range of nongenetic maternal effects in primates. Not surprisingly, maternal age, parity, dominance rank (in highly social species), and condition influence maternal investment and, therefore, offspring phenotype, including birth sex ratios, growth rate, reproductive development, hormonal physiology, and behavioral patterns. Many primate species experience long lives in stable environments, which facilitate the transmission of maternal condition or phenotype to their young, although the associated long developmental period can make it difficult to quantify such maternal effects. Like the hyenas discussed by Holekamp and Dloniak, in some primates with matrilineal dominance hierarchies, female offspring inherit their mother's rank, creating a nongenetic transmission of benefits (if any) of rank independent of an individual's actual condition. Offspring of high-ranking females have faster growth rates, wean sooner, and reproduce at an earlier age than those of lower-ranking females, although the generality of these effects varies by sex and species. Females differ in their mothering style, ranging from overprotective to rejecting, and given the typically long period of dependence on mothers, her style can have a significant impact on infant behavioral and neurohormonal development. Similar to rodents, cross-fostering studies in rhesus macaques (*Macaca mulatta*) have shown that mothering styles can be transmitted across generations through nongenetic mechanisms involving experience-induced long-term neuroendocrine and behavioral changes.

Most of the available literature on maternal effects in primates is on cercopithecine monkeys, yet our understanding of these phenomena would be broadened by studies on New World monkeys, monogamous, biparental species that twin, and by more studies on solitary species in which the confounds of rank and social competition are largely removed. Studies with provisioned macaque populations have been quite informative, but can be problematic because population densities are artificially high and species-typical life-history traits (e.g., male emigration and dispersal) are sometimes altered in these populations. The resulting stress and competition could influence maternal condition, and thus unintentionally create maternal effects. Despite decades of research with captive primates, the genetic bases of primate maternal effects, if any, are not well understood. However, cross-fostering and QTL studies could yield further insights into nongenetic maternal influences on offspring growth, physiology, and behavior.

Bjorklund, Grotuss, and Csinady turn our attention to our own species, humans (*Homo sapiens*), and to potential maternal effects on cognitive development and the evolution of intelligence. The authors summarize current theory on the evolution of human social intelligence, focusing on the evolutionary trends for increased brain size, delayed maturation, and increased social complexity in the primate order. For instance, species with larger executive brain ratios (the ratio of neocortex to brain stem) exhibit more tool use, social learning, and innovation (discovering novel solutions to environmental or social problems) than species with smaller ratios. Bjorklund et al. describe several types of cognitive abilities, such as self-awareness, theory of mind, language, and imitation, particularly those related to the social realm, and then discuss possible maternal effects in the development of social and cognitive abilities in children. Epigenetic theories of inheritance and evolution are presented, followed by a discussion of the potential influence that mothers have had in shaping social-cognitive changes over the course of human evolution. Bjorklund et al. also discuss the role of maternal emotional, chemical, and nutritional environments during gestation and lactation, and the effects of stress on offspring attachment and cognition. Large-scale cross-fostering studies are not feasible for quantifying maternal effects in humans, of course, but long-term twin studies as well as historical church records on births and deaths may illuminate some maternal effects similar to those described in other mammalian species, such as the influence of maternal age, condition, and parity on offspring sex, weight, and survival.

Throughout this volume, we have emphasized "maternal" effects because of the unique physiological and behavioral relationship between mammalian mothers and their offspring. Yet for some species paternal effects may provide a potent source of variation in offspring traits. Several species exhibit biparental care, such as California mice (*Peromyscus californicus*), Djungarian hamsters (*Phodopus campbelli*), Mongolian gerbils, most prairie voles (*Microtus ochrogaster*), some South American primate species (e.g., *Saguinus* and *Callithrix* spp.) and humans, with fathers building nests, retrieving, huddling over or carrying young, and, in some species, assisting in birth. There is every reason to expect postnatal paternal effects on offspring traits, although the strength of these effects may not be as robust as the cumulative pre- and postnatal maternal effects.

The chapters in this volume discuss maternal effects on a variety of offspring phenotypes in a variety of species, yet the proximate mechanisms producing these effects are not well understood. Long-term work with captive primates and rodents has pointed to effects at the molecular level (e.g., transcription, methylation) and the biochemical level (e.g., steroid hormones, neurotransmitters), but continued research into the genetic, physiological, and behavioral processes associated with maternal effects as the source of phenotypic variation will be crucial to fully understand these effects and their evolutionary significance in mammalian populations.

Some common threads are woven throughout this volume. First, the most robust maternal effects in mammals are a mother's influence on offspring survival and body weight. This is not surprising, given the defining feature of mammals—lactation—but these effects are similar in other taxonomic groups, including birds and insects. However, the extended interactions between mothers and offspring during lactation and weaning may promote maternal effects not seen to the same extent in other taxa, such as learning and memory, parenting style, dominance rank, food preferences, defensive behaviors, and reproductive strategy.

It has been noted that some of the maternal effects described in this volume are akin to data in search of theory. For example, why should animals that inherit low-ranking matrilines "accept" their position? To what extent are intrauterine-position effects adaptive? Why are the preferences of young rodents so affected by maternal diets, given the scant evidence that such preferences affect later foraging strategies? Functionally, why should maternal grooming of pups have effects on offspring physiology and behavior, and how robust are these effects taxonomically?

Long-term studies of wild mammals will allow us to observe variation in maternal effects within individuals across reproductive events as a function of, for example, climatic variables, anthropogenic effects, demographic shifts or senescence. Female age will increase linearly with each reproductive attempt, but parenting experience, condition, and perhaps dominance will not increase predictably. Future studies, in the lab or field, might further consider offspring sex differences in the degree of maternal effects, and in particular whether adaptive maternal effects are observed more often in the dispersing or philopatric sex. Finally, as noted in several chapters, it is unclear how long we should expect maternal effects to persist in offspring. With each new environment an offspring encounters, long after it leaves its mother, there could be new periods of gene expression. Effects could decline across the life span, or they could have a permanent influence, and both scenarios could lead to fitness costs or benefits. As detailed throughout this volume, there have been significant advances in maternal effects in the past decades, both conceptually and empirically, and in the field and laboratory. We hope that the data and ideas presented here inspire more interdisciplinary and integrated research in this rapidly evolving field.

ACKNOWLEDGMENTS

We wish to thank the many friends and colleagues who helped us during the editing process. Several contributors participated in the internal review process, most notably Bennett G. Galef, Jr.,, Andrew McAdam, and Alastair Wilson. External chapter reviews were kindly provided by Fred Bercovitch, Mathias Kölliker, Louis Lefebvre, Don Owings, Wendy Saltzman, Joanna Setchell, and Mike Tomasello, and two anonymous reviewers who provided very helpful, exhaustive reviews of the entire manuscript. Christie Henry at the University of Chicago Press provided encouragement and support at every stage.

INDEX

The letter f following a page reference denotes a figure.

337